BOUNDARY ELEMENTS VIII

Proceedings of the 8th International Conference, Tokyo, Japan, September 1986

Volume II

Editors:
M. Tanaka
C.A. Brebbia

Springer-Verlag Berlin Heidelberg GmbH

M. TANAKA

Department of Mechanical Engineering
Shinshu University
500 Wakasato
Nagano 380
Japan

C.A. BREBBIA

Computational Mechanics Institute
52 Henstead Road
Southampton
SO1 2DD
U.K.

 British Library Cataloguing in Publication Data

Boundary elements VIII:
 proceedings of the 8th international
 conference, Tokyo, Japan, September 1986
 1. Engineering mathematics   2. Boundar
 value problems
 I. Brebbia, C.A.   II. Keramidas, G.A.
 620'. 001'515353   TA347.B69

ISBN 978-3-662-22337-6   ISBN 978-3-662-22335-2 (eBook)
DOI 10.1007/978-3-662-22335-2

© Springer-Verlag Berlin Heidelberg 1986

Originally published by Computational Mechanics Publications in 1986.

Softcover reprint of the hardcover 1st edtion 1986

w9183120

CONTENTS

SECTION IV GEOMECHANICS

SECTION V ELECTROSTATICS & ELECTROMAGNETIC PROBLEMS

SECTION VI OPTIMIZATION

SECTION VII DYNAMICS AND VIBRATIONS

SECTION VIII STRUCTURAL PROBLEMS

VOLUME II

PREFACE

SECTION IX HYDRODYNAMICS

SECTION X NONLINEAR STRUCTURAL PROBLEMS

SECTION XI NONLINEAR MATERIAL PROBLEMS

SECTION XII COMPUTATIONAL ASPECTS

SECTION XIII NUMERICAL AND MATHEMATICAL ASPECTS

SECTION XIV FLUID MECHANICS

SECTION XV GROUNDWATER FLOW

EXTRA PAPER

PREFACE

The International Conference on Boundary Element Methods in
Engineering was started in 1978 with the following objectives:
i) To act as a focus for BE research at a time when the technique was just
emerging as a powerful tool for engineering analysis.
ii) To attract new as well as established researchers on Boundary Elements, in
order to maintain its vitality and originality.
iii) To try to relate the Boundary Element Method to other engineering
techniques in an effort to help unify the field of engineering analysis, rather
than to contribute to its fragmentation.

These objectives were achieved during the last 7 conferences and this meeting
– the eighth – has continued to be as innovative and dynamic as any of the
previous conferences.

Another important aim of the conference is to encourage the participation of
researchers from as many different countries as possible and in this regard it
is a policy of the organizers to hold the conference in different locations.

It is easy to forget when working on scientific projects that in science as well
as in other subjects, human relationships are as important as mathematical
equations. Science progresses not only as a result of laboratory and computer
experiments or abstract thinking but also by a process of personal interaction.
Furthermore if the advance of science were a simple linear process most
scientists would soon get tired of it. Fortunately, the pursuit of scientific
knowledge is a process as full of intuition, intelligent deduction and plain
guesses as any other human activity, including artistic endeavours.

The most direct and productive way of stimulating new ideas is to confront
scientists working on one problem with solutions produced by colleagues
working in a different field. For this reason it is important in these
conferences to bring together people working in different approaches, even
those which only marginally fall within the scope of the Boundary Element
Method.

The conference has changed over the years not only because more scientists are now using boundary elements but also as the range of applications of the method has expanded. This change is reflected in the contents of this meeting and in the emergence of some new topics of research. These changes are welcome, as a conference which is unable to develop will inevitably exhaust itself.

The success of this book is generated by the quality and originality of the papers presented at the conference. The success of the meeting itself depends on the degree to which the participants are prepared to listen to each other and it is indeed one of the most cherished characteristics of the Boundary Element Method Conference that participating scientists communicate and share their knowledge freely. Once we stop listening we will also stop learning and will negate the reason for being here.

In these days of international tension nobody seems to be prepared to understand another's arguments and the world is being fragmented into factions which have lost the capacity to learn from each other. It is the duty of scientists to resist this isolationism and to stress through our work that we are all part of the same community.

C.A. Brebbia
M. Tanaka
(Editors)

SECTION IX   HYDRODYNAMICS

A Hybrid Integral Equation Method for Evaluating Hydrodynamic Interaction Between Several Floating Bodies in Waves

T. Matsui
Department of Architecture, Nagoya University, Nagoya, Japan
K. Kato
Department of Architecture, Toyata College of Technology, Toyota, Japan

ABSTRACT

An efficient numerical technique is described for calculating the hydrodynamic interaction effects between adjacent floating bodies of arbitrary shape in waves. The hybrid integral equation method initially developed for a single body is extended for multiple bodies by introducing the concept of "multiple" fictitious vertical cylinders enclosing each body. The boundary element idealisations are thus used only in the localised regions close to the individual bodies, and these are coupled with the analytical solutions for the infinite region outside the fictitious cylinders. To demonstrate the validity and efficiency of the method, several numerical results are presented for two or three-body systems.

INTRODUCTION

If two or more bodies are floating in waves in close proximity to each other, the behaviour of the separate bodies will be influenced by the presence of the neighbouring bodies. The wave loads and responses will be different from those for a single body, due to the hydrodynamic interaction effects produced by the adjacent floating bodies. These interaction effects can be of particular relevance for multi-component offshore structures as well as for multi-body systems widely in use for offshore operations.

The analysis of the hydrodynamic interaction effects in the presence of closely-spaced bodies has been often performed using linear wave diffraction theory. Due to the complexity of the problem, only a few analytical solutions are available and are limited to problems with simple structural geometries. Spring and Monkmeyer [1] calculated wave forces on two neighbouring bottom-fixed vertical cylinders piercing the free sea surface by extending the eigen-series solution of MacCamy and Fuchs [2] for an isolated vertical cylinder. Ohkusu [3] extended the similar

approach to calculate wave forces and motion-induced hydrodynamic forces on groups of floating vertical cylinders. For
more complex structural geometries as widely used in offshore
fields, numerical solutions are available, e.g. using the 3-D
Green's function method [4], the finite element method [5] and
the coupled Green's function and finite element methods [6].
These numerical procedures, however, involve time-consuming
computations associated with the formation of influence coefficient matrices and equation solution for a large number of
unknowns. The computational efforts required will increase more
and more, as the number of the bodies increases. For this
reason, various efficient numerical procedures have been investigated for particular types of geometries. These include the
line integral equation Green's function methods of Isaacson [7]
for multiple vertical cylinders of arbitrary section and of
Matsui and Tamaki [8] for groups of vertical axisymmetric
bodies. The macro-element method of Kokkinnowrachos, Thanos and
Zibell [9] for groups of vertical bodies of revolution also
falls into this category.

This paper describes an efficient numerical technique for calculating the hydrodynamic interaction effects occuring between
several bodies of arbitrary shape floating in waves. The hybrid
integral equation method [10, 11] initially developed for a
single body is extended for multiple bodies by introducing the
concept of "multiple" fictitious vertical cylinders enclosing
each body. The boundary element idealisations to account for
arbitrary body geometries are thus needed to be used only in the
localised regions close to the individual bodies, and these are
coupled with the analytical eigen-series solutions for the infinite region outside the fictitious cylinders. The validity
and efficiency of the method are demonstrated by providing some
numerical results for the interaction effects between two or
three-body systems.

BASIC THEORY

Boundary value problem
M rigid bodies of arbitrary shape are considered to be floating
in waves in close proximity to each other, of which the typical
two are shown in Figure 1. The bodies are oscillating sinusoid-
ally about a mean position in response to excitation by a train
of regular plane waves. It is assumed that the amplitudes of
the motions of the bodies as well as of the waves are small, so
that the principle of linear superposition is valid. It is
further assumed that the fluid is invicid, incompressible and
irrotational, so that the fluid motion may be characterised by a
velocity potential function. The Cartesian coordinate system
oxyz is defined with the oxy plane in the mean free surface and
the oz axis measured positive vertically upwards.

The periodic motion of the body "b" oscillating in the k-th mode
may be expressed as

$$x_k^{(b)}(t)=\text{Re}[x_k^{(b)}e^{-i\omega t}], \qquad k=1,2,\ldots,6 \tag{1}$$

where ω is the frequency, t time, and $x_k^{(b)}$ is the complex amplitude of the motion.   The indices $k=1,2,\ldots,6$ designate   surge,
sway, heave, roll, pitch and yaw, respectively.

The irrotational fluid motion may be characterised by the velocity potential function which at the point (x,y,z) is written as

$$\Phi(x,y,z;t)=\text{Re}[\phi(x,y,z)e^{-i\omega t}] \tag{2}$$

where ϕ is the complex potential. This potential function
satisfies   the Laplace equation $\nabla^2\phi=0$ in the fluid,   the linearised free surface boundary condition, and the condition that
there is no flow through the body surface $S^{(b)}$ $(b=1,2,\ldots,M)$ and
the bottom $z=-h$.

The total potential ϕ may be separated into contributions from
the incident and scattered wave fields and from the waves radiated by the motions of all the M bodies:

$$\phi=\phi_I+\phi_S-i\omega\sum_{b=1}^{M}\sum_{k=1}^{6}\phi_k^{(b)}x_k^{(b)} \tag{3}$$

Here ϕ_I denotes the undisturbed incoming wave, ϕ_S the waves
scattered by the bodies when fixed held, and $\phi_k^{(b)}$ the waves radiated by unit-amplitude oscillation of the body "b" in the k-th
mode in otherwise still water.   Because of the linearisation
assumption,   each potential may be treated separately with the
appropriate boundary conditions on the body surface.

In a plane progressive wave with frequency ω, amplitude A and
incidence angle α relative to the ox axis, the incident potential is given by

$$\phi_I=-\frac{igA}{\omega}\frac{\cosh k(z+h)}{\cosh kh}e^{ik(x\cos\alpha+y\sin\alpha)} \tag{4}$$

where g is the acceleration of gravity, and k is the wave number
which is related to the frequency by the dispersion equation

$$\frac{\omega^2}{g}\equiv\nu=k\tanh kh \tag{5}$$

Because the incident potential is known, the hydrodynamic problem is reduced to determining the scattered potential ϕ_S and the
radiation potential $\phi_k^{(b)}$ which separately satisfy

$$\frac{\partial\phi_S}{\partial n^{(c)}}=-\frac{\partial\phi_I}{\partial n^{(c)}}, \qquad \frac{\partial\phi_k^{(b)}}{\partial n^{(c)}}=\delta_{bc}h_k^{(b)} \qquad \text{on } S^{(c)} \tag{6}$$

where $\partial/\partial n^{(c)}$ denotes the derivative with respect to the normal to the surface $S^{(c)}$ directed outward from the fluid, and δ_{bc} is the Kronecker delta. $h_k^{(b)}$ is written for each potential as

$$h_1^{(b)}=n_x^{(b)}, \qquad h_4^{(b)}=(y-y_G^{(b)})n_z^{(b)}-(z-z_G^{(b)})n_y^{(b)}$$

$$h_2^{(b)}=n_y^{(b)}, \qquad h_5^{(b)}=(z-z_G^{(b)})n_x^{(b)}-(x-x_G^{(b)})n_z^{(b)} \qquad (7)$$

$$h_3^{(b)}=n_z^{(b)}, \qquad h_6^{(b)}=(x-x_G^{(b)})n_y^{(b)}-(y-y_G^{(b)})n_x^{(b)}$$

where $n_x^{(b)}$, $n_y^{(b)}$ and $n_z^{(b)}$ denote the direction cosines of the outward normal to the surface $S^{(b)}$ in the x, y and z directions, respectively, and $(x_G^{(b)},y_G^{(b)},z_G^{(b)})$ denotes the coordinates of the point about which rotations of the body "b" are specified. Furthermore, ϕ_S and $\phi_k^{(b)}$ satisfy the radiation condition associated with outgoing waves at infinity.

Hydrodynamic loads and dynamic responses

Once the velocity potentials have been evaluated, the pressure at any point may be obtained from the linearised Bernoulli equation. The wave exciting force in the k-th mode on the body "b" is then obtained by integrating the pressure over the immersed surface of the body when fixed held in waves:

$$F_k^{(b)}(t)=Re[f_k^{(b)}e^{-i\omega t}]=Re\left[i\rho\omega e^{-i\omega t}\iint_{S(b)}(\phi_I+\phi_S)h_k^{(b)}dS\right] \qquad (8)$$

where ρ is the fluid density. The hydrodynamic force in the k-th mode on the body "b" due to the motion in the j-th mode of the body "c" may be described in terms of the added mass and damping coefficients:

$$F_{kj}^{(b)(c)}(t)=Re[(\omega^2 M_{kj}^{(b)(c)}+i\omega N_{kj}^{(b)(c)})x_j^{(c)}e^{-i\omega t}] \qquad (9)$$

where $M_{kj}^{(b)(c)}$ and $N_{kj}^{(b)(c)}$ designate the added mass and damping coefficients, respectively, given by

$$M_{kj}^{(b)(c)}+\frac{i}{\omega}N_{kj}^{(b)(c)}=\rho\iint_{S(b)}\phi_j^{(c)}h_k^{(b)}dS \qquad (10)$$

These coefficients satisfy the symmetry relations

$$M_{kj}^{(b)(c)}=M_{jk}^{(c)(b)}, \qquad N_{kj}^{(b)(c)}=N_{jk}^{(c)(b)} \qquad (11)$$

The motion of a freely-floating body "b" as depicted in Figure 1 may be described by Newton's law:

$$\sum_{j=1}^{6} m_{kj}^{(b)} \frac{d^2 x_j^{(b)}}{dt^2} = F^{(b)T}(t), \qquad k=1,2,\ldots,6 \qquad (12)$$

where $m_{kj}^{(b)}$ is the inertia matrix of the body "b", and $F_{kj}^{(b)T}$ is the total fluid force acting on the body "b", given by

$$F_k^{(b)T}(t) = F_k^{(b)}(t) + \sum_{c=1}^{M} \sum_{j=1}^{6} F_{kj}^{(b)(c)}(t) - \sum_{j=1}^{6} K_{kj}^{(b)} x_j^{(c)}(t) \qquad (13)$$

where $k_{kj}^{(b)}$ is the hydrostatic stiffness matrix of the body "b". Using Equations (1), (8), (9), (12) and (13), the complete equations of motion may be written as follows:

$$\sum_{c=1}^{M} \sum_{j=1}^{6} [-\omega^2 (m_{kj}^{(b)} \delta_{bc} + M_{kj}^{(b)(c)}) - i\omega N_{kj}^{(b)(c)} + K_{kj}^{(b)} \delta_{bc}] x_j^{(c)} = f_k^{(b)} \qquad (14)$$

from which the motions of the bodies are determined.

INTEGRAL EQUATION

Multiple fictitious vertical cylinders

Referring to Figure 1, the fluid is divided into a number of regions by introducing M fictitious vertical cylinders $S_R^{(b)}$ with radius R_b (b=1,2,...,M), which are just large enough to surround the body "b" but do not intersect with each other. The region inside $S_R^{(b)}$ is designated by $V^{(b)}$ with potential ϕ^b, and the remaining outside region by V* with potential ϕ*. (The subscripts and superscripts distinguishing the scattered potential and the radiation potentials are deleted hereafter.) Moreover, M local cylindrical coordinate systems (r_b, θ_b, z) (b=1,2,...,M) are defined with the origin on the axis of symmetry of $S_R^{(b)}$.

ϕ* may be represented as a superposition of the waves scattered or radiated from all the bodies, which are expressed by eigenfunction expansion in the cylindrical coordinates:

$$\phi* = \sum_{c=1}^{M} \sum_{m=0}^{\infty} \sum_{n=0}^{\infty} (\alpha_{mn}^c \cos n\theta_c + \beta_{mn}^c \sin n\theta_c) R_{mn}^c(r_c) Z_m(z) \qquad (15)$$

where

$$R_{mn}^c(r_c) = \begin{cases} H_n^{(1)}(kr_c)/H_n^{(1)}(kR_c) & \text{for } m=0 \\ K_n(\kappa_m r_c)/K_n(\kappa_m R_c) & \text{for } m \geq 1 \end{cases} \qquad (16)$$

$$Z_m(z) = \begin{cases} \cosh k(z+h)/\cosh kh & \text{for } m=0 \\ \cos \kappa_m(z+h)/\cos \kappa_m h & \text{for } m \geq 1 \end{cases} \qquad (17)$$

In these expressions, $H_n^{(1)}$ is Hankel function of the first kind of order n, K_n is the modified Bessel function of the second kind of order n, and κ_m is the positive real roots of the equation

$$\kappa_m \tan \kappa_m h + \nu = 0 \qquad (18)$$

$\phi*$ satisfies the free surface boundary condition, the bottom condition and the radiation condition at infinity. Furthermore, it must satisfy the following matching conditions:

$$\phi^b = \phi*, \quad \frac{\partial \phi^b}{\partial r_b} = \frac{\partial \phi*}{\partial r_b} \quad \text{on } S_R^{(b)} \qquad (19)$$

The expression for $\phi*$ given by Equation (15) is difficult to handle since each scattered or radiated wave is expressed in terms of different coordinates. In order to satisfy the matching conditions, Equation (19), it is convenient to express $\phi*$ in terms of one selected coordinates, say (r_b, θ_b, z). For this purpose, the Bessel "addition theorem" may be applied to give the relationships (e.g. Watson [12], pp.359-361)

$$\begin{Bmatrix} H_n^{(1)}(kr_c) \\ K_n(\kappa_m r_c) \end{Bmatrix} \begin{Bmatrix} \cos \\ \sin \end{Bmatrix} n\theta_c = \sum_{\ell=-\infty}^{\infty} (-1)^n \begin{Bmatrix} H_{n+\ell}^{(1)}(kL_{bc})J_\ell(kr_b) \\ K_{n+\ell}(\kappa_m L_{bc})I_\ell(\kappa_m r_b) \end{Bmatrix} \cdot$$

$$\cdot \left[\begin{Bmatrix} \cos \\ \sin \end{Bmatrix}(n+\ell)\Theta_{bc} \cos \ell\theta_b \begin{Bmatrix} + \\ - \end{Bmatrix} \begin{Bmatrix} \sin \\ \cos \end{Bmatrix}(n+\ell)\Theta_{bc} \sin \ell\theta_b \right] \qquad (20)$$

which is valid for $r_b < L_{bc}$. In the above expression, J_ℓ is Bessel function of the first kind of order ℓ, and I_ℓ is the modified Bessel function of the first kind of order ℓ. Using Equation (20), Equation (15) may be expressed in terms of the coordinates associated with $S_R^{(b)}$ as follows:

$$\phi* = \sum_{m=0}^{\infty} \sum_{n=0}^{\infty} (\alpha_{mn}^b \cos n\theta_b + \beta_{mn}^b \sin n\theta_b) R_{mn}^b(r_b) Z_m(z)$$

$$+ \sum_{c=1}^{M} (1-\delta_{bc}) \sum_{m=0}^{\infty} \sum_{n=0}^{\infty} \sum_{\ell=-\infty}^{\infty} \{ [\alpha_{mn}^c \cos(n+\ell)\Theta_{bc} + \beta_{mn}^c \sin(n+\ell)\Theta_{bc}] \cos \ell\theta_b$$

$$+ [\alpha_{mn}^c \sin(n+\ell)\Theta_{bc} - \beta_{mn}^c \cos(n+\ell)\Theta_{bc}] \sin \ell\theta_b \} S_{mn\ell}^{cb}(r_b) Z_m(z) \qquad (21)$$

where

$$S_{mn\ell}^{cb}(r_b) = \begin{cases} (-1)^n H_{n+\ell}^{(1)}(kL_{bc})J_\ell(kr_b)/H_n^{(1)}(kR_c) & \text{for } m=0 \\ (-1)^n K_{n+\ell}(\kappa_m L_{bc})I_\ell(\kappa_m r_b)/K_n(\kappa_m R_c) & \text{for } m \geq 1 \end{cases} \qquad (22)$$

Green's second identity

Application of Green's second identity to ϕ^b in $V^{(b)}$ and the fundamental solution, $1/R_{PQ}$, and making use of the free surface boundary condition, the bottom condition, the body surface condition (6) and the matching condition (19) leads to the integral equation

$$
\begin{aligned}
2\pi\phi^b(P) &+ \iint_{S(b)} \phi^b(Q)\, \frac{\partial}{\partial n_Q^{(b)}}\left(\frac{1}{R_{PQ}}\right) dS(Q) \\
&+ \iint_{S_F(b)} \phi^b(Q)\left[\frac{\partial}{\partial n_Q^{(b)}}\left(\frac{1}{R_{PQ}}\right) - \phi\left(\frac{1}{R_{PQ}}\right)\right] dS(Q) \\
&+ \iint_{S_B(b)} \phi^b(Q)\, \frac{\partial}{\partial n_Q^{(b)}}\left(\frac{1}{R_{PQ}}\right) dS(Q) \\
&+ \iint_{S_R(b)} \phi^b(Q)\left[\phi*(Q)\, \frac{\partial}{\partial n_Q^{(b)}}\left(\frac{1}{R_{PQ}}\right) - \frac{\partial\phi*}{\partial n^{(b)}}(Q)\left(\frac{1}{R_{PQ}}\right)\right] dS(Q) \\
&= \begin{cases} \displaystyle\iint_{S(b)} \delta_{bc} h_k^{(c)}(Q)\left(\frac{1}{R_{PQ}}\right) & \text{for } \phi_k^{(c)} \\[2ex] \displaystyle\iint_{S(b)} -\frac{\partial\phi_I}{\partial n^{(b)}}(Q)\left(\frac{1}{R_{PQ}}\right) & \text{for } \phi_S \end{cases}
\end{aligned}
\tag{23}
$$

where P designates a reference point on the boundary surface $\partial V^{(b)}$ which surrounds $V^{(b)}$, Q the integration point on $\partial V^{(b)}$, $dS(Q)$ is a differential area on $\partial V^{(b)}$, R_{PQ} denotes the distance between P and Q, and $S_F^{(b)}$ and $S_B^{(b)}$ denote the portions of the free surface and the bottom bounded by $S_R^{(b)}$, respectively. The subscript Q on $n^{(b)}$ designates that the normal derivative is to be evaluated at the point Q. The similar integral equation is obtained for the flow inside each fictitious cylinder ($b=1,2,\ldots,M$), thus constituting the complete set of integral equations which is sufficient to determine the potentials.

NUMERICAL SOLUTION

The set of integral equations such as given by Equation (23) may be solved numerically by dividing the boundary surfaces $S^{(b)}$, $S_F^{(b)}$ and $S_B^{(b)}$ into N_1, N_2 and N_3 small facets, ΔS_j, respectively, over which the potentials are assumed to be constant. If the integral equation is satisfied at one control point on each facet, say at the centroid, and truncating the double series of eigen-functions in Equation (15) at $m=\tilde{m}$ and $n=\tilde{n}$, Equation (23) is replaced by N^b algebraic equations of the form

$$\sum_{j=1}^{N_1} A_{ij}\phi^b(P_j) + \sum_{j=N_1+1}^{N_{12}} (A_{ij} - \nu B_{ij})\phi^b(P_j) + \sum_{j=N_{12}+1}^{N^b} A_{ij}\phi^b(P_j)$$

$$+ \sum_{c=1}^{M} \sum_{m=0}^{\tilde{m}} \sum_{n=0}^{\tilde{n}} (F_{imn}^{bc}\alpha_{mn}^c + G_{imn}^{bc}\beta_{mn}^c)$$

$$= \begin{cases} \sum_{j=1}^{N_1} B_{ij}\delta_{bc}h_k^{(c)}(P_j) & \text{for } \phi_k^{(c)} \\ \\ \sum_{j=1}^{N_1} -B_{ij}\frac{\partial\phi_I}{\partial n^{(b)}}(P_j) & \text{for } \phi_S \end{cases} \tag{24}$$

where $N_{12} = N_1 + N_2$ and $N^b = N_{12} + N_2$. The influence coefficients A_{ij}, B_{ij}, F_{imn}^{bc} and G_{imn}^{bc} in Equation (24) are given by

$$A_{ij} = \iint_{\Delta S_j} \frac{\partial}{\partial n_Q^{(b)}}\left\{\frac{1}{R_{P_iQ}}\right\}dS + 2\pi\delta_{ij} \tag{25}$$

$$B_{ij} = \iint_{\Delta S_j} \left\{\frac{1}{R_{P_iQ}}\right\}dS \tag{26}$$

$$\begin{pmatrix} F_{imn}^{bb} \\ G_{imn}^{bb} \end{pmatrix} = P_{imn}^b(r_{bi}, z_i)\begin{pmatrix} \cos \\ \sin \end{pmatrix}n\theta_{bi} \tag{27}$$

$$\begin{pmatrix} F_{imn}^{bc} \\ G_{imn}^{bc} \end{pmatrix} = \sum_{\ell=-\tilde{n}}^{\tilde{n}} Q_{imn\ell}^{bc}(r_{bi}, z_i)\left[\begin{pmatrix} \cos \\ \sin \end{pmatrix}(n+\ell)\Theta_{bc}\cos\ell\theta_{bi}\right.$$
$$\left.\begin{pmatrix} + \\ - \end{pmatrix}\begin{pmatrix} \sin \\ \cos \end{pmatrix}(n+\ell)\Theta_{bc}\sin\ell\theta_{bi}\right], \quad b \neq c \tag{28}$$

where

$$P_{imn}^b(r_{bi}, z_i) = \int_{-h}^{0}\left[\frac{\partial\rho^n}{\partial r_b}(r_{bi}, z_i; R_b, z)\right.$$
$$\left. - \frac{dR_{mn}^b}{dr_b}(R_b)\rho^n(r_{bi}, z_i; R_b, z)\right]R_b Z_m(z)dz$$
$$+\begin{cases} 0 & (1 \leq i \leq N^b) \\ 2\pi Z_m(z_i) & (N^b + 1 \leq i \leq N^b + N_R^b) \end{cases} \tag{29}$$

$$Q_{imn\ell}^{bc}(r_{bi}, z_i) = \int_{-h}^{0}\left[S_{mn\ell}^{cb}(R_b)\frac{\partial\rho^\ell}{\partial r_b}(r_{bi}, z_i; R_b, z)\right.$$
$$\left. - \frac{ds_{mn\ell}^{cb}}{dr_b}(R_b)\rho^\ell(r_{bi}, z_i; R_b, z)\right]R_b Z_m(z)dz$$

$$
+\begin{cases} 0 & (1\leq i\leq N^b) \\ 2\pi S_{mn\ell}^{cb}(R_b)Z_m(z_i) & (N^b+1\leq i\leq N^b+N_R^b) \end{cases} \tag{30}
$$

In these expressions, P_i and Q denote the control point on ΔS_i and the integration point on $\partial V^{(b)}$, respectively, whose coordinates are defined by (r_{bi},θ_{bi},z_i) and (r_b,θ_b,z) in terms of the cylindrical coordinates associated with $S_R^{(b)}$. In the derivation of the line integral expressions of Equations (29) and (30), use has been made of the Fourier series expansion of the fundamental solution:

$$
\frac{1}{R_{P_iQ}} = \frac{1}{2\pi}\sum_{n=-\infty}^{\infty}\rho^n(r_{bi},z_i;r_b,z)\cos n(\theta_{bi}-\theta_b) \tag{31}
$$

where

$$
\rho^n(r_{bi},z_i;r_b,z)=\frac{2}{\sqrt{r_{bi}r_b}}Q_{n-\frac{1}{2}}\left(\frac{r_{bi}^2+r_b^2+(z_i-z)^2}{2r_{bi}r_b}\right) \tag{32}
$$

$Q_{n-1/2}$ denoting the Legendre function of the second kind. The details of these evaluations are described by Matsui, Kato and Shirai [11].

The similar set of algebraic equations is obtained for the flow inside each fictitious cylinder ($b=1,2,\ldots,M$). Unknown variables involved in these equations are $N^1+N^2+\ldots+N^M$ potentials and $M(\tilde{m}+1)(2\tilde{n}+1)$ coefficients α_{mn}^b, β_{mn}^b. In order that the number of equations becomes equal to the number of unknowns, $N_R^b=$ $=(\tilde{m}+1)(2\tilde{n}+1)$ additional control points must be placed on each fictitious cylinder $S_R^{(b)}$. Solution of these algebraic equations leads to the numerical approximations to the scattered and radiation potentials.

EXAMPLES

A number of numerical examples have been studied to illustrate the applicability of the numerical procedure described in the foregoing, and to investigate the interaction effects between neighbouring floating bodies. The results are based on truncation of the double series of eigenfunctions for the flow outside each fictitious cylinder at $n=5$ and $m=9$, and neglect of the interaction effects between standing waves, which means that $S_{mn\ell}^{cb}=0$ for $m\geq 1$. Numerical integrations associated with the evaluation of the influence coefficient matrices A_{ij} and B_{ij} have been carried out using 16 points Gaussian quadrature for

each quadrilateral facet; while the line integrals in the expressions for the coefficients F_{imn}^{bc} and G_{imn}^{bc} have been evaluated
by dividing the water depth into 10 subsegments of equal size,
for each of which 4 points Gaussian quadrature has been applied.

Two fixed vertical cylinders

The first example studied is two bottom-fixed vertical cylinders
piercing the free sea surface. This case has been studied
analytically by many other investigators, including Spring and
Monkmeyer [1] and Ohkusu [3], and serves as a useful test case
for the numerical technique described herein. The cylinders
analysed here have equal depth to radius ratios of h/a=4 and
alined in the incident wave direction with axes five radii
apart. In view of the double symmetry of the body geometries,
only one quadrant of each cylinder has been analysed adopting
the boundary element idealisations shown in Table 1. Results
for the horizontal wave forces are presented in Figure 2 and
compared with the analytical solutions. Agreement between the
analytical and numerical results is seen to be excellent over
the frequency range examined.

Three floating vertical cylinders

The second example illustrated is three vertical cylinder configurations floating at the sea surface, for which analytical
results have been presented by Ohkusu [3]. The analysis was
based on the boundary element idealisations shown in Table 1,
and the double symmetry of the body geometries was again exploited. Results for the horizontal wave forces are presented
in Figure 3 and compared with the corresponding results obtained by the line integral equation method for vertical bodies
of revolution [8]. Agreement is again seen to be close for the
frequency range studied.

Floating cylinder and box

As illustration of use of the present method for groups of
bodies of arbitrary three-dimensional shape, a two-body system
composed of floating vertical cylinder and rectangular box, as
depicted in Figure 4, was selected. In view of the double
symmetry of each body configuration, only one quadrant of the
cylinder and the box has been analysed using the boundary
element idealisations shown in Table 1. Only selected results
are presented in Figures 5 to 8 for the motion-induced hydrodynamic interaction forces and the dynamic responses. These
results are compared with the corresponding results obtained by
the 3-D Green's function method, where the same element idealisations were adopted for the body surfaces. Agreement between
the two numerical methods is observed to be satisfactory over
the frequency range of interest.

The central processing (CPU) times for the hybrid integral
equation method as well as for the line integral equation method
and the 3-D Green's function method are listed in Table 2.

Here T_0 represents the time taken for the computations independent of frequency, whereas T_F the time per frequency for the frequency dependent computations. These CPU times were obtained on the FACOM M-382 computer at Nagoya University Computation Center.

CONCLUSION

The hybrid integral equation method initially developed for a single body has been extended to calculate the hydrodynamic interaction effects in the presence of neighbouring floating bodies in waves. The validity and efficiency of the method have been confirmed by comparing the results from the present method with those from other analytical and numerical approaches. The present hybrid integral equation method is based on the use of multiple fictitious vertical cylinders enclosing each body, which minimises the size of the boundary element regions to account for arbitrary body geometries. The computational effort required was found to be equivalent to that for the line integral equation method for vertical bodies of revolution and be much less than that for the 3-D Green's function method. The method will provide an effective means to evaluate the hydrodynamic interaction effects between multi-component bodies frequently in use offshore.

REFERENCES

1. Spring B.H. and Monkmeyer P.L. (1974), Interaction of Plane Waves with Vertical Cylinders, Proc. 14th Int'l Coast. Engng. Conf., Copenhagen, Vol.III, pp.1828-1847.
2. MacCamy R.C. and Fuchs R.A. (1954), Wave Forces on Piles: A Diffraction Theory, Beach Erosion Board Tech. Memo. No.69.
3. Ohkusu M. (1974), Hydrodynamic Forces on Multiple Cylinders in Waves, Proc. Int'l Symp. Dyn. Marine Vehicles and Struct. in Waves, London, pp.107-112.
4. Oortmerssen G. van (1979), Hydrodynamic Interaction between Two Structures Floating in Waves, Proc. 2nd Int'l Conf. Behav. Offsh. Struct., London, Vol.1, pp.339-356.
5. Huang M.C., Hudspeth, R.T. and Leonard, J.W. (1985), FEM Solution of 3-D Wave Interference Problems, J. Wat. Port Coast. and Ocean Engng., Proc. A.S.C.E., Vol.111, No.4, pp.661-677.
6. Eatock Taylor R. and Zietsman J. (1982), Hydrodynamic Loading on Multi-Component Bodies, Proc. 3rd Int'l Conf. Behav. Offsh. Struct., Massachusetts, Vol.1, pp.424-443.
7. Isaacson M. de St.Q. (1978), Vertical Cylinders of Arbitrary Section in Waves, J. Wat. Port Coast. and Ocean Div., Proc. A.S.C.E., Vol.104, No.WW4, pp.309-324.
8. Matsui T. and Tamaki T. (1981), Hydrodynamic Interaction between Groups of Vertical Axisymmetric Bodies Floating in Waves, Proc. Int'l Symp. Hydrodyn. Ocean Engng., Trondheim, Vol.2, pp.817-836.

9. Kokkinnowrachos K. Thanos I. and Zibell H.G. (1986), Hydro-
    dynamic Interaction between Several Vertical Bodies of Revo-
    lution in Waves, Proc. 5th Int'l Offsh. Mech. and Arc.
    Engng. Symp., Tokyo, Vol.1, pp.194-205.
10. Matsui T., Kato K. and Shirai T. (1985), A Hybrid Boundary
    Element Method for Ocean Wave Diffraction and Radiation
    Problems, in Boundary Elements VII (Ed. Brebbia C.A. and
    Maier G.), Vol.I, pp.5/37-5/50, Proc. 7th Int'l Conf.
    B. E. M. in Engng., Villa Olmo, Lake Como, Italy, 1985,
    Springer-Verlag, Berlin Heidelberg New York Tokyo.
11. Matsui T., Kato K. and Shirai T. (1986), A Hybrid Integral
    Equation Method for Diffraction and Radiation of Water Waves
    by Three-Dimensional Bodies, Computational Mechanics, to
    appear.
12. Watson G.N. (1966), A Treatise on the Theory of Bessel
    Functions, 2nd Edn., Cambridge Univ. Press, Cambridge.

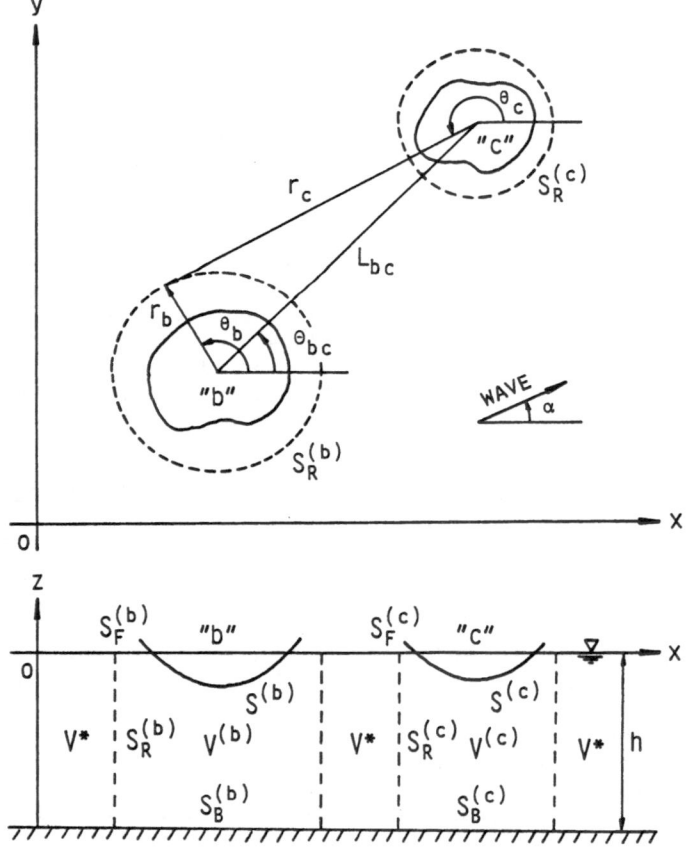

Figure 1   Coordinates for multi-body system

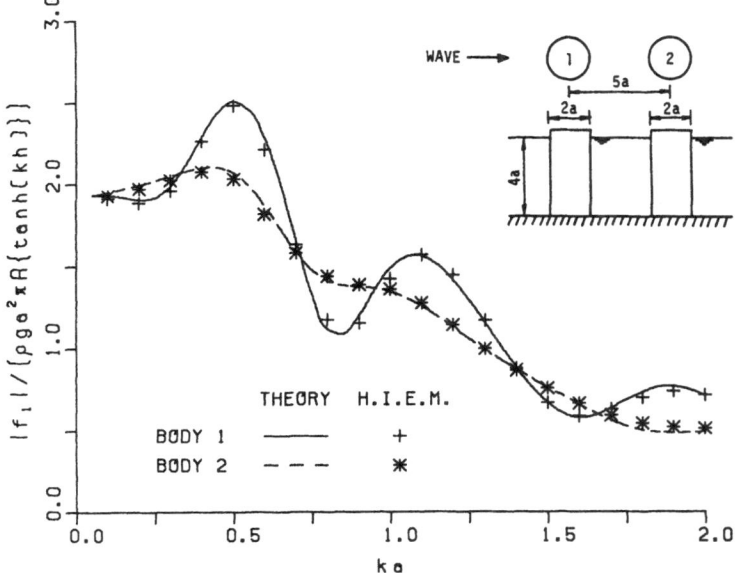

Figure 2   Horizontal wave forces on the two fixed vertical
cylinders

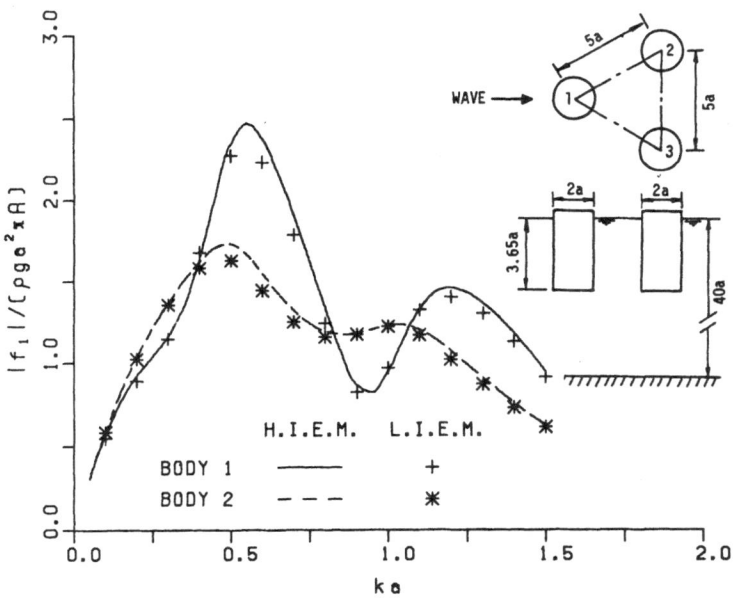

Figure 3   Horizontal wave forces on the three floating vertical
cylinders

Table 1   Boundary element idealisations for quarter of the body

Configuration	Body	R	N_1	N_2	N_3	N_R	N_{Total}
2 fixed cylinders	Each cylinder	1.2a	60	6	6	30	102
3 floating cylinders	Each cylinder	1.2a	38	4	2	30	74
Floating cylinder and box	Cylinder	60 m	52	8	4	30	94
	Box	85 m	55	30	5	30	120

Table 2   Central processing times (sec)

Configuration	Method	T_0	T_F
2 fixed cylinders	H.I.E.M.	4.4	7.5
3 floating cylinders	L.I.E.M.	0.02	5.3
	H.I.E.M.	2.2	10.4
Floating cylinder and box	G.F.M.	0.06	37.8
	H.I.E.M.	9.7	8.2

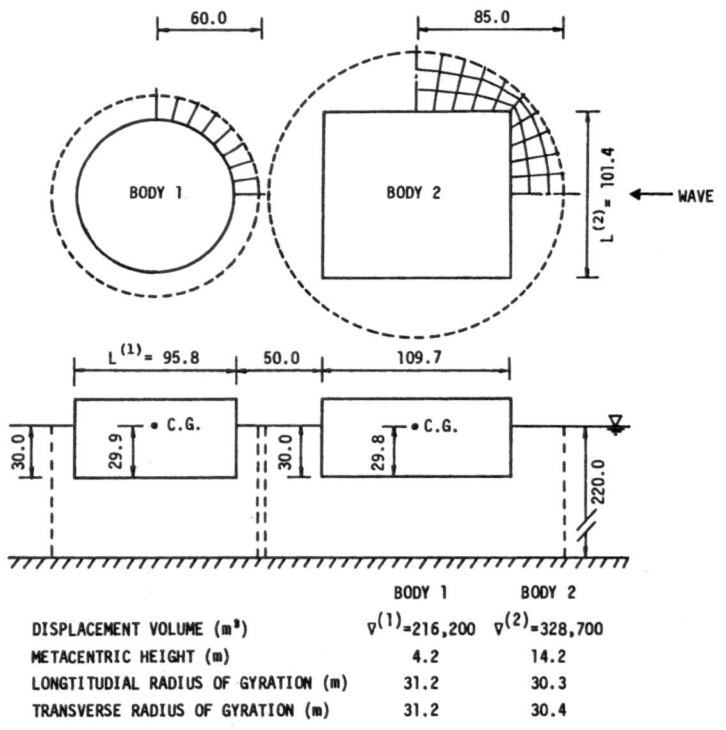

	BODY 1	BODY 2
DISPLACEMENT VOLUME (m³)	$V^{(1)}$=216,200	$V^{(2)}$=328,700
METACENTRIC HEIGHT (m)	4.2	14.2
LONGTITUDIAL RADIUS OF GYRATION (m)	31.2	30.3
TRANSVERSE RADIUS OF GYRATION (m)	31.2	30.4

Figure 4   Dimensions and mesh subdivision on the free surface
for the floating cylinder and box (in metre)

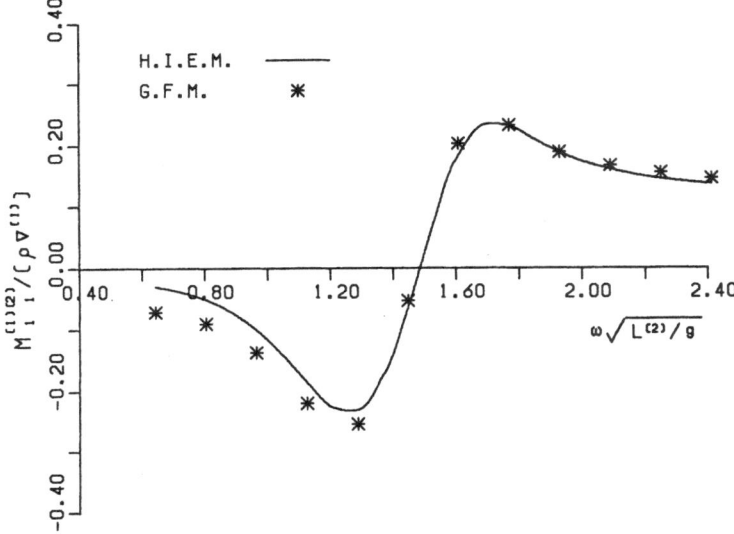

Figure 5   Hydrodynamic interaction force (added mass) on the cylinder due to the motion of the box

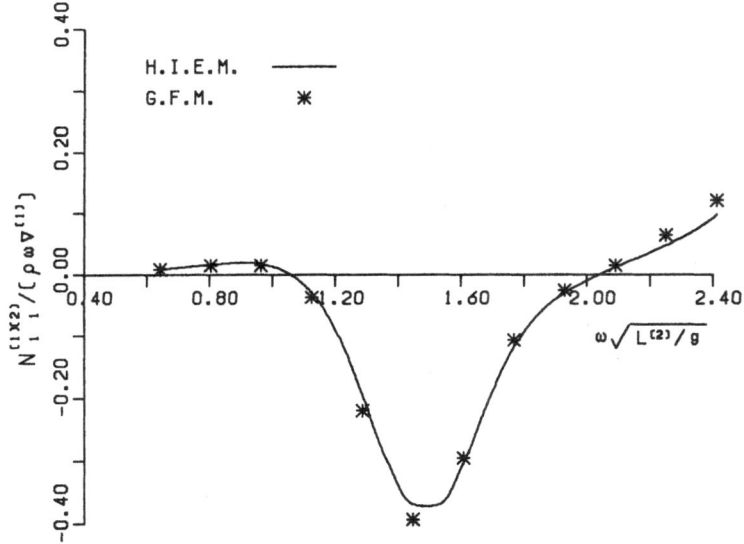

Figure 6   Hydrodynamic interaction force (added damping) on the cylinder due to the motion of the box

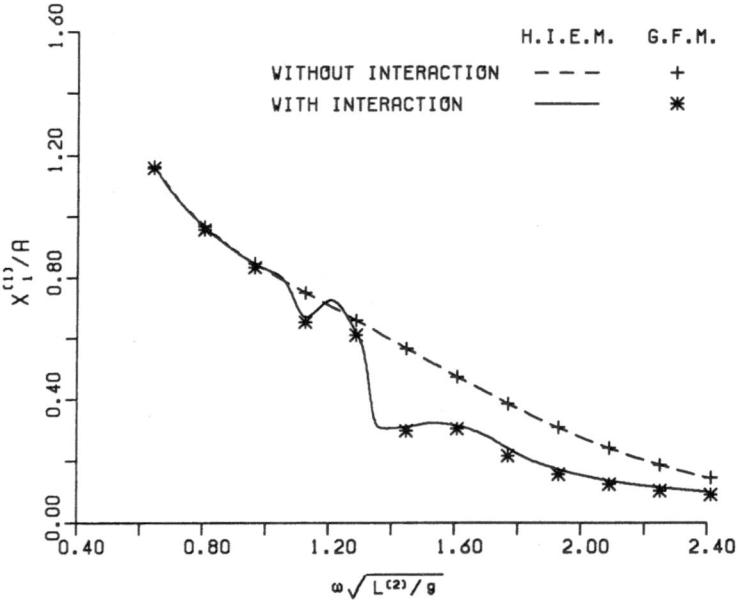

Figure 7   Surge amplitude ratio of the cylinder

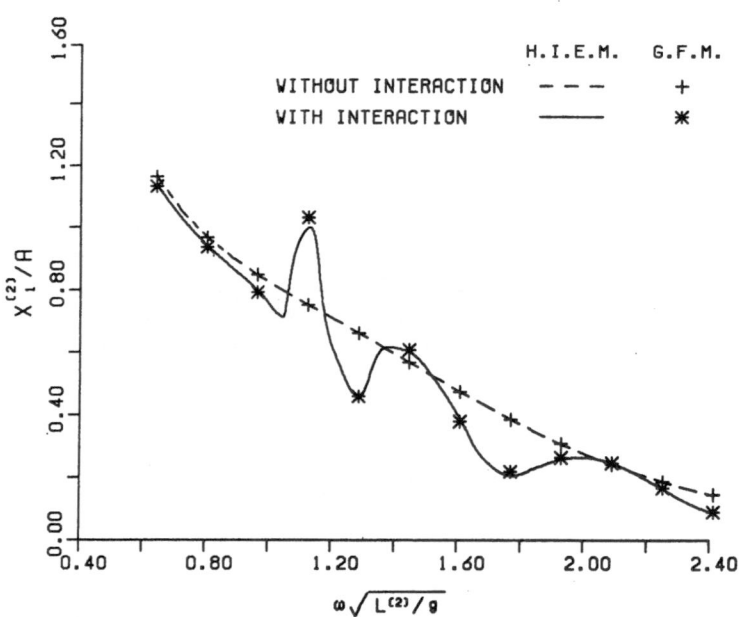

Figure 8   Surge amplitude ratio of the box

Analysis of Structure – Fluid Dynamic Interaction Problems by Boundary Integral Equation Methods

T. Kawakami, M. Kitahara
Department of Marine Civil Engineering, Tokai University, 3-20-1 Orido, Shimizu, Shizuoka, Japan

ABSTRACT

The structure-fluid interaction problems are studied by the Boundary Integral Equation (BIE) method. The structure is assumed to be linear elastic body and the fluid is assumed to be incompressible inviscid fluid. The integral equations are formulated for the elastic body and the fluid domain, respectively, by means of Green's formula. The systems of boundary integral equations can be obtained by considering the interface conditions on the common boundary of the elastic body and the fluid.

In the numerical examples, the characteristics of deformations and stresses of the elastic structure are discussed.

INTRODUCTION

The boundary integral equation (BIE) method has been applied to the analysis of various types of structure - fluid interaction problems in water waves such as wave scatterings, wave forces, and the motion of floating structures. In most of these studies, the structures have been assumed to be rigid body (F. John(1950), Ijima & Chou(1976), Kiyokawa & Ohyama (1984)). On the other hand, there are few studies on the deformable elastic body and fluid interaction problems. The ·summary and useful methodology in this field are given by Zienkiewicz & Bettess(1978). The eigenvalue problems for the elastic body and fluid interaction problems have been analysed by Kakuda & Tosaka(1983).

In this paper, we apply the BIE method to the

"elastic" structure-fluid dynamic interaction problems. The structure is assumed to be homogeneous, isotropic and linear elastic body. The fluid is assumed to be incompressible inviscid fluid and the waves are assumed to be small amplitude wave. The integral equations are formulated by Green's formula on the boundary surfaces of the elastic and fluid domains, respectively. Then, the problem is reduced to the coupling problem of the elastic body and the fluid in terms of the displacement vector of the elastic body and the velocity potential of the fluid. The interface conditions on the common boundary are expressed as the equilibrium condition of the normal component of the surface traction and the continuity condition of the normal component of the velocity.

After the verification of the accuracy of numerical results for wave scattering problems, we show the characteristics of deformations and stresses in the tow dimensional elastic structure.

GOVERNING EQUATIONS AND BOUNDARY CONDITIONS

Governing equations in the elastic domain

The fundamental equations in the homogeneous, isotropic, and linear elastic body are expressed as follows:
- equation of motion:

$$\nabla \cdot \bar{\tau} + \rho \bar{b} = \rho \ddot{\bar{u}}, \tag{1}$$

- constitutive equation:

$$\bar{\tau} = \lambda (tr\bar{\varepsilon}) 1 + 2\mu\bar{\varepsilon} , \tag{2}$$

- strain-displacement relation:

$$\bar{\varepsilon} = 1/2 (\nabla\bar{u} + \bar{u}\nabla), \tag{3}$$

where $\bar{\tau}$ and $\bar{\varepsilon}$ are stress and strain tensors; $\bar{u}$, $\bar{b}$ and $\ddot{\bar{u}}$ are displacement, body force and acceleration vectors; λ , μ and ρ are Lamé constants and the mass density; ∇ and 1 denote the gradient and the unit tensor; $tr\varepsilon$ means the trace of ε . If eqs. (2) and (3) substituted into eq. (1), we have the following governing equation

$$\mu\Delta\bar{u} + (\lambda + \mu)\nabla\nabla \cdot \bar{u} = \rho\ddot{\bar{u}} \tag{4}$$

in the absence of the body force, where Δ denotes the Laplacian. Here we consider the case that the displacement vector $\bar{u}(x,t)$ is harmonic in time with an angular frequency ω , i.e.,

$$\bar{u}(x,t)=\text{Re}\{u(x)e^{-i\omega t}\}, \tag{5}$$

where Re means the real part of the $u(x)^{-i\omega t}$. In this steady elastodynamic state, the governing equation (4) reduces to

$$Lu \equiv \mu \Delta u + (\lambda + \mu) \nabla \nabla \cdot u + \rho \omega^2 u = 0 , \tag{6}$$

where the time factor $\exp(-i\omega t)$ and the symbol "Re" have been suppressed.

Governing equations in the fluid domain
We confine ourselves to the case of the incompressible inviscid fluid. Furthermore, we assume that the flow is irrotational. Then, the velocity vector $\bar{V}(x,t)$ has to satisfy the following equations:

$$\nabla \cdot \bar{V}(x,t)=0, \tag{7}$$

$$\nabla \times \bar{V}(x,t)=0. \tag{8}$$

Eq. (8) is satisfied if the velocity is derived from the potential $\bar{\Phi}(x,t)$ by

$$\bar{V}(x,t)=\nabla\bar{\Phi}(x,t). \tag{9}$$

From eqs. (7) and (9), the velocity potential $\bar{\Phi}(x,t)$ satisfies

$$\Delta\bar{\Phi}(x,t)=0. \tag{10}$$

In the time-harmonic case such that

$$\bar{\Phi}(x,t)=\text{Re}\{\Phi(x)e^{-i\omega t}\} , \tag{11}$$

eq. (10) reduces to

$$\Delta\Phi(x)=0. \tag{12}$$

Boundary conditions in the elastic domain
Let D be an elastic body enclosed by its boundary $\partial D_u \cup \partial D_t$. The boundary conditions for this domain are given in the following form:

$$u(x)=f(x) \qquad x \in \partial D_u , \tag{13}$$

$$t(x)=\overset{n}{T} u(x)$$

$$=\lambda n\{\nabla \cdot u(x)\}+\mu n \cdot [\nabla u(x)+\{\nabla u(x)\}^T] =g(x) \quad x \in \partial D_t, \tag{14}$$

where n is the unit outward normal vector on the boundary and t is the traction vector.

Boundary conditions in the fluid domain

In the fluid domain, the condition for the boundary which has no flow through the surface is given by

$$\frac{\partial \phi(\mathbf{x})}{\partial n} = 0 \quad . \tag{15}$$

Also the boundary condition on the free surface is expressed as

$$\frac{\partial \phi(\mathbf{x})}{\partial x_2} = \frac{\omega^2}{g} \phi(\mathbf{x}) \quad . \tag{16}$$

COUPLING CONDITIONS FOR THE ELASTIC AND FLUID REGIONS

Now we consider the elastic body which is interacting with the fluid motion, as shown in Fig. 1. In this figure, the elastic body has been denoted as the region IV. The boundary ∂D_t is the traction free and the boundary ∂D_c is the common boundary with the fluid. The fluid region is separated into three regions, I, II, and III. In the regions I and III which have an infinite extent, the water depth is assumed to be constant. The region II can be considered as the region where the scattering effect is predominant. The boundary conditions on the free surface ∂D_s and the bottom ∂D_r have been given in eqs. (16) and (15), respectively. On the artificial boundaries $\partial D_i^{\pm}$, of course, suitable continuity conditions must be satisfied.

Interface conditions on the common boundary

On the common boundary ∂D_c of the elastic body and the fluid, the interface conditions are given as

$$\mathbf{t} = -n p = -n i \bar{\rho} \omega \phi \quad , \tag{17}$$

$$-i \omega \mathbf{u} \cdot \mathbf{n} = \mathbf{V} \cdot \mathbf{n} = (\nabla \phi) \cdot \mathbf{n} \quad , \tag{18}$$

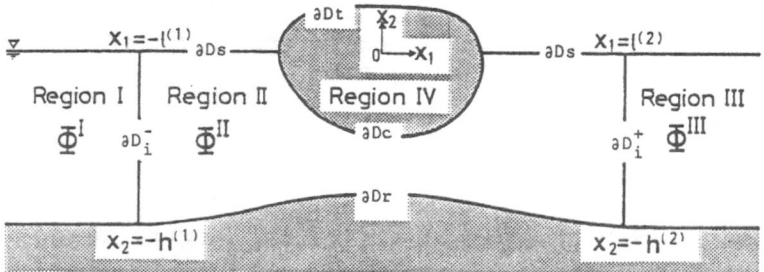

Fig.1 The elastic body (IV) in the fluid regions (I,II,and III)

where P is the fluid pressure and $\bar{\rho}$ is the fluid density. Eq. (17) is the equilibrium condition of the surface traction and eq. (18) is the continuity condition of the normal component of the velocity.

Continuity conditions on the artificial boundary

Since the water depth in the regions I and III is constant and the regions extend to infinity, we can easily obtain the solution which satisfies the free surface (eq. (16)) and bottom (eq. (15)) conditions in these regions. If we denote the velocity potentials in the regions I and III as ϕ^{I} and ϕ^{III}, respectivety, these ϕ^{I} and ϕ^{III} which satisfy the radiation condition at infinity can be expressed as

$$\phi^{I} = \phi^{i} + \phi^{r} + \phi^{sI}$$

$$= -\frac{ig}{\omega}\{a^{i}e^{ik(x_1+\ell^{(1)})} + a^{r}e^{-ik(x_1+\ell^{(1)})}\}\frac{\cosh k(x_2+h^{(1)})}{\cosh kh^{(1)}}$$

$$-\sum_{m=1}^{\infty} a_m^{sI}\frac{ig}{\omega}\frac{\cos k_m(x_2+h^{(1)})}{\cos k_m h^{(1)}} e^{k_m(x_1+\ell^{(1)})} \tag{19}$$

$$\phi^{III} = \phi^{t} + \phi^{sIII} = -a^{t}\frac{ig}{\omega}\frac{\cosh k(x_2+h^{(2)})}{\cosh kh^{(2)}} e^{ik(x_1-\ell^{(2)})} \tag{20}$$

$$-\sum_{m=1}^{\infty} a_m^{sIII}\frac{ig}{\omega}\frac{\cos k_m(x_2+h^{(2)})}{\cos k_m h^{(2)}} e^{-k_m(x_1-\ell^{(2)})} .$$

In eqs. (19) and (20), velocity potentials ϕ^{i}, ϕ^{r} and ϕ^{t} correspond to the incident, reflected and transmitted waves, respectively; ϕ^{sI} and ϕ^{sIII} are velocity potentials for the scattered waves in the regions I and III, respectively. The coefficients a^{i}, a^{r}, a^{t}, a^{sI}, and a^{sIII} mean the amplitudes for each potentials ϕ^{i}, ϕ^{r}, ϕ^{t}, ϕ^{sI} and ϕ^{sIII}. Furthermore, the angular frequency ω and wave numbers k and km in eqs.(19) and (20) have to satisfy the following dispersion relation:

$$\omega^2 = gk \tanh kh = -gk \tan kmh . \tag{21}$$

Now we can write down the continuity conditions on the artificial boundaries $\partial D_i^{\pm}$:

$$\phi^{I} = \phi^{II} , \quad \frac{\partial\phi^{I}}{\partial n} = \frac{\partial\phi^{II}}{\partial n} , \quad \text{on } \partial D_i^{-} , \tag{22}$$

$$\phi^{II} = \phi^{III} , \quad \frac{\partial\phi^{II}}{\partial n} = \frac{\partial\phi^{III}}{\partial n} , \quad \text{on } \partial D_i^{+} , \tag{23}$$

where ϕ^{II} is the velocity potential which is to be determined in the region II .

BOUNDARY INTEGRAL EQUATIONS

In the elastic body, Green's second identity can be expressed as

$$\int_D \{U(x,y;\omega)\cdot Lu(y) - LU(x,y;\omega)\cdot u(y)\}dV_y$$

$$=\int_{\partial D}\{U(x,y;\omega)\cdot\overset{n_y}{T}u(y)-\overset{n_y}{T}U(x,y;\omega)\}\cdot u(y)\ dS_y\ ,\qquad(24)$$

where D means the elastic region Ⅳ and ∂D means $\partial D_t \cup \partial D_c$ in Fig.1. In this equation, the displacement vector **u** satisfies eq. (6) and the fundamental solution **U** satisfies the following inhomogeneous equation:

$$LU=(x,y;\omega)=-1\delta(x-y),\qquad(25)$$

where δ is the Dirac measure. By taking into account the eqs. (6) and (25), the second identity (24) is reduced to the following form on the boundary:

$$C_d^+u(x)=\int_{\partial D}[U(x,y;\omega)\cdot(\overset{n_y}{T}u(y))-\{\overset{n_y}{T}U(x,y;\omega)\}\cdot u(y)]dS_y$$

$$\equiv(St)(x)-(Du)(x),\qquad(26)$$

where **S** and **D** are the simple and double layer operators. C_d^+u is the free term of the exterior limit of the double layer potential. If the boundary is smooth, C_d^+u reduces to

$$C_d^+u=\frac{1}{2}u\ .\qquad(27)$$

Furthermore, the fundamental solution **U** is given as

$$U=\frac{i}{4\mu}[H_0^{(1)}(k_Tr)1+\frac{1}{k_T^2}\nabla\nabla\{H_0^{(1)}(k_Tr)-H_0^{(1)}(k_Lr)\}]\ ,(28)$$

where $H_0^{(1)}$ is the zeroth order Hankel function of the first kind; k_T and k_L are respectively transverse and longitudinal wave numbers; and r= $|x-y|$.

In the same way, the integral equation for the fluid region Ⅱ in Fig. 1 can be expressed in the following form on the boundary:

$$C_d^+\phi(x)=\int_{\partial D}\{G(x,y)\frac{\partial\phi(y)}{\partial n} - \frac{\partial G(x,y)}{\partial n}\phi(y)\}dS_y$$

$$\equiv(G\tilde{\phi})(x)-(H\phi)(x),\qquad(29)$$

where ∂D means $\partial D_c \cup \partial D_s \cup \partial D_i^\pm \cup \partial D_r$. The fundamental solution G is defined as

$$\Delta G(x,y)=-\delta(x-y)\qquad(30)$$

and has the following form:

$$G=\frac{1}{2\pi}\ log\frac{1}{r}\ .\qquad(31)$$

In eq. (29), $C_d^+\phi$ is the free term of the exterior limit of the double layer potential and has the value $\phi/2$ if the boundary is smooth.

NUMERICAL PROCEDURES

The system of boundary integral equations for the structure-fluid interaction problem can be obtained from eqs. (26) and (29) by considering the interface conditions in eqs. (17) and (18) and the continuity conditions in eqs. (22) and (23). As for the discretization of this system, we employ the quadratic isoparametric element. The hoop stress on the boundary of the elastic structure is obtained by taking the covariant derivative of the displacement on the boundary surface of the structure. The final form of the system of boundary integral equations can be expressed as follows:

$$
\begin{bmatrix}
-\bar{D}_t \mid -\bar{D}_c \mid S^*_c \mid & 0 & \mid 0 \mid & 0 & \mid 0 \\
-\!-\!\mid\!-\!-\!\mid\!-\!-\!\mid\!-\!-\!-\!+\!-\!-\!+\!-\!-\!-\!+\!-\!-\!- \\
0 \mid G^*_c \mid -\bar{H}_c \mid \frac{\omega^2}{g} G_s - \bar{H}_s \mid -\bar{H}_r \mid G^*_- - \bar{H}^*_- \mid G^*_+ - \bar{H}^*_+
\end{bmatrix}
\begin{Bmatrix}
u_t \\ u_c \\ \Phi_c \\ \Phi_s \\ \Phi_r \\ a_- \\ a_+
\end{Bmatrix}
$$

$$
= \begin{bmatrix}
-S_t \mid & 0 \\
-\!-\!+\!-\!-\!-\!-\! \\
0 \mid -G^*_- + H^*_-
\end{bmatrix}
\begin{Bmatrix}
g_t \\ a^i
\end{Bmatrix}
\tag{32}
$$

in the matrix form, where $\bar{D}=C^+_d + D$ and $\bar{H}=C^+_d 1 + H$. In this equation, the subscript c in u_c, for example, means that the displacement vector u is evaluated on the boundary ∂D_c in Fig. 1. The vectors a_- and a_+ are amplitudes of the potentials in eq. (19) and (20), i.e., $a_-=\{a^r, a^{sI}_1, a^{sI}_2, \cdots a^{sI}_M\}^T$ and $a_+=\{a^t, a^{sIII}_1, a^{sIII}_2 \cdots a^{sIII}_M\}^T$, where the series in eqs. (19) and (20) is truncated at M. Moreover, the symbol $*$ in S^*_c, for example, means that the simple layer matrix S_c on ∂D_c is rearranged by taking into account the interface conditions (17) and (18).

NUMERICAL EXAMPLES

In order to check the accuracy of our present formulation, we consider two types of steel structures as shown in Fig. 2 (a) and (b). The material properties of these structures are ν (Poisson's ratio)=0.30, $\mu=0.857\times10^{10}$ kgw/m^2, and $\rho=0.786\times10^4$ kgw/m^3. The reflection (KR) and transmission (KT) coefficients for these structures are shown in Fig. 3, where KR and KT are defined as

$$
KR=|a^r/a^i| , \quad KT=|a^t/a^i| . \tag{33}
$$

Since there is no energy loss in the system, these KR and KT should satisfy the following relation

$$
KR^2 + KT^2 = 1 \tag{34}
$$

at infinity ($x_1 \to \pm\infty$). Our results in Fig. 3 satisfy this relation with the accuracy less than 0.1% for

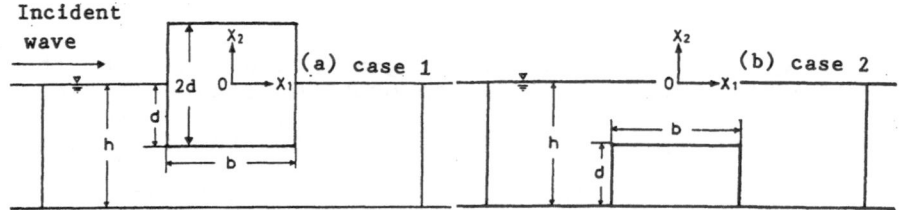

Fig. 2 Numerical model

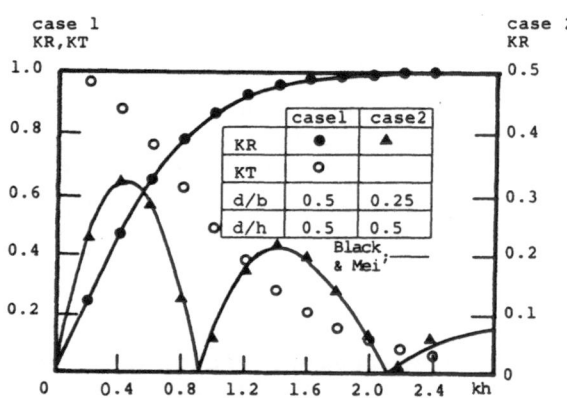

Fig. 3 Reflection and transmission
coefficients

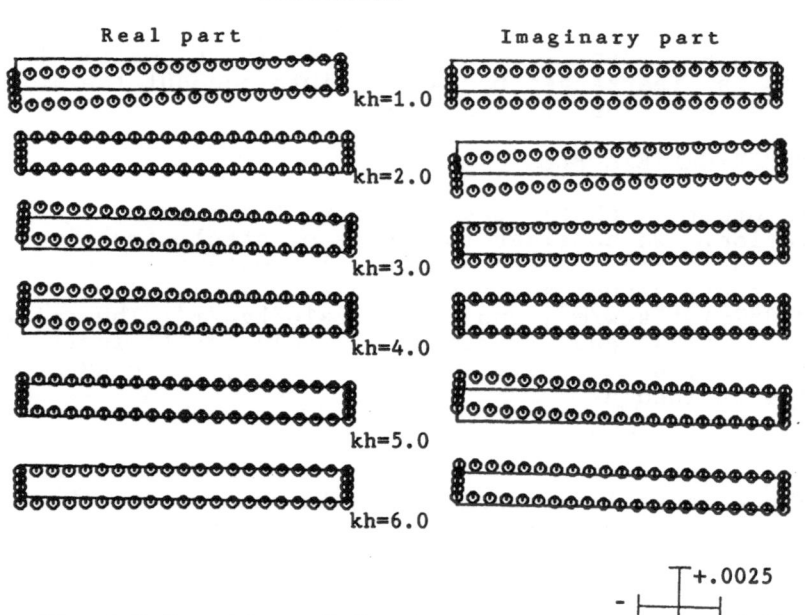

Fig. 4 Deformations of the structure
$(2d/a^i=4.0, \ 2d/b=0.1, \ d/h=0.1)$

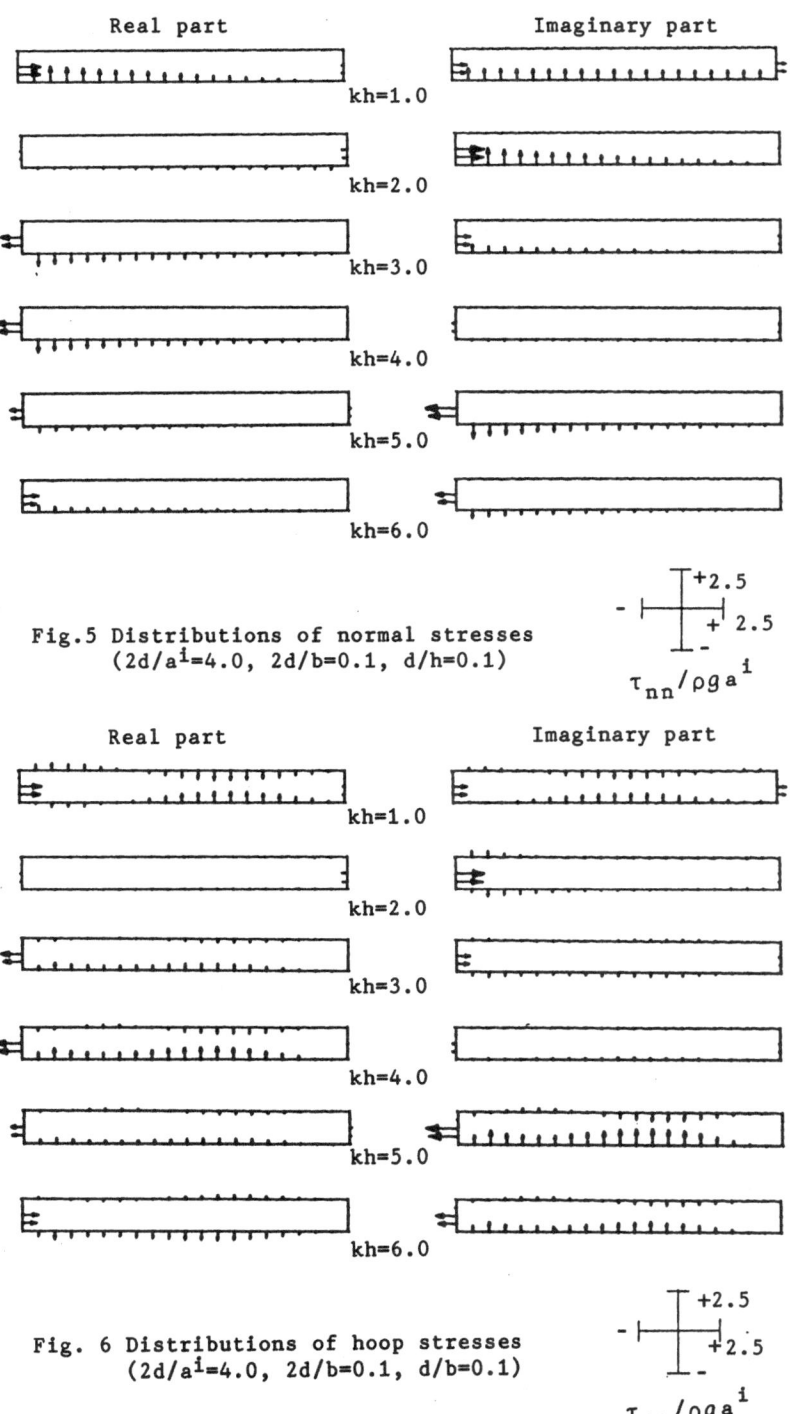

Fig.5 Distributions of normal stresses
$(2d/a^i=4.0,\ 2d/b=0.1,\ d/h=0.1)$

Fig. 6 Distributions of hoop stresses
$(2d/a^i=4.0,\ 2d/b=0.1,\ d/b=0.1)$

all values of nondimensional wave number kh's. For comparison, the results by Black & Mei (1970) for the rigid body are also shown in this Fig. 3.

As an example of the flexible elastic structure, we consider the rubber type of structure which has the material properties of $\nu=0.46$, $\mu=0.5 \times 10^5 \, \text{kgw/m}^2$, and $\rho = 0.91 \times 10^3 \, \text{kgw/m}^3$. Figs. 4, 5, and 6 show the deformation of the structure, the normal stress on the boundary, and the hoop stress on the boundary, respectively. From these figures, we can observe the following:
1) The deformation of the structure decreases as the wave number kh increases.
2) The normal stress on the boundary has a large value at the front (the side of the incident wave) of the structure. At the bottom of the structure, also, the front side has a large value of the normal stress. Moreover, the absolute value of the normal stress decreases as the wave number increases.
3) The hoop stress has a peculiar distribution pattern along the boundary for each wave number and there seems to be no special tendency for every wave numbers.

REFERENCES

Black, J. L. and Mei, C. C. (1970) Scattering and Radiation of Water Waves, Report No. 121, Dpt. of Civil Engineering, Mass. Inst. Technol.

Ijima, T. and Chou, C. R. (1976) Analyses of Two-Dimensional Wave Problems by Means of Green's Identity Formula, Proc. JSCE, No. 252, pp. 57-71.

John, F. (1950) On the Motion of Floating Bodies. 2., Communications of Pure and Applied Mathematics, 3, pp. 45-101.

Kiyokawa, K. and Ohyama, T. (1984) Analysis of Scattered Waves and Induced Forces on Axisymmetric Body by Means of Hybrid Method, Proc. JSCE, No.345, pp. 131-140.

Kakuda, K. and Tosaka, N. (1983) Numerical Analysis of Coupled Fluid -Elasticity Systems Using the Boundary Element Method, Boundary Elements, Springer-Verlag, pp.1005-1016.

Zienkiewicz, O. C. and Bettess, P. (1978) Fluid-structure Dynamic Interaction and Wave Forces, An Introduction to Numerical Treatment, International Journal for Num. Meth. Engng., Vol. 13, pp. 1-16.

Nonlinear Wave Transformation

K. Mizumura
Department of Civil Engineering, Kanazawa Institute of Technology, 7-1, Ogigaoka, Nonoichimachi, Ishikawa Pref., Japan

INTRODUCTION

Recently, the boundary element method has been developed and applied to many engineering problems. The problems on the free surface profiles due to wave action (Nakayama et al[1]; Mizumura[2]) or running water (Mizumura[3]; Mizumura[4]) become easier through the merit of the boundary element method. The difficulty in solving these problems exists in the nonlinearity of the dynamic boundary condition (Bernoulli equation). The nonlinearity appears remarkably in the case of the shallow water. Herein, the boundary element method is applied to analyze the wave transformation over simple coastal structures and the iteration procedure is employed to deal with the nonlinear dynamic boundary condition. The problems on the transformation of a solitary wave over a uniform slope are numerically solved by using different methods (Hauguel[5]; Madsen[6]). The linear wave transformation over a semi-circular cylinder (Bird[7]) and a rectangular obstacle are calculated by different methods (Mei et al[8]). The problems on the wave interactions with submarine trenches are also analytically computed (Lee et al[9,10]) for the linearized case. In this study, the nonlinear wave transformations over a uniform slope, a semi-circular cylinder and a submarine semi-circular trench are numerically solved by using the boundary element method.

PROBLEM FORMULATION

To apply the boundary element formulation, we consider the motion of a fluid in a two-dimensional wave tank as shown in Fig.1. A fluid region V is bounded by a free surface S_1, a piston plate S_2, and a bottom and a side wall of the wave tank S_3. A co-ordinate system is represented in Fig.1. The x-axis coincides with the stationary free surface and the y-axis coincides with a piston plate on the left and is directed upwards. Assuming that the fluid is inviscid, incompressible, and irrotational, the governing equation is given by

$$\frac{\partial^2\phi}{\partial x^2} + \frac{\partial^2\phi}{\partial y^2} = 0 \tag{1}$$

in which ϕ is the velocity potential. The boundary conditions on S_1, S_2, and S_3 are shown as follows:

$$\frac{\partial\phi}{\partial t} + \frac{1}{2} \{(\frac{\partial\phi}{\partial x})^2 + (\frac{\partial\phi}{\partial y})^2\} + g\eta = 0 \quad ; \text{ on } S_1 \tag{2}$$

$$\frac{\partial\phi}{\partial n} = n_y \frac{\partial\eta}{\partial t} \qquad\qquad ; \text{ on } S_1 \tag{3}$$

$$\frac{\partial\phi}{\partial n} = -f(t) \qquad\qquad ; \text{ on } S_2 \tag{4}$$

$$\frac{\partial\phi}{\partial n} = 0 \qquad\qquad ; \text{ on } S_3 \tag{5}$$

in which $\eta(x, t)$ is the displacement of the free surface from the stationary free surface and g is the gravitational acceleration. n is the co-ordinate normal to the boundary and n_x and n_y are the direction cosines of the normal with respect to the x-axis and y-axis, respectively. Next, the two-dimensional Green's formula is transformed by selecting the velocity potential ϕ and the fundamental solution $\ln(1/r)$ in it. The variable r is the distance between the source point and the observation point. If the source point is inside the region V,

$$2\Pi\phi_i + \int_S \phi \frac{\partial}{\partial n}(\ln\frac{1}{r})ds - \int_S \frac{\partial\phi}{\partial n}\ln\frac{1}{r}ds = 0 \tag{6}$$

and if it lies on the boundary S,

$$\alpha_i\phi_i + \int_S \phi \frac{\partial}{\partial n}(\ln\frac{1}{r})ds - \int_S \frac{\partial\phi}{\partial n} \ln\frac{1}{r} ds = 0 \tag{7}$$

in which the subscript i shows the source point and α_i denotes the angle between two tangents at the source point. By using Eqs.(2), (3), (4), and (5), Eq.(7) becomes as:

$$\alpha_i\phi_i + \int_S \phi \frac{\partial}{\partial n}(\ln\frac{1}{r}) ds + \int_{S_2} f(t)\ln\frac{1}{r} ds - \int_{S_1}\frac{\partial\phi}{\partial n}\ln\frac{1}{r} ds = 0 \tag{8}$$

The above integral equation is nonlinear.

DYNAMIC BOUNDARY CONDITION

To obtain the free surface location, a relation between ϕ and $\partial\phi/\partial n$ at the free surface nodal points is derived. The kinematic and dynamic boundary conditions are rewritten as follows:

$$\frac{\partial \eta}{\partial t} = \frac{1}{\cos \beta} \frac{\partial \phi}{\partial n} \tag{9}$$

$$(\frac{\partial \phi}{\partial t})_x = -\frac{1}{2} [(\frac{\partial \phi}{\partial s})^2 - 2\tan\beta \frac{\partial \phi}{\partial n} \frac{\partial \phi}{\partial s} - (\frac{\partial \phi}{\partial n})^2] - \frac{\eta}{F_r^2} \tag{10}$$

in which β = the surface angle; $\partial \phi/\partial s$ = tangential velocity; and F_r = the Froude number. Denoting Δt as a time step between time level k and k+1, we can rewrite the above equation in the finite difference forms as:

$$\eta^{k+1} = \eta^k + \Delta t \{ \frac{(1-\theta_1)}{\cos\beta^k}(\frac{\partial \phi}{\partial n})^k + \frac{\theta_1}{\cos\beta^{k+1}}(\frac{\partial \phi}{\partial n})^{k+1} \} \tag{11}$$

$$\phi^{k+1} = \phi^k - \Delta t \{ (1-\theta_2) [\frac{\eta^k}{F_r^2} + \frac{1}{2}\{(\frac{\partial \phi}{\partial s})^2 - 2\tan\beta \ (\frac{\partial \phi}{\partial n})(\frac{\partial \phi}{\partial s}) - (\frac{\partial \phi}{\partial n})^2\}^k$$

$$+ \theta_2 [\frac{\eta^{k+1}}{F_r^2} + \frac{1}{2}\{(\frac{\partial \phi}{\partial s})^2 - 2\tan\beta(\frac{\partial \phi}{\partial n})(\frac{\partial \phi}{\partial s}) - (\frac{\partial \phi}{\partial n})^2\}^{k+1}] \} \tag{12}$$

in which θ_1 and θ_2 = weighting factors. The tangential velocity component, $\partial \phi/\partial s$, at the nodal point can be approximated by the central difference equation as (Liu el al[11,12]):

$$(\frac{\partial \phi}{\partial s})_j = \frac{1}{\ell_j \ell_{j-1}^2 + \ell_{j-1} \ell_j^2} [\ell_{j-1}^2 \phi_{j+1} + (\ell_j^2 - \ell_{j-1}^2)\phi_j$$

$$- \ell_j^2 \phi_{j-1}] \tag{13}$$

To calculate nonlinear terms in Eqs.(11) and (12), the following iteration equations are employed:

$$\frac{1}{\cos\beta^{k+1}}(\frac{\partial \phi}{\partial n})^{k+1} \cong \frac{1}{\cos\bar\beta^{k+1}} (\frac{\partial \phi}{\partial n})^{k+1} \tag{14a}$$

$$[(\frac{\partial \phi}{\partial s})^{k+1}]^2 \cong 2 (\frac{\partial \bar\phi}{\partial s})^{k+1} (\frac{\partial \phi}{\partial s})^{k+1} - [(\frac{\partial \bar\phi}{\partial s})^{k+1}]^2 \tag{14b}$$

$$[(\frac{\partial \phi}{\partial n})^{k+1}]^2 \cong 2 (\frac{\partial \bar\phi}{\partial n})^{k+1} (\frac{\partial \phi}{\partial n})^{k+1} - [(\frac{\partial \bar\phi}{\partial n})^{k+1}]^2 \tag{14c}$$

$$\tan\beta^{k+1}(\frac{\partial \phi}{\partial n})^{k+1} (\frac{\partial \phi}{\partial s})^{k+1} \cong \tan\bar\beta \{(\frac{\partial \bar\phi}{\partial n})^{k+1} (\frac{\partial \phi}{\partial s})^{k+1} + (\frac{\partial \phi}{\partial n})^{k+1} (\frac{\partial \bar\phi}{\partial s})^{k+1}$$

$$- (\frac{\partial \bar\phi}{\partial n})^{k+1} (\frac{\partial \bar\phi}{\partial s})^{k+1} \} \tag{14d}$$

in which quantities with an overbar are obtained from the previous iteration and quantities without the overbar are to be determined at the current iteration. Both quantities are on the time level k+1. Substitution of Eqs.(11) and (14) into

Eq.(12) yields a relation among ϕ, $\partial\phi/\partial s$, and $\partial\phi/\partial n$ at the jth iteration on the (k+1) time level.

$$\phi^{k+1} + a \left(\frac{\partial\phi}{\partial n}\right)^{k+1} + b \left(\frac{\partial\phi}{\partial s}\right)^{k+1} = r \tag{15}$$

NUMERICAL PROCEDURE

To discretize Eqs.(8) and (10), the functions ϕ and η are assumed to be linear functions with respect to s, which is a local co-ordinate along each element. By using the above assumption, Eq.(8) becomes as follows:

$$\alpha_i\phi_i + \Sigma\int_S \int_0^\ell \underline{N}^T\frac{\partial}{\partial n}(\ell n\frac{1}{r})ds \ \underline{\phi} - \Sigma\int_{S_1}\int_0^\ell \underline{N}^T\ell n\frac{1}{r} \ \underline{\phi}_n + \Sigma\int_{S_2}\int_0^\ell f(t)\ell n\frac{1}{r}ds = 0 \tag{16}$$

for i = 1, 2, , M, in which ℓ is the length of the line element, M is the number of the elements on the boundary, $\underline{\phi}^T = [\phi_j, \phi_{j+1}]$, and $\underline{n}^T = [n_j, n_{j+1}]$. $\underline{N}^T$ denotes $[\ell-s, s]/\ell$. Assuming that the fluid is entirely at rest at t=0, we have the following initial conditions:

$$\phi = \eta = \frac{\partial\phi}{\partial n} = 0 \text{ at } t = 0 \tag{17}$$

The piston plate on the left side of the tank is assumed to be located by the following function:

$$f(t) = \frac{d}{dt} [X_0 \tanh \omega(t-t_c)] \tag{18}$$

in which X_0 = the semistroke of the piston plate, and ω and t_c = parameters to characterize the motion of the piston plate.

INSTABILITY, SMOOTHING AND ACCURACY

In the computations, the instability appears after a long time. The cause of the instability is rounding errors or the growth of short gravity-waves. They are partly damped by the existence of viscosity. To remove the instability, the following equation is used (Longuet-Higgins[14]).

$$f_j = (-f_{j-2} + 4f_{j-1} + 10f_j + 4f_{j+1} - f_{j+2})/16 \tag{19}$$

The accuracy of numerical solution was tested by calculating the mean level of water surface profile on each time.

NUMERICAL RESULTS

Fig.3 shows the wave progress in the case of h/h' = 1.5, namely h' = 0.67 m. The solitary wave generated by the motion of the piston plate advances without my transformation. Fig.4 represents the wave transformation in the case of h/h' = 2.0. The indication of wave transformation is found at the end of

the slope when t = 4.4 sec and the disturbance grows. The wave
transformation over a submerged semi-circular cylinder of which
radius is 0.8 m is given in Fig.5. The wave amplitude gradually increases as the wave approaches to the semi-circular cylinder. Then, as the crest passes the semi-circular cylinder, it
decreases its amplitude and disintegrates into two crests. One
crest progresses over the semi-circular cylinder and the other
crest goes back to the opposite direction. The other wave is
recognized by the phenomenon of wave reflection. The wave
transformation due to the semi-circular trench of which radius
is 3 m is represented in Fig.6. The amplitude of the prog-
ressing wave decreases as it approaches the trench. Then, the
amplitude increases as the water depth decreases. The reflected
wave is also formed at t = 7.2. It is generated by the sudden
decrease of the water depth between the semi-circular trench
and the uniform water depth.

CONCLUSIONS

Applying the boundary element method to the wave transformation
in the shallow water, the following conclusions are obtained
for the given solitary wave:
(1) The shape of the initial solitary wave becomes unstable on
the uniform slope, if the ratio of the uniform water depth to
the water depth on the shelf is greater than a fixed value.
The occurrence of the instability is considered as the wave
breaking.
(2) In the existence of the semi-circular cylinder or the semi-
circular trench the wave crest disintegrates into two crests as
its amplitude increases. One crest advances and the other goes
back to the opposite direction and forms the reflected wave.

REFERENCES

1. Nakayama T. and Washizu K. (1981). The Boundary Element
    Method Applied to the Analysis of Two-dimensional Nonlinear
    Sloshing Problems, Int. Jour. Num.Meth.Engrg., 17, pp.1631-
    1646.
2. Mizumura K. (1985), Nonlinear Water Waves Developed by an
    Accelerated Circular Cylinder, Boundary Elements Vll, pp.
    9.49-59.
3. Mizumura K. (1983), Free-Surface Flow over a Channel with
    Rectangular Obstruction, Boundary Elements V, pp.301-309.
4. Mizumura K. (1985), Nonlinear Water Waves over Wavy Bed,
    Boundary Elements Vll, pp.5.61-70.
5. Hauguel A. (1980), A Numerical Model of Storm Waves in
    Shallow Water, 17th Int.Coastal Engrg.conf., Vol.1, pp.746-
    762.
6. Madsen O.S. and Mei C.C. (1969), The Transformation of a
    Solitary Wave over an Even Bottom, Jour. of Fluid Mech.,
    Vol.39, pp.781-791.
7. Bird H.W. and Shepherd R. (1982), Wave Interaction with
    Large Submerged Structures, Jour. of WW Div., ASCE, Vol.108,

No.WW2, pp.146-162.

8. Mei C.C. and Black J.L. 1969), Scattering of Surface Waves by Rectangular Obstacles in Waves of Finite Depth, Jour. of Fluid Mech., Vol.38, pp.499-511.

9. Lee J.J., Ayer R.M. and Chiang W.L. (1980), Interactions of Waves with Submarine Trenches, 17th Int.Coastal Engrg.Conf., Vol.1, pp.812-822.

10. Lee J.J. and Ayer R.M. (1981), Wave Propagation over a Rectangular Trench, Jour. of Fluid Mech., Vol.110, pp.335-347.

11. Liu P.L.-F. and Liggett J.A. (1982). Applications of boundary element methods to problems of water waves. In:Developments in Boundary Element Methods-2, Applied Science Publishers Ltd, London, pp.37-67.

12. Liu P.L.-F. and Liggett J.A. (1983). Boundary element formulations and solutions for some nonlinear water wave problems. In:Developments in Boundary Element Methods-3, Applied Science Publishers Ltd, London, pp.171-190.

13. Nakayama T. (1983), Boundary Element Analysis of Nonlinear Water Wave Problems, Int.Jour.Num.Meth.Engrg., 19, pp.953-970.

14. Longuet-Higgins F.R.S. and Cokelet E.D. 1976), The Deformation of Steep Surface Waves on Water, A Numerical Method of Computation, Proc.R.Soc.Lond. A350, pp.1-26.

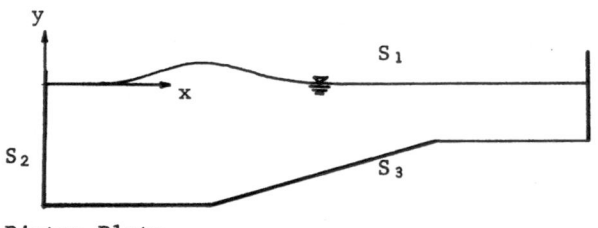

Fig. 1   Definition Sketch

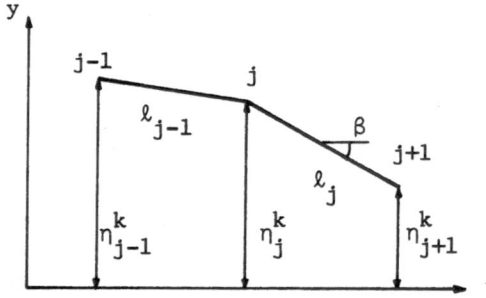

Fig. 2 Nodal Points for Central Difference

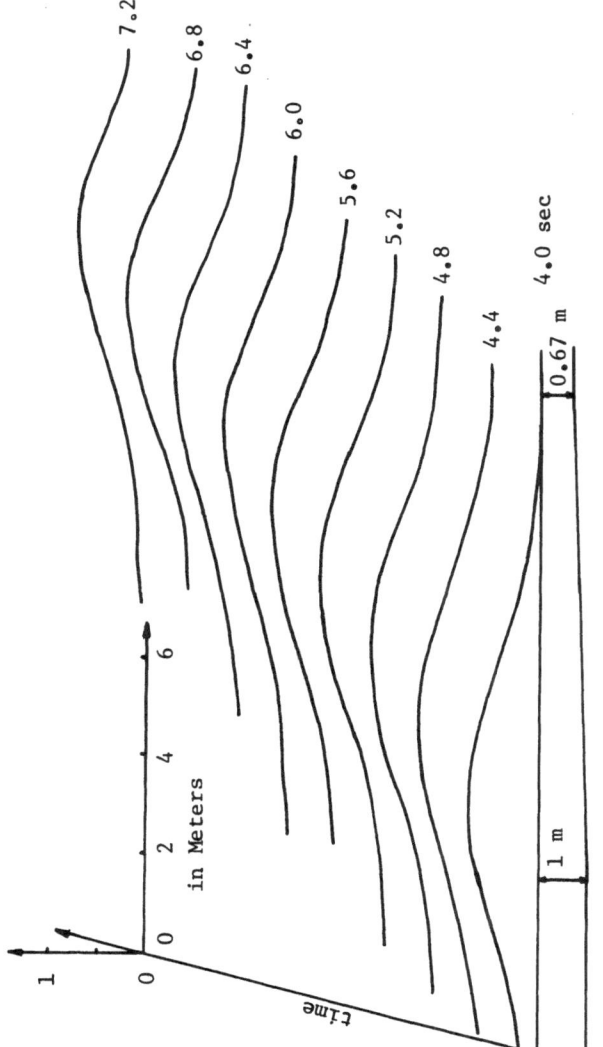

Fig. 3 Wave Transformation over Uniform Slope (h/h' = 1.5)

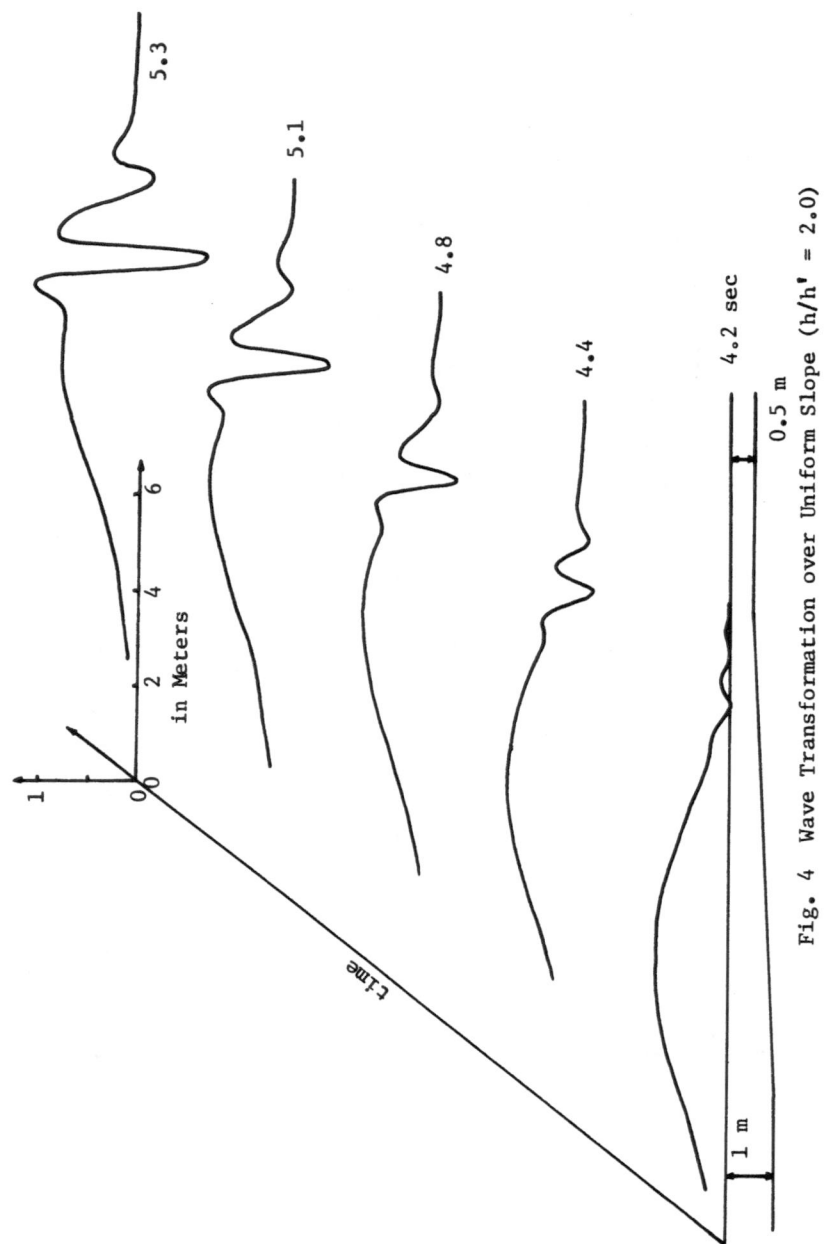

Fig. 4 Wave Transformation over Uniform Slope (h/h' = 2.0)

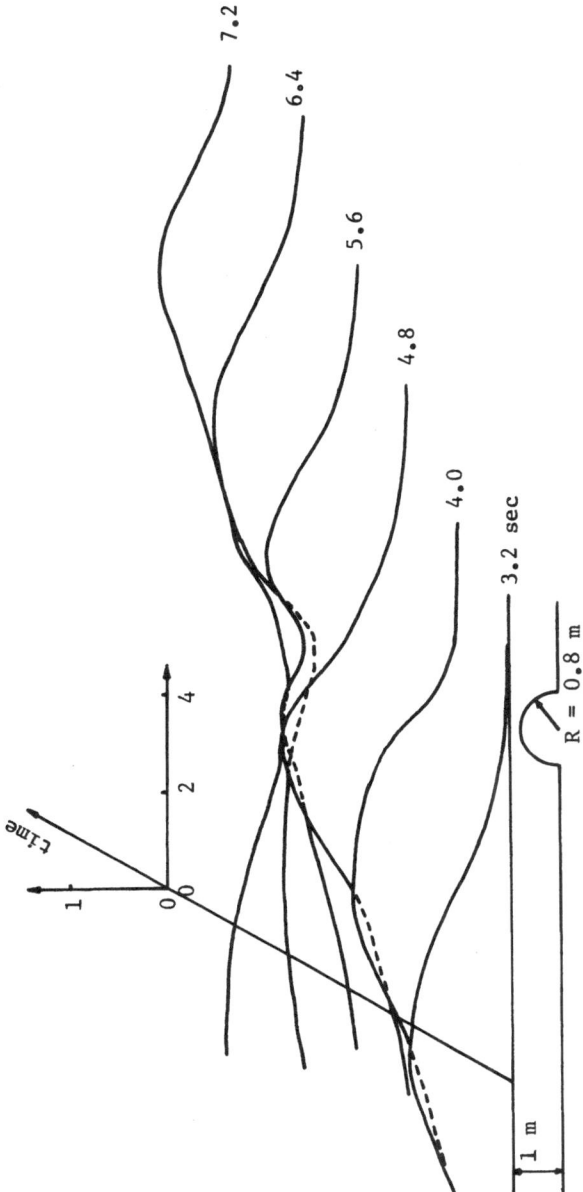

Fig. 5   Wave Transformation over Semi-Circular Cylinder (R = 0.8 m)

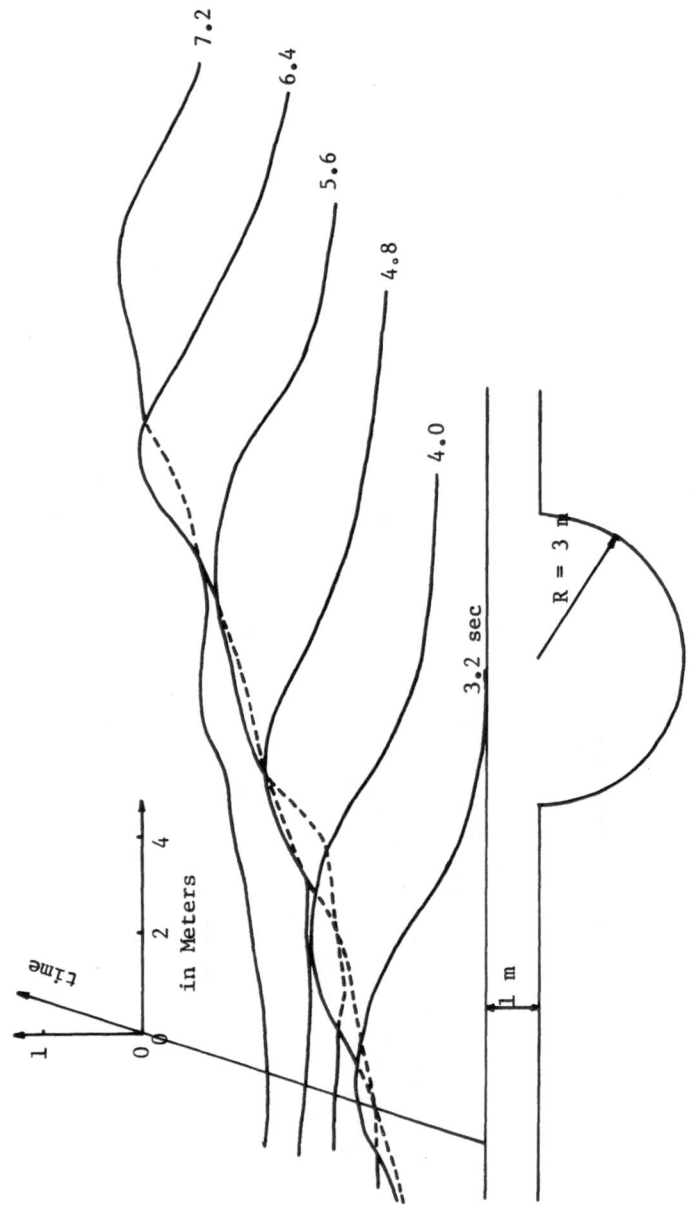

Fig. 6 Wave Transformation over Semi-Circular Trench (R = 3 m)

SECTION X   NONLINEAR STRUCTURAL PROBLEMS

Geometrically Nonlinear Analysis of Shallow Spherical Shell Using an Integral Equation Method

N. Tosaka, S. Miyake
Department of Mathematical Engineering, College of Industrial Technology, Nihon University, Narashino-shi, Chiba 275, Japan

INTRODUCTION

Recently, numerical solution techniques have been developed and widely used in various fields of natural science and engineering.

In most recent years, the developments of the boundary element method[1] is very notable among the various numerical solution methods. The boundary element method for linear boundary-value problems is recognized to be an efficient numerical approximation method in comparison with the finite element method or the finite difference one. But, the analysis of nonlinear problem by the boundary element method is very few at present stage[1,6]. Especially, the effective application to geometrically nonlinear bending problems of shallow shells has been just started[2,3,4,5,6].

The general integral equation formulations for the solutions of elastic shallow shell bending problems were presented with the stress function approach and the displacement one in Ref.[7] for the linear problem and in Ref.[5] for the geometrically nonlinear one. In Ref.[2], we obtained the very accurate numerical solutions with few constant boundary elements for linear bending problems of shallow spherical shells by using the stress function approach from among two types of integral equation formulation. In Refs.[3,4,6], we succeed in tracking the completely load-deflection curves with snap-through phenomena by using not the conventional boundary method but the hybrid type integral equation method.

The purpose of the present investigation is to develop the above approach to nonlinear analysis of shallow spherical shells and then to show its usefulness and potentiality through a number of illustrative numerical results. In the first place, we indicate the fundamental equations of our shell bending problem and its integral equation expression is derived with the aid of the weighted residual method and the fundamental solution for a linear differential operator. The derived nonlinear integral equations are discretized by using not only boundary elements but also interior cells. In order to track completely the load-deflection curves with high nonlinearity which characterize the snap-through phenomena of shallow spherical shells, the corresponding nonlinear system of algebraic equation is solved effectively by means of the Riks-Wempner method[8] instead of the simple iteratve procedure or the incremental Newton-Raphson method.

FUNDAMENTAL EQUATIONS

Let us consider an elastic isotropic shallow spherical shell of radius R and constant thickness h subjected to a uniformly distributed normal load p. Under these conditions, the well-known partial differential equations governing the nonlinear behaviour of the shell are given as follows[2,3,4,6]:

$$\frac{E h^3}{12(1-\eta^2)} \nabla^2 \nabla^2 w = [\phi, w] - \frac{1}{R} \nabla^2 \phi + P, \tag{1}$$

$$\frac{1}{E h} \nabla^2 \nabla^2 \phi = -\frac{1}{2} [w, w] + \frac{1}{R} \nabla^2 w, \tag{2}$$

where the nonlinear terms $[\widetilde{\phi}, \widetilde{w}]$ and $[\widetilde{w}, \widetilde{w}]$ are represented by the following notations

$$[\phi, w] = \frac{\partial^2 w}{\partial x^2} \frac{\partial^2 \phi}{\partial y^2} - 2\frac{\partial^2 w}{\partial x \partial y} \frac{\partial^2 \phi}{\partial x \partial y} + \frac{\partial^2 w}{\partial y^2} \frac{\partial^2 \phi}{\partial x^2}, \tag{3}$$

$$[w, w] = 2\frac{\partial^2 w}{\partial x^2} \frac{\partial^2 w}{\partial y^2} - 2\left(\frac{\partial^2 w}{\partial x \partial y}\right)^2. \tag{4}$$

Here w is the normal displacement of middle surface, ϕ the Airy-type stress function, E the Young's modulus, η the Poisson's ratio of shell material, and ∇^2 the Laplacean.

In this place, we intend to introduce the following new functions and convenient parameters defined by

$$\widetilde{\phi} = \frac{\sqrt{12(1-\eta^2)}}{R^2 h} \phi \quad , \quad \widetilde{w} = \frac{w}{R^2/Eh} \quad , \quad \widetilde{\Psi} = \widetilde{w} + i\widetilde{\phi} \quad ,$$

$$\epsilon^2 = \frac{\sqrt{12(1-\eta^2)}}{Rh} \quad , \quad \alpha = \frac{1}{2}\frac{R^3 \epsilon^2}{E h} \quad , \quad \beta = \frac{-\epsilon^4 R^4}{E\sqrt{12(1-\eta^2)}}. \tag{5}$$

With using these new functions and constants to equations (1) and (2), we can rewrite the fundamental equation set as the following single complex-valued expression:

$$\nabla^2 \nabla^2 \widetilde{\psi} - i\,\epsilon^2 \nabla^2 \widetilde{\psi} + \beta\,[\,\widetilde{\phi}\,,\,\widetilde{w}\,] + i\,\alpha\,[\,\widetilde{w}\,,\,\widetilde{w}\,] = p\,\epsilon^4. \tag{6}$$

INTEGRAL EQUATION FORMULATION

Let us transform preceding single complex-valued equation (6) into the integral equations Following the previously proposed formulation [2,3,4,6], we can get the following nonlinear integral equation expression in terms of fundamental solution ω^* and the complex unknown function $\widetilde{\psi}$:

$$c\,\widetilde{\psi}(\xi) = \int_{\Omega} \left\{ p\epsilon^4 - \beta\,[\,\widetilde{\phi}\,,\,\widetilde{w}\,] - i\,\alpha\,[\,\widetilde{w}\,,\,\widetilde{w}\,] \right\} \omega^* d\,\Omega$$

$$+ \int_{\Gamma} \left\{ \widetilde{\psi}\,(\,\nabla^2 \omega^*),_\nu - (\,\nabla^2 \widetilde{\psi}),_\nu \cdot \omega^* \right\} d\,\Gamma$$

$$+ \int_{\Gamma} \left\{ (\,\nabla^2 \widetilde{\psi}) \cdot \omega^*,_\nu - \widetilde{\psi},_\nu \cdot (\,\nabla^2 \omega^*) \right\} d\,\Gamma$$

$$+ i\,\epsilon^2 \int_{\Gamma} (\,\widetilde{\psi},_\nu \cdot \omega^* - \widetilde{\psi} \cdot \omega^*,_\nu\,) d\,\Gamma \tag{7}$$

where ξ and c denote the source point and the shape coefficient.

Further more, differentiating equation (7) in normal direction ν_0 at the source point on the smooth boundary Γ , we can get the second equation such that

$$\frac{1}{2}\,\frac{\partial \widetilde{\psi}}{\partial \nu_0}(\xi) = \int_{\Omega} \left\{ p\epsilon^4 - \beta\,[\,\widetilde{\phi}\,,\,\widetilde{w}\,] - i\,\alpha\,[\,\widetilde{w}\,,\,\widetilde{w}\,] \right\} \omega^*,_{\nu_0} d\,\Omega$$

$$+ \int_{\Gamma} \left\{ \widetilde{\psi}\,(\,\nabla^2 \omega^*),_{\nu\nu_0} - (\,\nabla^2 \widetilde{\psi}),_\nu\,\omega^*,_{\nu_0} \right\} d\,\Gamma$$

$$+ \int_{\Gamma} \left\{ (\,\nabla^2 \widetilde{\psi})\,\omega^*,_{\nu\nu_0} - \widetilde{\psi},_\nu\,(\,\nabla^2 \omega^*),_{\nu_0} \right\} d\,\Gamma$$

$$+ i\,\epsilon^2 \int_{\Gamma} (\,\widetilde{\psi},_\nu \cdot \omega^*,_{\nu_0} - \widetilde{\psi} \cdot \omega^*,_{\nu\nu_0}\,) d\,\Gamma. \tag{8}$$

The fundamental solution ω^* which is the sum of the real part θ^* and the imaginary one χ^* is given in Refs.[2,3,4,6]. Substituting the fundamental solution into equations (7) and (8), and decomposing the nonlinear integral equations into the real part and the imaginary one, we can easily obtain the coupled system of nonlinear integral equations in terms of the real unknown functions $\widetilde{w}$ and $\widetilde{\phi}$ [2,3,4,6].

DISCRETIZATION OF INTEGRAL EQUATIONS

The nonlinear integral equations (7) and (8) is generally difficult to solve analytically. Therefore the derived nonlinear integral equations must be solved approximately by means of some numerical procedure. We consider such simple discretization that the boundary Γ is divided in to a series of straight line segments and the domain Ω is discretized into a series of triangular cells. The boundary nodes are taken to be in the middle on each segment and the nodes in the domain are taken to be the center of gravity in each triangular cells.

Calculation of both boundary and cell integrals can be computed numerically using the standard Gaussian quadrature except for the one corresponding to the node under consideration. For the singular integrals corresponding to both the boundary elements and cell ones, we evaluate its in the sense of Cauchy principle values. The singular integrals on boundary are estimated by the previously proposed scheme[2,3,6]. The singular integrals on cell can be estimated by applying the semi-analytical integration proposed in Refs.[3,4,6]. Adopting of the above mentioned discretized scheme for the drived nonlinear integral equation set, we arrive at the following nonlinear system of algebraic equations:

$$
\begin{bmatrix} \mathscr{A}_1 & \mathscr{N}_1(W_\Omega, \phi_\Omega) \\ \mathscr{A}_2 & \mathscr{N}_2(W_\Omega, \phi_\Omega) \end{bmatrix} \begin{Bmatrix} W_\Gamma \\ \phi_\Gamma \\ W_\Omega \\ \phi_\Omega \end{Bmatrix} = \begin{Bmatrix} P_1 \\ P_2 \\ P_3 \\ P_4 \end{Bmatrix} \tag{9}
$$

where $\mathscr{A}_i\ (i=1,2)$ and $\mathscr{N}_i\ (i=1,2)$ are coefficient matrices obtained from the boundary integral terms and the domain integral terms due to nonlinearlity of governing equations, respectively. And $P_i\ (i=1 \sim 4)$ is the load vector, and W and ϕ denote the unknown nodal vectors.

SOLUTION PROCEDURE

Since the derived matrix equation(9) is highly nonlinear, we must solve this by making use of some linearized solution procedure. In general, the incremental Newton-Raphson method is valid for the nonlinear algebraic equations. However, this method is not always efficient for the problems with the limit point appearing in the theory of elastic stability.

In this paper, in order to pursue the complete load-deflection curve with high nonlinearity, we wish to apply effectively the Riks-Wempner method[8] to derive nonlinear system of algebraic equation (9). By using this arc-length type solution procedure, we can track easily the curve beyond the limit point(the so-called snap-through point).

NUMERICAL EXAMPLES

In order to demonstrate the potentiality and applicability of our integral equation method developed in this investigation, we study the nonlinear problem of shallow spherical shells with rectangular plan form with the lengths L_1, L_2 and the rise H as shown in Fig.1.

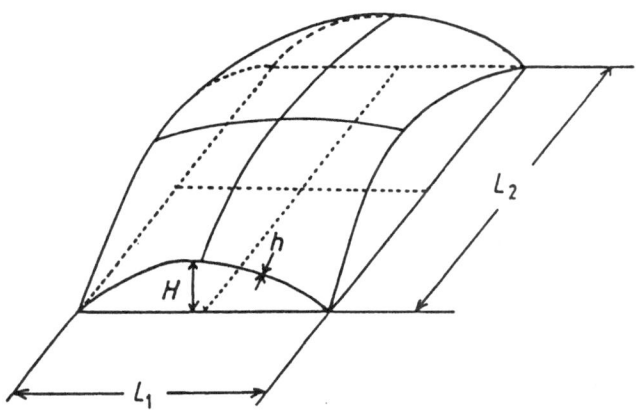

Fig.1 Shallow spherical shell

In this investigation, the boundary conditions are assumed as the simply supported boundary conditions given by

$$w = \phi = 0,$$
$$\nabla^2 w = \nabla^2 \phi = 0.$$
$$(10)$$

For convenience, the results are presented in terms of nondimensional uniform pressure $\bar{P}$, and the average deflection $\bar{W}$, and central deflection $\bar{W_c}$, which are defined as follows:

$$\bar{P} = \frac{R^2 p}{E h^2} \qquad \bar{W} = \frac{1}{L_1 \cdot L_2} \int_0^{L_1} \int_0^{L_1} \frac{w}{h} dx dy, \qquad \bar{W_c} = \frac{w(L_1/2, L_2/2)}{h}. (11)$$

A value of the Poisson's ratio of the shell material is taken $\eta = 0.3$. The numerical integration for boundary are performed with the six-point Gaussian quadrature. The calculation of integrals over each cell are evaluated by the seven-point Gaussian quadrature for triangles. The boundary and internal discretizations employed to run the problem are shown in Fig.2.

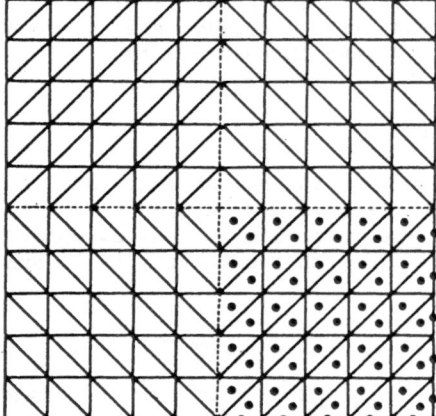

Fig.2 Element mesh

Figs.3,4 and 5 show the load-average deflection curve for the rise to thickness ratio H/h=1.25,1.75 and 5.0 through a comparison with linear Fourier series solutions, respectively. Especially, the variation of deflection modes with increasing of load is also given in Figs.4 and 5. Points A, B, C, D in the curves denote the each equilibrium state. Figs.6 and 7 show the load-central deflection curves for H/h=0.5,1.0,1.75 and 5.0.

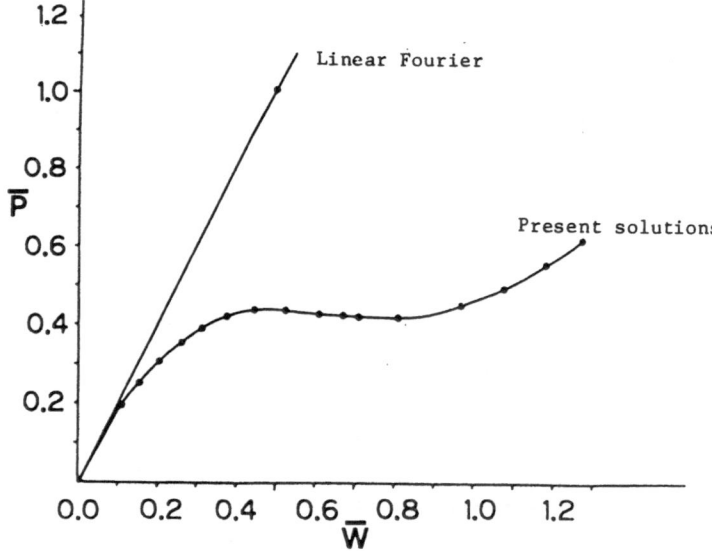

Fig.3 Load-average deflection curve for H/h=1.25

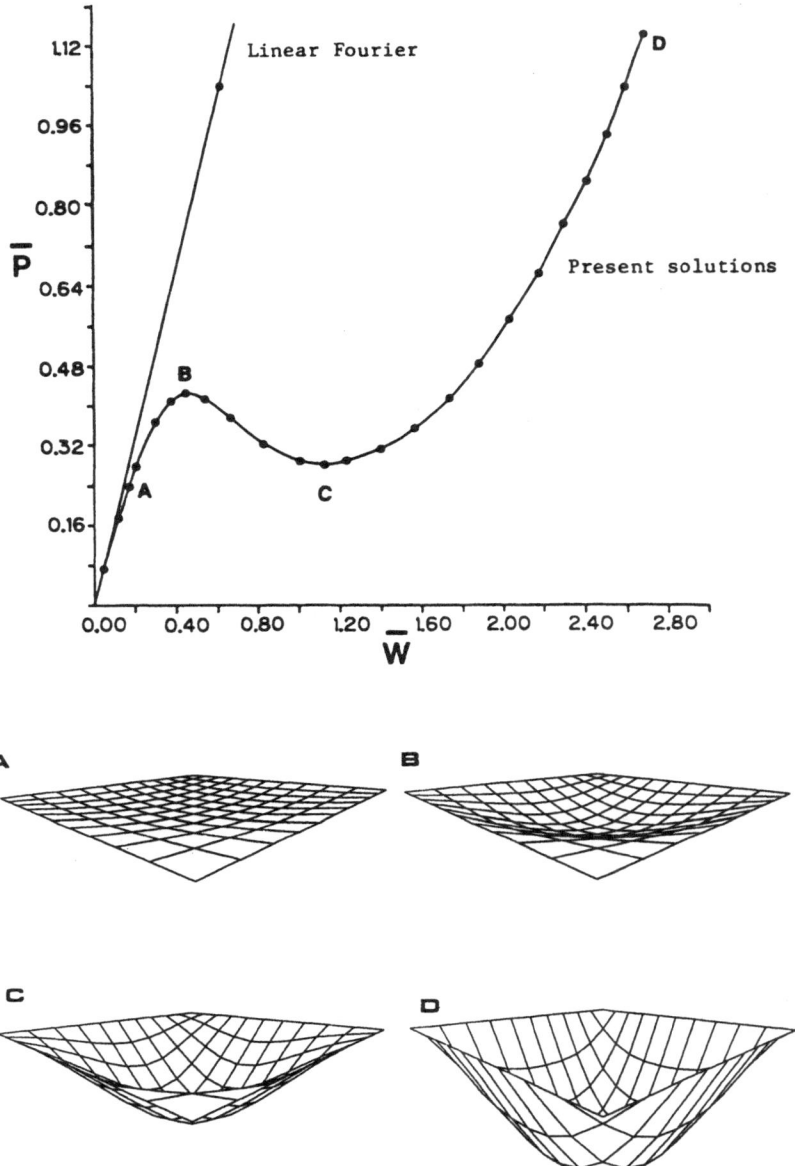

Fig.4 Load-average deflection curve and deflection modes
for H/h=1.75

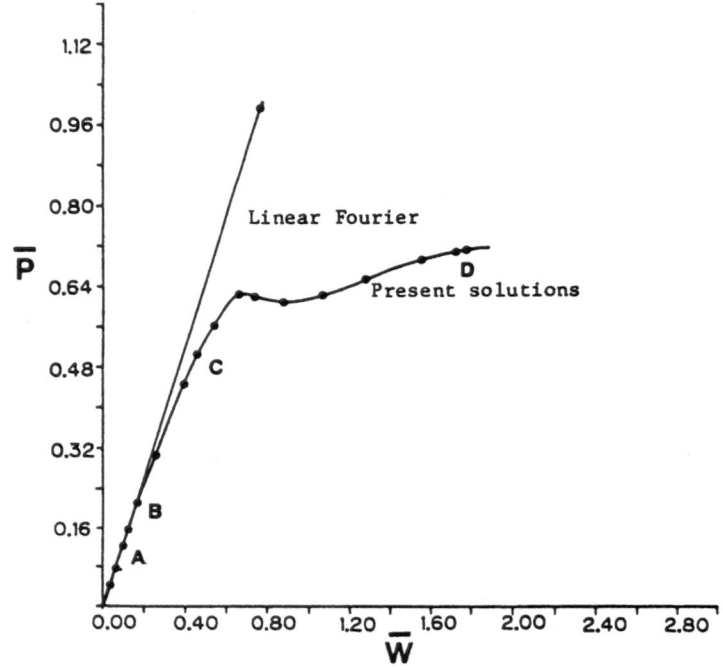

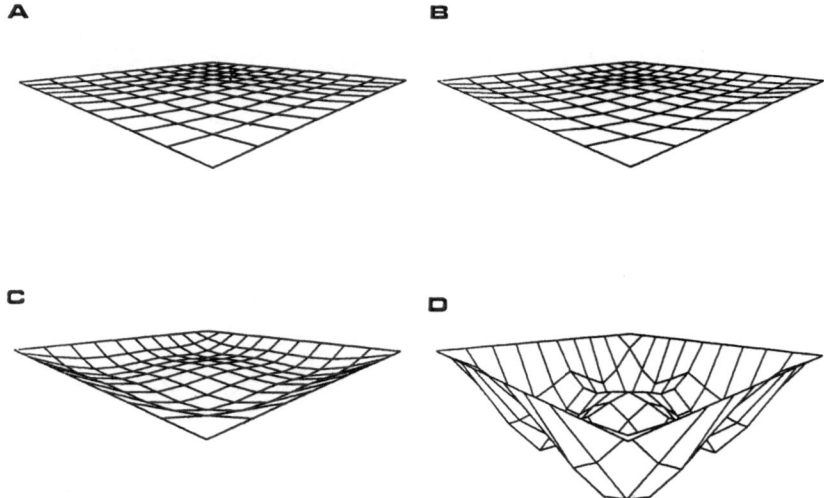

Fig.5 Load–average deflection curve and deflection modes
for H/h=5.0

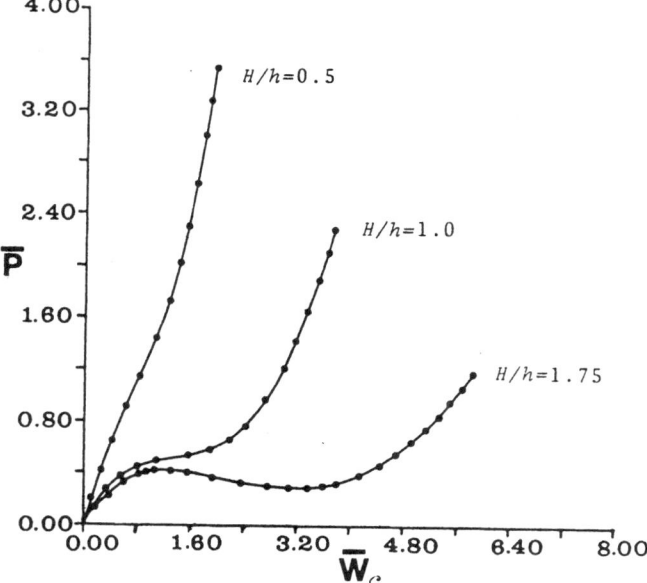

Fig.6 Load-central deflection curves

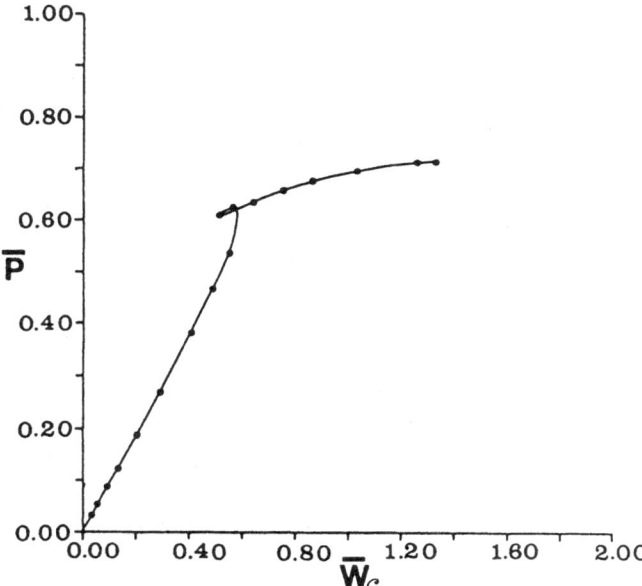

Fig.7 Load-central deflection curve for H/h=5.0

CONCLUSIONS

An integral equation method is applied to the geometrically
nonlinear analysis of shallow spherical shell bending problems.
Our integral equation formulation is based on the normal
displacement-stress function approach.   Nonlinear algebraic
equations obtained from the nonlinear integral equations are
solved by using the Riks-Wempner method.   Throughout numerical
calculations, we can track completely the load-deflection
curves manifesting the well-known snap-through phenomena in
shell bending problems.   From the performance of numerical
experiments for each value of the rise to thickness ratio H/h,
we can verify that there are various deflection patterns
associated with nonlinear bahaviour of shallow spherical
shell.

REFERENCES

1.   Brebbia C.A. Telles J.C.F. and Wrobel L.C. (1984). Boundary
     Element Techniques, Springer-Verlag, Berlin, Heidelberg,
     New York and Tokyo.
2.   Tosaka N. and Miyake S. (1985). Bending Analysis of
     Shallow Spherical Shells Using the Boundary Element Method,
     Theoretical and Apllied Mechanics, Vol.33, pp.195-204.
3.   Miyake S. and Tosaka N. (1985). Nonlinear Bending Analysis
     of Shallow Spherical Shell by the Integral Equation Methods
     Proceedings of the 2nd Japan National Symp. on B.E.M.,
     JASCOME, pp.257-262,(in Japanese).
4.   Tosaka N. and Miyake S. (1986). Large Deflection Analysis
     of Shallow Spherical Shell Using an Integral Equation
     Method, Proceeding of Int. Conf. on Boundary Elements,
     Beijing, China,(in press).
5.   Tosaka N. and Miyaka S. (1985). Nonlinear Analysis of
     Elastic Shallow Shells by Boundary Element Method,
     Boundary Elements VII (Ed. Brebbia C.A. and Maier G.),
     pp.4-43 to 4-52, Proceedings of the 7th Int. Conf., Villa
     Olmo, Lake Como, Italy, Springer-Verlag, Berlin, Heidelberg
     New York and Tokyo.
6.   Miyake S. (1986). Integral Equation Analyses for
     Geometrically Nonlinear Problems, Dr. Thesis, College of
     Industrial Technology, Nihon University,(in Japanese).
7.   Tosaka N. (1984). Fundamental Formulations in Elastic
     Shallow Shell Bending Problem by a Boundary Integral
     Equation, Proceedings of The 1st Japan National Symp. on
     B.E.M., JASCOME, pp.295-300,(in Japanese).
8.   Ram E. (1982). The Riks/Wempner Approach-An Extension of
     the Displacement Control Method in Nonlinear Analyses,
     Recent Advances in Nonlinear Computational Mechanics
     (Ed. Hinton E., Owen D.R.J. and Taylor C.), pp.63-86,
     Pineridge Press Limited, Newton Road, Mumbles, Swansea,
     U.K..

Elastic Buckling Analysis of Assembled Plate Structures by Boundary Element Method

M. Tanaka
Department of Mechanical Engineering, Shinshu University, 500 Wakasato, Nagano 380, Japan
K. Miyazaki
Seiko-Epson, Co., Shiojiri City, Nagano Prefecture, Japan

ABSTRACT

In this paper, we first discuss the integral equation formulation for the buckling problem of a single plate, using the biharmonic fundamental solution for the plate bending problems. The so called boundary-volume element method previously proposed by the senior author is applied to the numerical solution of the resulting set of integral equations. The total set of simultaneous equations are derived for nodal unknowns included in the whole domain, and reduced to eigenvalue equations. The method of solution is extended to the solution of elastic buckling of assembled plate structures. A few examples are computed and the results obtained are compared with other solutions to demonstrate the potential usefulness of the proposed method.

INTRODUCTION

In the last decades the boundary element methods (BEM) have attracted the attention of engineers because of their numerical efficiency in a wide variety of applications[1-5]. Several attempts have been also made for the bending problem of plates and assembled plate structures[6-9]. For the buckling problem of plates, however, there are still few investigations and no generalized boundary element solution procedures are available.

In the buckling problem we must take into account a

combined action of in-plane and out-of-plane deformation.
The governing differential equation of this problem includes
the biharmonic differential operator and also second-order
partial differential operators. Therefore, some innovative
treatment should be made for the integral equation formulation
and its numerical implementation.

In this paper, we discuss first the integral equation
formulation for the buckling problem of a single plate,
using the biharmonic fundamental solution for the plate
bending problem. The so called boundary-volume element
method previously proposed by the senior author[10] is applied
to the numerical solution to the resulting set of integral
equations. Namely, the plate boundary is divided into
boundary elements and also the inner domain is subdivided
to finite elements. The total set of simultaneous equations
are derived for nodal unknowns included in the whole domain.
Then, these equations are reduced to an algebraic set of
eigenvalue equations by taking account of the prescribed
homogeneous boundary conditions.

After the power of the proposed method is demonstrated
through several sample computations, the method is extended
to the elastic buckling of assembled plate structures.
Emphasis is placed on the rigorous numerical implementation
of the compatibility and equilibrium equations along the
jointed edges of plate components. A few examples are
computed by the computer program developed in this study
and compared with other solutions available in the literature.

BUCKLING OF A SINGLE PLATE

The boundary and the inner domain of the plate under
consideration are denoted by Γ and Ω, respectively. The
cartesian coordinate systen $0-x_1x_2x_3$ is introduced in which
the plane x_1-x_2 lies on the middle plane of the plate as
shown in Fig. 1. It is assumed that the plate material is
isotropic and homogeneous.

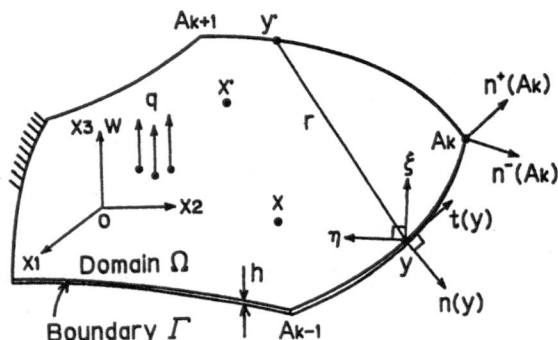

Fig. 1 Notation and coordinate system

To reduce the buckling problem of plates to an eigenvalue problem, the distributions of membrane forces N_{11}, N_{12} and N_{22} should be known when the plate subjected to the reference in-plane load. It is assumed that the membrane forces in the buckling state is λ times the values under the reference in-plane load. Thus, the governing differential equation of the buckling problem can be expressed as follows[11]:

$$\nabla^4 w = \frac{\lambda}{D} (N_{11}w_{,11} + N_{22}w_{,22} + 2N_{12}w_{,12}) \tag{1}$$

where w denotes the lateral deflection of the plate, the biharmonic differential operator and D the flexural rigidity of the plate. In addition, partial differentiation is denoted such that $w_{,11} = \partial w / \partial x_1^2$, etc. If equation (1) is solved under the prescribed boundary conditions, the buckling load can be obtained as λ times the reference in-plane load.

For the integral equation formulation of the buckling problem, the pseudo lateral load q defined by

$$q = \lambda (N_{11}w_{,11} + N_{22}w_{,22} + 2N_{12}w_{,12}) \tag{2}$$

is assumed to act in the lateral direction of the plate. Then, we can formulate equation (1) in the same manner as in the plate bending problem. Namely, making use of the well known biharmonic fundamental solution, we finally arrive at the following set of boundary integral equations:

$$\kappa(y)w(y) + \int_\Gamma [V_{n_1}{}^*(y,y')w(y')$$

$$-M_{n_1}{}^*(y,y')T_n(y') + T_{n_1}{}^*(y,y')M_n(y') - w_1{}^*(y,y')V_n(y')]d\Gamma(y').$$

$$+ \sum_{k=1}^{\kappa} [M_{nt_1}{}^*(y,A_k)w(A_k) - w_1{}^*(y,A_k)M_{nt}(A_k)]$$

$$= \lambda \int_\Omega w_1{}^*(y,x')[N_{11}(x')w_{,11}(x') + N_{22}(x')w_{,22}(x')$$

$$+ 2N_{12}(x')w_{,12}(x')]d\Omega(x') \tag{3}$$

$$\kappa_\xi(y)\frac{\partial w(y)}{\partial \xi} + \kappa_\eta(y)\frac{\partial w(y)}{\partial \eta}$$

$$+ \int_\Gamma [V_{n_2}{}^*(y,y')\widehat{w}(y') - M_{n_2}{}^*(y,y')T_n(y')$$

$$+ T_{n_2}{}^*(y,y')M_n(y') - w_2{}^*(y,y')V_n(y')]d\Gamma(y')$$

$$+ \sum_{k=1}^{\kappa} [M_{nt_2}{}^*(y,A_k)\widehat{w}(A_k) - w_2{}^*(y,A_k)M_{nt}(A_k)]$$

$$= \lambda \int_\Omega w_2{}^*(y,x')[N_{11}(x')w_{,11}(x') + N_{22}(x')w_{,22}(x')$$

$$+ 2N_{12}(x')w_{,12}(x')]d\Omega(x') \tag{4}$$

where $\hat{w}(y') = w(y') - w(y)$, and κ , κ_ξ and κ_η are the coefficients depending only on the geometrical property of the plate boundary. It is noted that T_n , M_n , M_{nt} and V_n are the slope, the bending moment, the twisting moment and the effective shearing force, respectively, while $[\cdot]$ means the discontinuity jump at a corner point of the plate boundary. The two-point functions denoted by the asterisk can be derived from the biharmonic fundamental solution.

The right-hand sides of equations (3) and (4) include the unknowns $w,_{11}$, $w,_{12}$ and $w,_{22}$ in the inner domain Ω . To express these unknowns in terms of the boundary variable, we will use the integral equations which can be obtained by differentiation of the integral equation (3) with the coefficient $\kappa = 1$ for an internal point. That is,

$$w,_{ij}(x) + \int_\Gamma [V_{n_1}^*,_{ij}(x,y')w(y')$$

$$-M_{n_1}^*,_{ij}(x,y')T_n(y') + T_{n_1}^*,_{ij}(x,y')M_n(y')$$

$$-w_1^*,_{ij}(x,y')V_n(y')]d\Gamma(y')$$

$$+\sum_{k=1}^{\kappa}[M_{nt_1}^*,_{ij}(x,A_k)w(A_k) - w_1^*,_{ij}(x,A_k)M_{nt}(A_k)]$$

$$=\lambda\int_\Omega w_1^*,_{ij}(x,x')[N_{11}(x')w,_{11}(x') + N_{22}(x')w,_{22}(x')$$

$$+2N_{12}(x')w,_{12}(x')]d\Omega(x') \tag{5}$$

where i, j = 1, 2. For the discontinuity jump $[M_{nt}]$ arising at a corner point which can be interpreted as a concentrated force, we can use the following relationship:

$$M_{nt}(y) = -D(1-\nu)\frac{\partial T_n(y)}{\partial t} \tag{6}$$

where t denotes the unit tangent along the plate boundary. Making use of these relations, we can reduce the number of unknowns to w, T_n , M_n , V_n on the plate boundary as well as $w,_{11}$, $w,_{12}$ and $w,_{22}$ in the inner domain Ω , in totality 8 variables.

Now, the boundary conditions can be expressed such that

$w = 0$, $T_n = 0$ (clamped)

$w = 0$, $M_n = 0$ (simply supported) (7)

$M_n = 0$, $V_n = 0$ (free)

The buckling problem can be thoroughly analyzed if the integral equations (3), (4) and (5) can be solved in an appropriate manner under the boundary conditions.

The so called boundary-volume element method[10] can be applied to a numerical solution of the integral equations

(3), (4) and (5). That is, the plate boundary is divided into boundary elements and also the inner domain is subdivided into regular finite elements. In order to maintain good numerical accuracy care should be taken for the evaluation of integrals including the singularities inherent to the asterisked two-point functions. It is recommended that singular integrals should be evaluated in an analytical manner, while the other integrals can be computed by means of the usual Gaussian numerical integration formula.

Taking account of the boundary conditions (7), we can reduce the discretized set of the integral equations (3), (4) and (5) to the following system of simultaneous equations:

$$[A_{BB}]\{x_B\} + \lambda[M_{BD}]\{\phi_D\} = \{0\} \tag{8}$$

$$\{\phi_D\} + [A_{DB}]\{x_B\} + \lambda[M_{DD}]\{\phi_D\} = \{0\} \tag{9}$$

where $[A_{BB}]$ to $[M_{DD}]$ are the coefficient matrices whose components are known, while $\{X_B\}$ and $\{\phi_D\}$ denote the column vectors of unknown nodal values on the plate boundary and in the inner domain, respectively. Eliminating $\{X_B\}$ in equations (8) and (9), we finally obtain the algebraic set of eigenvalue equations as follows:

$$([A_{DB}][A_{BB}]^{-1}[M_{BD}] - [M_{DD}])\{\phi_D\} = \frac{1}{\lambda}\{\phi_D\} \tag{10}$$

It is an easy matter to find the minimum value of λ from equation (10) by using the usual solution procedure for eigenvalue problems, which is at present available at almost every computing center.

APPLICATION TO ASSEMBLED PLATE STRUCTURES

Now, we shall extend the solution procedure developed for a single plate to the buckling problem of assembled plate structures as shown in Fig.2. In this problem, a combined action of the in-plane (membrane) and the out-of-plane (bending) deformation should be exactly taken into account.

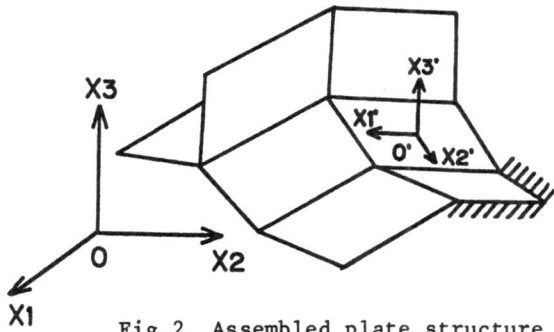

Fig.2 Assembled plate structure

It can be assumed that the in-plane (membrane) problem is
in a plane stress state. The boundary integral equation
for this problem can be expressed as

$$c_{ki}(y)u_i(y) + \int_\Gamma p_{ki}^*(y,y')u_i(y')d\Gamma(y')$$
$$= \int_\Gamma u_{ki}^*(y,y')p_i(y')d\Gamma(y') \qquad (11)$$

where the asterisked functions correspond to Kelvin's
fundamental solution, while u_j and p_j are the displacement
and traction components in the middle plane of the plate.
Equation (11) can be solved numerically through the usual
boundary element method.

Next, we consider the solution procedure for the buckling
problem of assembled plate structures. The local cartesian
coordinate system is used for each plate component as shown
in Fig. 2. The necessary set of integral equations for every
plate component can be obtained in terms of the variable
expressed in the local coordinate system. It is noted again
that the distributions of membrane forces N_{ij} are known
under the reference in-plane load. The discretization of
the integral equations (3), (4), (5) and (11) through the
boundary-volume element method leads to

$$[H_\theta]\begin{Bmatrix} w \\ T_n \end{Bmatrix} + [G_\theta]\begin{Bmatrix} V_n \\ M_n \end{Bmatrix} + \lambda[J_\theta]\begin{Bmatrix} w_{,11} \\ w_{,22} \\ w_{,12} \end{Bmatrix} = \{0\} \qquad (12)$$

$$\begin{Bmatrix} w_{,11} \\ w_{,22} \\ w_{,12} \end{Bmatrix} + [H_D]\begin{Bmatrix} w \\ T_n \end{Bmatrix} + [G_D]\begin{Bmatrix} V_n \\ M_n \end{Bmatrix} + \lambda[J_D]\begin{Bmatrix} w_{,11} \\ w_{,22} \\ w_{,12} \end{Bmatrix} = \{0\} \qquad (13)$$

$$[h_\theta]\begin{Bmatrix} u_1 \\ u_2 \end{Bmatrix} + [g_\theta]\begin{Bmatrix} p_1 \\ p_2 \end{Bmatrix} = \{0\} \qquad (14)$$

where the coefficient matrices from $[H_B]$ to $[g_B]$ are
known. Combining equations (12) and (13) we obtain

$$\begin{bmatrix} h_\theta & 0 \\ 0 & H_\theta \end{bmatrix}\begin{Bmatrix} u_1 \\ u_2 \\ w \\ T_n \end{Bmatrix} + \begin{bmatrix} g_\theta/h & 0 \\ 0 & G_\theta \end{bmatrix}\begin{Bmatrix} p_1 h \\ p_2 h \\ V_n \\ M_n \end{Bmatrix} + \lambda\begin{bmatrix} 0 \\ J_\theta \end{bmatrix}\begin{Bmatrix} w_{,11} \\ w_{,22} \\ w_{,12} \end{Bmatrix} = \{0\} \qquad (15)$$

It is noted that the nodal vectors of the traction $\{p_1\}$ and
$\{p_2\}$ are multiplied by the factor h (plate thickness)
to identify their dimension with that of the effective
shearing force $\{V_n\}$. Equation (13) can be rewritten as
follows:

$$\begin{Bmatrix} w_{,11} \\ w_{,22} \\ w_{,12} \end{Bmatrix} + [0 \ H_D] \begin{Bmatrix} u_1 \\ u_2 \\ w \\ T_n \end{Bmatrix} + [0 \ G_D] \begin{Bmatrix} p_1 h \\ p_2 h \\ V_n \\ M_n \end{Bmatrix} + \lambda [J_D] \begin{Bmatrix} w_{,11} \\ w_{,22} \\ w_{,12} \end{Bmatrix} = \{0\} \qquad (16)$$

In order to derive the set of linear equations for the assembled plate structure, the nodal values $\{u_1\}$, $\{u_2\}$, $\{p_1 h\}$, $\{p_2 h\}$ and $\{V_n\}$ in the local coordinate system should be transformed into the global coordinate system. The variables in the global coordinate system are the displacement $_0 u_i$ and the resulting force $_0 f_i$ $(i = 1,2,3)$ as well as the slope T_n and the moment M_n , in totality 8 variables. Furthermore, the variables $w_{,11}$, $w_{,12}$ and $w_{,22}$ should be taken into account. Using these variables we can obtain the set of equations in the global coordinate system for each plate component of the assembled structure.

On the jointed edge between the plate components, we have to consider the compatibility conditions of the displacement and the slope as well as the equilibrium conditions of the resulting forces and moments. Provided that n plate components are jointed together along the edge under consideration, the compatibility of the displacement leads to

$$_0 u_i^1 = _0 u_i^2 = \cdots\cdots = _0 u_i^n \qquad (17)$$

and the equilibrium of the resulting force yields

$$_0 f_i^1 + _0 f_i^2 + \cdots\cdots + _0 f_i^n = 0 \qquad (18)$$

where the superscript denotes the number of the plate component. Supposed that the angle between the plate components is unaltered before and after deformation of the assembled plate structure, the compatibility conditions of the slope can be expressed by

$$\beta^1 T_n^1 = \beta^2 T_n^2 = \cdots\cdots = \beta^n T_n^n \qquad (19)$$

The equilibrium of the bending moment on the jointed edge yields

$$\beta^1 M_n^1 + \beta^2 M_n^2 + \cdots\cdots + \beta^n M_n^n = 0 \qquad (20)$$

In equations (19) and (20) β^i $(i = 1,2,\ldots,n) = 1$ when the tangential direction s of the jointed edge is the same as that of the reference plate component, and otherwise $\beta^i = -1$. For example, if the reference plate is assumed to be plate 1 in Fig.3, β's values become such that

$$\begin{aligned} \beta^1 &= 1 \ , \ \beta^2 = -1 \quad \text{for Fig.3(a)} \\ \beta^1 &= 1 \ , \ \beta^2 = 1 \quad \text{for Fig.3(b)} \end{aligned} \qquad (21)$$

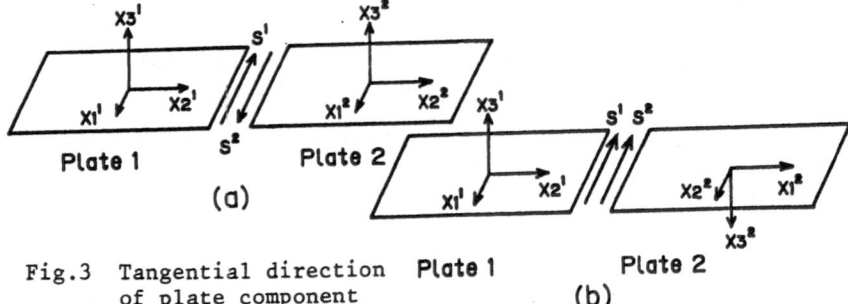

Fig.3 Tangential direction **Plate 1**   **Plate 2**
of plate component   **(b)**

It is noted that the boundary conditions on the periphery
of the assembled plate structure can be expressed as

$$_0u_i = 0 \;,\; T_n = 0 \qquad \text{(clamped)}$$
$$_0u_i = 0 \;,\; M_n = 0 \qquad \text{(simply supported)} \qquad (22)$$
$$M_n = 0 \;,\; _0f_i = 0 \qquad \text{(free)}$$

Through the above-mentioned treatment together with
the homogeneous boundary conditions, we finally obtain the
set of linear simultaneous equations for the assembled
structures in the following form:

$$[A]\{x\}+\lambda[M]\{\phi\}=\{0\} \tag{23}$$

$$\{\phi\}+[A']\{x\}+\lambda[M']\{\phi\}=\{0\} \tag{24}$$

where [A] to [M'] are the coefficient matrices whose
components are known, while $\{X\}$ denotes the column vector
of nodal unknowns on the periphery of the structure and $\{\phi\}$
the column vector of nodal unknowns on the jointed edges
as well as in the inner domains of the plate components.

Eliminating $\{X\}$ from equations (23) and (24), we can
arrive at the following algebraic set of eigenvalue equations.
That is,

$$([A'][A]^{-1}[M]-[M'])\{\phi\}= \frac{1}{\lambda} \{\phi\} \tag{25}$$

The minimum value of λ gives the buckling load, which can
be obtained by multiplying the in-plane reference load by
the factor λ .

NUMERICAL RESULTS AND DISCUSSION

On the basis of the theoretical considerations mentioned
above, a new computer program has been developed in this
study. Computational results obtained by this program for

several sample problems will be shown in the following. In all the computations the material constants are assumed such that

Poisson's ratio $\nu = 0.3$
Young's modulus $E = 2.058 \times 10^2$ GPa

The outer boundary as well as the jointed edges of the assembled structure are divided into constant boundary elements and the inner domain is subdivided into constant volume elements in which the variables $w_{,11}$, $w_{,12}$ and $w_{,22}$ are assumed to be piece-wise constant.

First, attempt is made to check numerical accuracy of the present solution procedure. The buckling problem of a square plate with the edge length a and the flexural rigidity D is investigated for the following three kinds of boundary conditions:

(1) All the edges are clamped and the plate is subjected to uniform compressive in-plane load from the four edges.
(2) Two counter edges are clamped and the other two edges are simply supported. Uniform compressive in-plane load is applied at the simply supported edges.
(3) All the edges are simply supported and the plate is subjected to uniform in-plane shearing force along the four edges.

The whole boundary is equally divided into 48 constant boundary elements and the inner domain is also equally subdivided into 32 triangular constant volume elements. The aspect ratio is assumed as $a/h = 240$ (h is the plate thickness). In Table 1 comparison is made between the present solution and the analytical solution available in the literature[11]. The results obtained for the cases (1) and (2) are in excellent agreement with the analytical ones. A relatively large error in the case (3) seems to be improved by modification of the volume element subdivision.

Table 1 Results on buckling of square plate under various boundary conditions

Boundary condition and in-plate Force	$\dfrac{N_{cr}a^2}{\pi^2 D}$		Error (%)
	BEM	Analytical	
(1)	5.43	5.33	1.95
(2)	0.95	0.95	0.17
(3)	9.86	9.34	5.59

As the first example of the assembled plate structure, we analyze the buckling problem of the plate structure which consists of two equal rectangular plates jointed at the right angle along the one edge as shown in Fig.4(a). The top and bottom boundaries of this structure are simply supported, while the other sides are free. We also analyze a single plate with three edges clamped and the other one free as shown in Fig.4(b). The aspect ratios are assumed such that b/a = 0.5 and a/h = 240. The element subdivision is shown in Fig.5 in which Fig.(a) corresponds to Fig.4(a) and Fig.(b) to Fig.4(b). The results obtained for the critical in-plane load N_{cr} are shown in Table 2 where comparison is made with the analytical solutions[11]. It can be seen that both the results are in good agreement. More excellent agreement could be obtained if the element subdivision is improved.

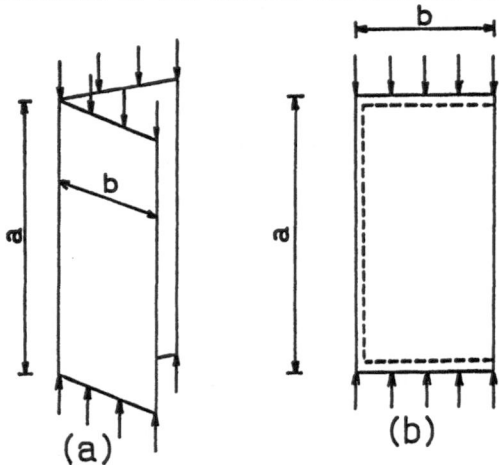

Fig.4 Buckling problems of assembled structure and its equivalent single plate

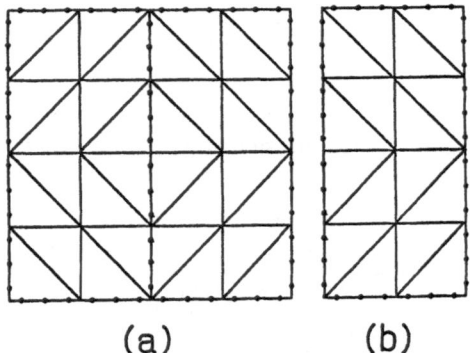

Fig.5 Discretization of plate structures shown in Fig.4

Table 2   Results on plate sturctures of Fig.4

	$\dfrac{N_{cr} b^2}{\pi^2 D}$		Error (%)
	BEM	Analytical	
Assembled Plate Structure	0. 67	0. 70	4. 06
Thin Plate	0. 68	0. 70	2. 11

Next, we consider the assembled plate structure where four equal square plates are jointed to produce a hollow cylinder as shown in Fig.6.   The structure is subjected to the in-plane load applied at the top and bottom boundaries in the direction of the cylinder axis.   It is assumed that the top and bottom edges are all simply supported and that a/h = 240.   It is well known[11] that the lowest buckling mode of this structure will be like Fig.6(b).   For the sake of comparison we also analyze the square plate with four edges simply supported and subjected to the same in-plane load as that of the hollow boxed structure.   The element discretization is shown in Fig.7.   The results obtained are compared with the analytical solutions[11], together with those of single square plate, in which N_{cr} means the buckling load.   Good agreement can be recognized between both the results.   Further improvement of numerical accuracy can be expected if the element subdivision is made in a more appropriate manner.

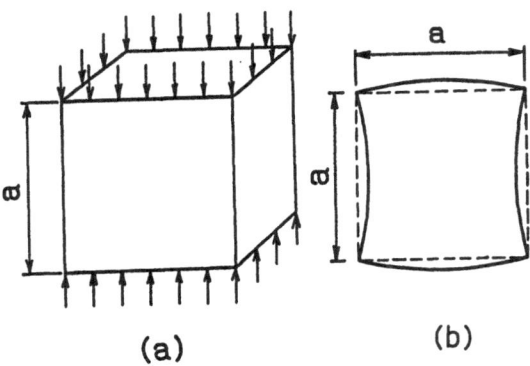

(a)   **(b)**

Fig.6   Buckling problem of four-plate boxed structure and its buckling mode

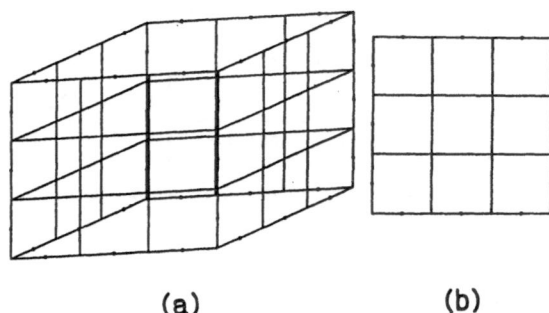

(a)   **(b)**

Fig.7   Discretization of plate structure
of Fig.6 and its equivalent single plate

Table 3   Results on plate structure of Fig.6
and its equivalent single plate

	$\dfrac{N_{cr}a^2}{\pi^2 D}$		Error (%)
	BEM	Analytical	
Assembled Plate Structure	3.76	4.00	5.99
Thin Plate	3.76	4.00	5.97

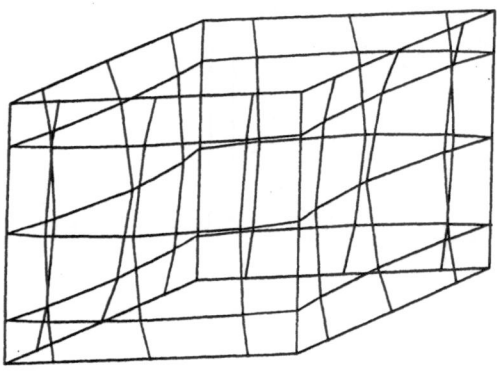

Fig.8   Buckling mode output for plate structure
of Fig.6

CONCLUDING REMARKS

The integral equation approach has been presented for the
buckling problem of assembled plate structures. After the
problem of a single plate was investigated by means of the
boundary-volume element method, the method was extended to
the problem of assembled plate structures. The potential
usefulness of the proposed method has been demonstrated
through the numerical computation of several sample
problems.

Since the proposed method is in principle not limited
by the number of plate components of the assembled plate
structure, it is easily applicable to more complicated plate
structures. Such practical applications of the method could
be recommended as the future research work.

REFERENCES

1.   Brebbia, C.A.(ed.) (1983). Progress in Boundary Element
      Methods, Vols. 1 & 2, Pentech Press, London.
2.   Banerjee, P.K. and Butterfield, R. (1981). Boundary
      Element Methods in Engineering Science, McGraw-Hill,
      London-New York.
3.   Brebbia, C.A. Futagami, T. and Tanaka, M.(eds.) (1983).
      Boundary Elements, Springer-Verlag, Berlin-New York-
      Tokyo.
4.   Brebbia, C.A. Telles, J.C.F. and Wrobel, L.C. (1984).
      Boundary Element Techniques, Springer-Verlag, Berlin-
      New York-Tokyo.
5.   JASCOME(ed.) (1986). Boundary Element Methods - Theory
      and Applications, Corona, Tokyo.
6.   Stern, M. (1979), A General Boundary Integral Formulation
      for the Numerical Solution of Plate Bending Problems,
      Int. J. Solids and Structures, Vol.15, pp.769-782.
7.   Bezine, G. (1978), Boundary Integral Formulation for
      Plate Flexure with Arbitrary Conditions, Mech. Res.
      Comm., Vol.5, pp.197-206.
8.   Tanaka, M. and Miyazaki, K. (1985), A Direct Boundary
      Element Method for Elastic Bending Analysis of Plates,
      Trans. Japan Soc. Mech. Engrs., Ser.A, Vol.51, pp.1636-
      1641.
9.   Tanaka, M. and Miyazaki, K. (1985), A Direct BEM for
      Elastic Plate-Structures Subjected to Arbitrary Loadings,
      Boundary Elements VII, ed. by C.A. Brebbia and G. Maier,
      Springer-Verlag, Berlin-New York-Tokyo, pp.4/3-4/16.
10.   Tanaka, M. and Tanaka, K. (1981), On a New Boundary
      Element Solution Scheme for Elastoplasticity, Ingenieur-
      Archiv, Vol.50, pp.289-295.
11.   Timoshenko, S.P. and Woinowsky-Krieger, S. (1959). Theory
      of Plates and Shells, 2nd ed., McGraw-Hill, New York-
      Tokyo.

An Alternative Boundary Element Analysis of Plates Resting on Elastic Foundation

N. Kamiya, Y. Sawaki
Department of Mechanical Engineering, Mie University, Kamihamacho, Tsu 514, Japan

Key words: Boundary element method, Thin plate, Elastic
foundation, Biharmonic differential equation

Abstract

An alternative boundary element deflection analysis of thin
elastic plates resting on Winkler-type elastic foundation
is presented in this paper. Integral equations formulated
using conventional fundamental solution to the biharmonic
differential operator include domain integrals relating to
the reaction term of the original differential equation.
The present study attempts to overcome this shortcoming by
the following scheme. First, distribution of the reaction
from the elastic foundation is approximated by the
polynomials in terms of its value at some internal and
boundary points of the plate. Next, the above-mentioned
domain integrals are transformed into equivalent boundary
integrals which is estimated on the boundary alone. The
transformed reaction is considered as inhomogeneous term
and successive iteration scheme is employed for the
solution of the integral equation. The proposed method,
combination of transformation and iteration, suggests a
possibility of the analogous scheme to nonlinear problems.
The effectiveness of the method is examined through some
numerical examples for which analytical solutions are
available.

INTRODUCTION

In recent years, the boundary element method (BEM) is
increasingly getting popularity among numerical methods.
One of the main reasons of its development is a
possibility of reduced dimensionality of the problem, which
leads to reduced set of equations and to a smaller amount
of data required for computation. For the linear and
homogeneous problems, the boundary element method is
especially effective because integral equations are
formulated only on the boundary of the domain under
consideration. When, however, nonlinear or inhomogeneous
problems are formulated by the BEM, the integral equations
usually include domain integrals. Their transformation
into equivalent boundary integrals increases surely
computational advantage of the BEM.

Integral equation formulation is generally done by using
the respective fundamental solutions to a whole or a main
part of linear differential operator of the governing
equation. For the former case, integral equation is
formulated only with respect to the boundary unknowns while
the fundamental solution is generally more complicated than
that for the latter. For the latter case, the fundamental
solution relates only a part of the differential operator,
and therefore the remaining terms are thought to be
additional inhomogeneous terms which appear as domain
integral in the integral equation formulation. Two
different approaches to the estimation of the such domain
integrals are possible: i) we solve integral equations
formulated for the domain unknowns as well as for the
boundary simultaneously, or ii) we estimate the domain
integrals through successive iteration as same as we do
when solving the nonlinear problem by the BEM.

In this paper, we will try to consider the transformation
of the such domain integrals into equivalent boundary
integrals and employ them during successive iteration
calculation. In order to examine the usefulness of the
above-mentioned coupled method of transformation and
iteration, we consider boundary element deflection analysis
of thin elastic plates resting on Winkler-type elastic
foundation as an example problem. Integral equations are
formulated by using the fundamental solution to the
biharmonic differential equation which is a part of the
differential operator of the original governing equation.
Therefore they include the domain integral relating to the
reaction term of the original differential equation. The
distribution of reaction from the foundation is supposed to
be represented by the linear combination of the
functions of distance between the point on which the
magnitude of reaction is evaluated and the other arbitrary
points. Some of the such points are located both on the

plate boundary and inside the domain. And then the above-
mentioned domain integrals are transformed into equivalent
boundary integrals. Analogous transformation was proposed
and employed in the elastodynamics by Brebbia and Nardini
[1]. Numerical results for clamped circular plates are
obtained and compared with the corresponding analytical
solution to demonstrate the effectiveness of the proposed
alternative approach.

GOVERNING DIFFERENTIAL EQUATION AND ITS INTEGRAL
FORMULATION

The governing differential equation for a thin elastic
plate of constant thickness h rested on the Winkler-type
elastic foundation with stiffness k and subjected to
arbitrary lateral load $q(x, y)$ is represented, with
reference to the rectangular cartesian coordinate system x,
y, z, as follows:

$$D\nabla^4 w + kw = q \tag{1}$$

where w, D and ∇^4 denote the deflection, the bending
rigidity of the plate and the biharmonic differential
operator, respectively. The middle plane of the plate
occupies a two-dimensional domain S bounded by a smooth
curve C.

We can formulate the governing equation (1) as integral
equation in two ways by selecting one of distinct
fundamental solutions v_1 and v_2, defined by

$$D\nabla^4 v_1(P, Q) = \delta(P, Q) \tag{2}$$

and

$$D\nabla^4 v_2(P, Q) + kv_2(P, Q) = \delta(P, Q) \tag{3}$$

where $\delta(P, Q)$ is Dirac delta function and P and Q arbitrary
points on the x-y plane. The both solutions have been
obtained analytically. The advantage of employing the
fundamental solution v_2 is that the boundary integral
equation for the deflection is formulated only on the
boundary of the plate [2-4]. But v_2 and its related
kernels defined later are more complicated than those for
v_1. On the other hand, if we employ the solution v_1,
integral equation includes the domain integral due to the
unknown reaction term kw multiplied by v_1.

For a smooth boundary C, the following generalized Green
integral identity holds for functions w and v:

$$- \int_S (v\nabla^4 w - w\nabla^4 v) \, dS$$

$$= \frac{1}{D} \int_C [vK(w) - \frac{\partial v}{\partial n} M(w) + \frac{\partial w}{\partial n} M(v) - wK(v)] \, ds \tag{4}$$

where differential operators K and M are defined as follows:

$$K = D\{(1-\nu)\frac{\partial}{\partial s} [\ell m(\partial^2/\partial x^2 - \partial^2/\partial y^2) - (\ell^2 - m^2)\partial^2/\partial x \partial y]$$
$$- \partial/\partial n \nabla^2\} \tag{5}$$

$$M = - D[\nu\nabla^2 + (1-\nu)(\ell^2 \partial^2/\partial x^2 + m^2 \partial^2/\partial y^2 + 2\ell m \partial^2/\partial x \partial y)] \tag{6}$$

In the above equations, ℓ and m are the direction cosines of the outward normal n and $\partial/\partial s$ denotes the directional derivative along the tangent to the boundary. Substituting eq. (1) into eq. (4) and employing v_1 defined by eq. (2) as v , we finally obtain the following integral equation [5, 6]:

$$c(P)w(P) = \int_S [q(Q) - kw(Q)]v(P, Q) \, dS(Q)$$

$$+ \int_C \{v(P, Q)K[w(Q)] - \frac{\partial v}{\partial n}(P, Q)M[w(Q)] + \frac{\partial w}{\partial n}(Q)M[v(P, Q)]$$

$$- w(Q)K[v(P, Q)]\} \, ds(Q) \tag{7}$$

where $v(P, Q)$ can be expressed as follows:

$$v(P, Q) = \frac{1}{8\pi D} r^2(P, Q) \log r(P, Q) \qquad (r(P, Q) = |\overline{PQ}|) \tag{8}$$

and the coefficient $c(P)$ is

$$c(P) = \begin{cases} 1 & (P \in S) \\ \\ 1/2 & (P \in C) \end{cases} \tag{9}$$

If the point P is taken on the boundary C, we have to estimate the singular boundary integrals of eq. (7) as Cauchy principal value integrals. By differentiating eq. (7) with respect to the outward normal n_0 at the point P taken on the boundary, we have

$$\frac{1}{2}\frac{\partial w}{\partial n_0}(P) = \int_S [q(Q) - kw(Q)]\frac{\partial v}{\partial n_0}(P, Q)dS(Q)$$

$$+ \int_C \{\frac{\partial v}{\partial n_0}(P, Q)K[w(Q)] - \frac{\partial^2 v}{\partial n_0 \partial n}(P, Q)M[w(Q)]$$

$$+ \frac{\partial w}{\partial n}(Q)\frac{\partial M}{\partial n_0}[v(P, Q)] - w(Q)\frac{\partial K}{\partial n_0}[V(P, Q)]\}ds(Q) \quad (10)$$

Equations (7) and (10) constitute a system of simultaneous integral equations in terms of the four basic boundary values, i.e., deflection w, normal slope $\partial w/\partial n$, normal bending moment M(w) and equivalent shear force K(w). The first terms of the right-hand-side of eqs. (7) and (10) are the domain integrals over S, which include the lateral load and the reaction from the foundation.

EVALUATION OF DOMAIN INTEGRALS

We consider the transformation of the domain integrals appearing in eqs. (7) and (10) into the equivalent boundary integrals. The domain integral terms due to the lateral load of relatively simple distribution multiplied by the fundamental solution (8) are conveniently transformed into the corresponding boundary integrals [7, 8]. Hence in what follows, we restrict ourselves to the evaluation of the domain integrals due to the reaction from the foundation:

$$I(P) = \int_S k\ w(Q)\ v(P, Q)dS(Q) \quad (11)$$

$$I'(P) = \int_S k\ w(Q)\ \frac{\partial v}{\partial n_0}(P, Q)dS(Q) \quad (12)$$

Following the idea of Brebbia and Nardini [1] proposed for the elastodynamic problem, the deflection w is supposed to be represented, for example, in the following linear equation:

$$w(Q) = \sum_{j=1}^{M} \alpha_j f_j(R, Q) = \sum_{j=1}^{M} \alpha_j f_j[\rho(R, Q)] \quad (13)$$

where f_j is the function of the distance ρ between the point Q for which the deflection is estimated and the arbitrary point R in the x-y plane, and α_j ($j = 1,2,---,M$) is a set of unknown coefficients. If we can find a function F_j corresponding to f_j satisfying

$$\nabla^4 F_j = f_j \quad (14)$$

eq. (11) may be transformed into the following boundary integral with the help of eq. (13) and the Rayleigh-Green integral identity:

$$I(P) = \sum_{j=1}^{M} k\alpha_j \{c(P)F_j(R, P) + \int_C [v(P, Q)\frac{\partial}{\partial n} \nabla^2 F_j(R, Q)$$

$$- \frac{\partial v}{\partial n}(P, Q)\nabla^2 F_j(R, Q) + \frac{\partial F_j}{\partial n}(R, Q)\nabla^2 v(P, Q)$$

$$- F_j(R, Q) \frac{\partial}{\partial n} \nabla^2 v(P, Q)]ds(Q)\} \tag{15}$$

where coefficient $c(P)$ is identical to the one defined in eq. (9). Similarly, eq. (12) is transformed into the following boundary integral form:

$$I'(P) = \sum_{j=1}^{M} k\alpha_j \{c(P)\frac{\partial F_j}{\partial n_0}(R, P) + \int_C [\frac{\partial v}{\partial n_0}(P, Q)\frac{\partial}{\partial n} \nabla^2 F_j(R, Q)$$

$$- \frac{\partial^2 v}{\partial n_0 \partial n}(P, Q)\nabla^2 F_j(R, Q) + \frac{\partial F_j}{\partial n}(R, Q)\frac{\partial}{\partial n_0} \nabla^2 v(P, Q)$$

$$- (F_j(R, Q) - F_j(R, P)) \frac{\partial^2}{\partial n_0 \partial n} \nabla^2 v(P, Q)]ds(Q)\} \tag{16}$$

We take here the following quadratic expression as the function f_j:

$$f_j(R, Q) = 1 - \rho(R, Q) - \rho^2(R, Q) \tag{17}$$

Equation (13) is represented in matrix form as follows:

$$\{w\} = [f]\{\alpha\} \tag{18}$$

If matrix $[f]$ has its inverse $[f]^{-1}$, a set of the unknown coefficients $\{\alpha\}$ is obtained by

$$\{\alpha\} = [f]^{-1}\{w\} \tag{19}$$

Some of arbitrary points R appearing in the eq. (13) are placed on the boundary (not necessarily to coincide with the boundary nodes) and some of them in the domain. The reason why the such points are located in the domain as well as on the boundary is that if they are taken only on the boundary of the plate with clamped or supported edges, for example, the deflection vector $\{w\}$ of eq. (19) vanishes identically, and therefore the corresponding coefficient vector $\{\alpha\}$ disappears, which is not appropriate to the present purpose. The deflection at the point R taken in the domain is calculated from eq. (7) with $c(P) = 1$.

BOUNDARY ELEMENT ANALYSIS

The simultaneous integral equations (7) and (10) can not be

solved analytically and hence their numerical solutions are
obtained using the boundary element method.   The boundary
of the plate is discretized into N straight line elements
with a node at their center.   Then equations (7) and (10)
are reduced to a following system of discretized algebraic
equations, represented in the matrix form:

$$[A]\{W\} + [B]\{T\} + [C]\{M\} + [D]\{K\} = \{H\} + \{I\}$$

(20)

$$[A']\{W\} + [B']\{T\} + [C']\{M\} + [D']\{K\} = \{H'\} + \{I'\}$$

where $[A]$, $[B]$, $---$, $[D']$ are N × N coefficient matrices,
$\{W\}$, $\{T\}$, $\{M\}$ and $\{K\}$ are the boundary value vectors
constructed of nodal values of w , $\partial w/\partial n$, M(w) and
K(w), respectively, $\{H\}$, $\{H'\}$ are the transformed boundary
integral terms resulting from the lateral load and $\{I\}$, $\{I'\}$
are the vectors due to the reaction term and represented as
follows:

$$\{I\} = [F]\{\alpha\}$$

(21)

$$\{I'\} = [F']\{\alpha\}$$

In the above equations, $[F]$, $[F']$ are N × M matrices
calculated numerically from eqs. (15) and (16) by
discretizing the boundary into N elements which can be same
as those employed for the calculation of matrices $[A]$, $[B]$,
etc. and $\{\alpha\}$ is a vector composed of a set of unknown
coefficients.

Two of the above-mentioned four boundary values w, $\partial w/\partial n$,
M(w) and K(w) are prescribed by the boundary condition and
the remaining two unknowns are determined from eqs. (20).
Since the vectors $\{I\}$, $\{I'\}$ contain the unknown quantities
as shown in eqs. (21), we employ numerical iteration to get
converged solution.

NUMERICAL EXAMPLES

Numerical solutions for a clamped circular plate subjected
to a concentrated load at the center (radius a,
concentrated load P), resting on an elastic foundation, are
obtained and compared in Table 1 with the corresponding
analytical solutions.   Figure 1 shows the distribution of
the points at which the deflections are calculated in order
to estimate eq. (19).   These points are also taken on the
boundary nodes, where the deflections are identically zero
in the present example problems.

In Table 1 (a) and (b), the center deflection obtained
respectively by using different boundary elements and the

various distribution of domain points as shown in Fig. 1 are presented for different magnitudes of the dimensionless foundation parameter λ ($\equiv ka^4/D$). The results obtained by the present BEM show relatively good agreement with those obtained analytically.

Another example is demonstrated for a clamped circular plate subjected to a eccentric concentrated load. The deflection distributions on the diameter including the loading point are shown for various λ's in Fig. 2.

The advantages of the present method are summarized as follows:
(i) to evaluate the domain integrals including unknown quantities, we take merely some domain points but need not to discretize the domain into cells,
(ii) it is enough only to descretize the boundary, and
(iii) this makes the input data structure simpler.

REFERENCES

1. Brebbia C. A. and Nardini D., Dynamic Analysis in Solid Mechanics by an Alternative Boundary Element Procedures, Int. J. Soil Dynamics and Earthquake Eng., 2, 228, 1983.

2. Katsikadelis J. T. and Armenakas A. E., Analysis of Clamped Plates on Elastic Foundation by the Boundary Integral Equation Method, J. Appl. Mech., Trans. ASME, 51, 574-580, 1984.

3. Katsikadelis J. T. and Armenakas A. E., Plates on Elastic Foundation by BIE Method, J. Eng. Mech., Proc. ASCE, 110, 1086-1105, 1984.

4. Costa J. C. A. and Brebbia C. A., Bending of Plates on Elastic Foundations Using the Boundary Element Method, Proc. Second Int. Conf. Variational Meth. Eng., Springer-Verlag, Brebbia C. A. ed., 5-23-33, 1985.

5. Bezine G., Boundary Integral Formulation for Plate Flexure with Arbitrary Boundary Conditions, Mech. Res. Comm., 5, 197-206, 1978.

6. Stern M., A General Boundary Integral Formulation for the Numerical Solution of Plate Bending Problems, Int. J. Solids Struct., 15, 769-782, 1979.

7. Kamiya N. and Sawaki Y., The Boundary Integral Formulation for Some Inhomogeneous Differential Equations, Eng. Anal., 1, 188-194, 1984.

8. Costa J. A. C. and Brebbia C. A., Plate Bending Problem Using BEM, Proc. Sixth Int. Conf. Boundary Elements, Springer-Verlag, Brebbia C. A. ed., 3-43-63, 1984.

Table 1.   Center deflection Dw_c/Pa^2 ($\times$ 10)

(a)   Number of domain points = 25

λ	EXACT	Number of boundary elements		
		4 8	3 6	2 4
0	0.1989	0.1982	0.1977	0.1961
1	0.1973	0.1963	0.1957	0.1940
2	0.1958	0.1945	0.1938	0.1920
3	0.1942	0.1927	0.1919	0.1900
4	0.1927	0.1909	0.1901	0.1881
5	0.1912	0.1891	0.1883	0.1862
6	0.1897	0.1874	0.1865	0.1843
7	0.1882	0.1858	0.1848	0.1825
8	0.1868	0.1841	0.1831	0.1808
9	0.1854	0.1825	0.1814	0.1780
1 0	0.1839	0.1810	0.1798	0.1773

(b) Number of boundary elements = 48

λ	EXACT	Number of domain points			
		4 9	2 5	1 3	7
0	0.1989	0.1982	0.1982	0.1982	0.1982
1	0.1973	0.1963	0.1963	0.1964	0.1966
2	0.1958	0.1944	0.1945	0.1946	0.1950
3	0.1942	0.1925	0.1927	0.1929	0.1934
4	0.1927	0.1906	0.1909	0.1912	0.1918
5	0.1912	0.1889	0.1891	0.1895	0.1903
6	0.1897	0.1871	0.1874	0.1878	0.1888
7	0.1882	0.1854	0.1858	0.1862	0.1874
8	0.1868	0.1837	0.1841	0.1846	0.1859
9	0.1854		0.1825	0.1831	0.1845
1 0	0.1839		0.1810	0.1816	0.1831

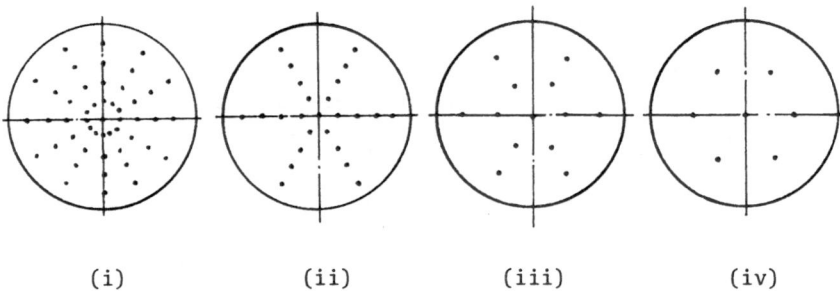

(i)   (ii)   (iii)   (iv)

Figure 1.   Distribution of domain points
(Number of domain points: (i) 49
(ii) 25, (iii) 13, (iv) 7).

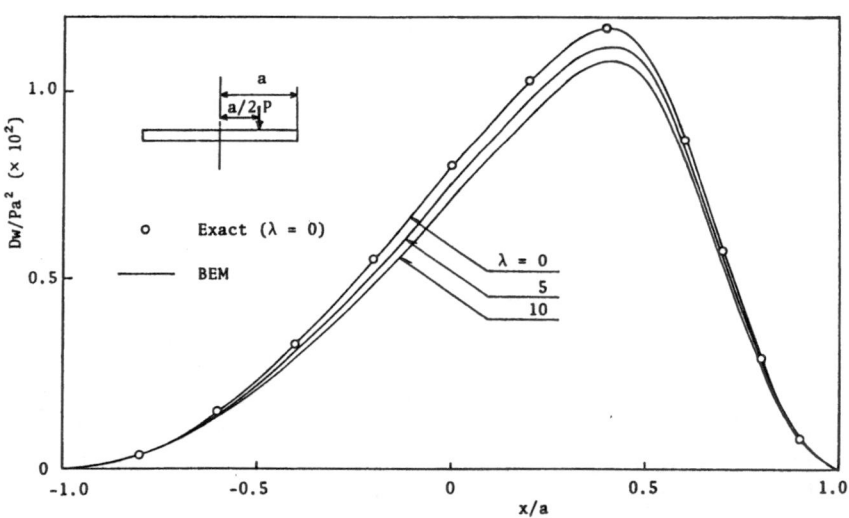

Figure 2.   Distribution of deflection of circular plate
subjected to eccentric concentrated load
(Number of boundary elements = 48,
Number of domain points = 25).

SECTION XI   NONLINEAR MATERIAL PROBLEMS

Application of BEM to Calculation of the Work Roll Deformation and Residual Stress in the Rolled Sheet

J. Kihara
Department of Metallurgy, Faculty of Engineering, University of Tokyo, 7-3-1 Hongo, Bunkyo-ku, Tokyo 113, Japan
M. X. Peng
Graduate School of University of Tokyo, 7-3-1 Hongo, Bunkyo-ku, Tokyo 113, Japan

1. INTRODUCTION

The increase of rolling forces due to rolling of high strength materials or in the low temperature range causes severe large displacement of rolls , so that a finer control of sheet shape is urgently needed. For such purposes, Boundary Element Method is found to be an efficient tool to estimate displacement and stresses in the rolls or sheets with sufficient accuracy.

In the present paper, we further develop this BEM approach to cope with an evaluation of the residual stresses increase which is observed when the applied bending force become larger at the work roll, or the thickness of sheet is severely reduced. In addition, since the deformation in the roll is dependent on stress redistribution due to the rolling pressure, we need iterative procedure to determine the above residual stresses.

Here, we will first present the formulation of three dimensional elastic BEM analysis for rolling system in the cylindrical coordinates and some numerical results for thick wall cylinder under uniform pressure will show the effectiveness of using the cylindrical coordinates in such problems. Furthermore,the iterative procedure will be described to calculate the residual stress, and then we will make some sensitivity analysis to classify the influences of both the reduction in the thickness and the bending forces of work roll on the variation of residual stress distributions.

2. FORMULATION OF 3-D ELASTIC BEM IN CYLINDRICAL COORDINATES

When body force is not so significant and negligible, the integral equation for 3-D elastic problems can be written as

$$c_{ki} u(P) + \int p^*_{ki}(P,Q) u_i(Q) d\Gamma = \int u^*_{ki}(P,Q) p_i(Q) d\Gamma \quad (1)$$

the $u^*_{ki}(P,Q)$ and $p^*_{ki}(P,Q)$ is the
fundamental solutions for the 3-D elastic isotropic bodies and
written as follows:

$$U^*_{ki}(P,Q) = \frac{1}{16\pi G(1-\nu)r} \left\{ (3-4\nu)\delta_{ki} + r_k r_i \right\} \quad (2)$$

$$P^*_{ki}(P,Q) = -\frac{1}{8\pi(1-\nu)r^2} \left[n_m r_m \left\{ (1-2\nu)\delta_{ki} + 3 r_k r_i \right\} - (1-2\nu)(r_k n_i - r_i n_k) \right] (3)$$

Where r is the distance from the point Q of application of
load to the point P under consideration, δ_{ki} is Kroneker's
delta, nj is the direction cosine component of the normal vector
on the surface, r_i is the direction cosines of the connecting
vector between the two points of Xi direction, G and ν is the
shear modulus and the poisson's ratio respectively. The integral
equation can be now expressed in matrix form as follows:

$$[C]\{U\} + \int [P^*]\{u(Q)\} d\Gamma = \int [U^*]\{p(Q)\} d\Gamma \quad (4)$$

where the vector $\{u(Q)\}$ and $\{p(Q)\}$ are defined in the cartesian
coordinates as follows:

$$u(Q) = [\begin{array}{ccc} u_x & u_y & u_z \end{array}]^T, \quad p(Q) = [\begin{array}{ccc} p_x & p_y & p_z \end{array}]^T \quad (5)$$

and in the cylindrical coordinates, both of the displacements and
tractions can be expressed in form as follows:

$$\hat{u}(Q) = [\begin{array}{ccc} u_R & u_T & u_z \end{array}]^T, \quad \hat{p}(Q) = [\begin{array}{ccc} p_R & p_T & p_z \end{array}]^T \quad (6)$$

So that the relationship of both displacements and tractions
between the two coordinates system can be written as

$$\{u(Q)\} = [T]\{\hat{u}(Q)\}, \quad \{p(Q)\} = [T]\{\hat{p}(Q)\} \quad (7)$$

where [T] is a transformation matrix such as below

$$[T] = \begin{bmatrix} \cos\theta & -\sin\theta & 0 \\ \sin\theta & \cos\theta & 0 \\ 0 & 0 & 1 \end{bmatrix} \quad (8)$$

Now, substitute the equation(7) into the equation(4), and the
formulation of three dimensional elastic BEM in the cylindrical
coordinates can be written as the following equations.

$$[c] \; [T] \; \{\hat{U}\} + \int [P^{\cdot}] \; [T] \; \{\hat{U}\} \, d\Gamma$$

$$= \int [U^{\cdot}] \; [T] \; \{\hat{U}\} \, d\Gamma \qquad (9)$$

It is easy to write the fundamental solution $p_{\kappa i}^{*}(P,Q)$ and $u_{\kappa i}^{*}(P,Q)$ in the cylindrical coordinates, but it will make the calculation complicated. Therefore in the present paper, the expression of fundamental solution in the cartesian coordinates is employed. The boundary of the domain is discretized into certain number of elements, the integral equation (9) is approximated by element numerical integrations. , and then the boundary element integral equations are obtained. In the numerical integration, the real coordinates should be transformed into the general coordinates system. By use of the transformation, a differential of area can be given by

$$d\Gamma = J \, d\xi \, d\eta \qquad J = (g_1{}^2 + g_2{}^2 + g_3{}^2)^{1/2} \qquad (10)$$

where
$$g_1 = \frac{r\,\partial\theta}{\partial\xi} \cdot \frac{\partial z}{\partial\eta} - \frac{r\,\partial z}{\partial\xi} \cdot \frac{\partial\theta}{\partial\eta} \qquad g_3 = \frac{r\,\partial r}{\partial\xi} \cdot \frac{\partial\theta}{\partial\eta} - \frac{r\,\partial\theta}{\partial\xi} \cdot \frac{\partial r}{\partial\eta}$$

$$g_2 = \frac{\partial z}{\partial\xi} \cdot \frac{\partial r}{\partial\eta} - \frac{\partial r}{\partial\xi} \cdot \frac{\partial z}{\partial\eta} \qquad \{\vec{n}\} = \frac{1}{J} \begin{Bmatrix} g_1 \\ g_2 \\ g_3 \end{Bmatrix}$$

the normal vector $\vec{n}$ is expressed in the cylindrical coordinates system, and then it must be transformed to the cartesian system when it is used in the numerical integration of the fundamental solution.

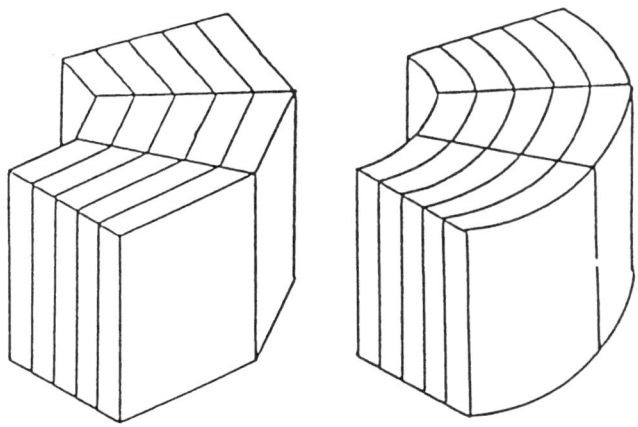

in cartesian coordinates   in cylindrical coordinates

Fig. 1. Boundary element subdivisions for thick-
wall cylinder

3. AN EXAMPLE FOR THICK WALL CYLINDER

In order to compare the accuracy of the numerical results in using the cylindrical coordinates with that of the cartesian coordinates system, the cylinder under uniform pressure is taken for an example .The calculation conditions are as follows:

External radius of the cylinder	R	= 200 mm
Internal radius of the cylinder	r	= 100 mm
Internal pressure	p	= 10 kg / mm
Young's modulus	E	= 21000 kg / mm
Poisson's ratio		= 0.30

In this problem, because of symmetry, only a quarter of the cylinder segment is considered as the analysed domain.The boundary element subdivision for the quarter in the cartesian coordinates and in the cylindrical coordinates are shown in Fig.1. In the cartesian system,the subdivided quarter of cylinder is just like a polyhedron. But in the cylindrical system, it is shown that even though the subdivisions may be rougher , the exact shape of cylinder is well kept. Here some numerical results can show the effectiveness of using the cylindrical system in such problems. The plane strain condition and the plane stress condition are used to make the comparison of the residual stress between these two coordinates system.

The numerical results for the radial displacement and the circumferential stress of the cylinder in the two coordinates

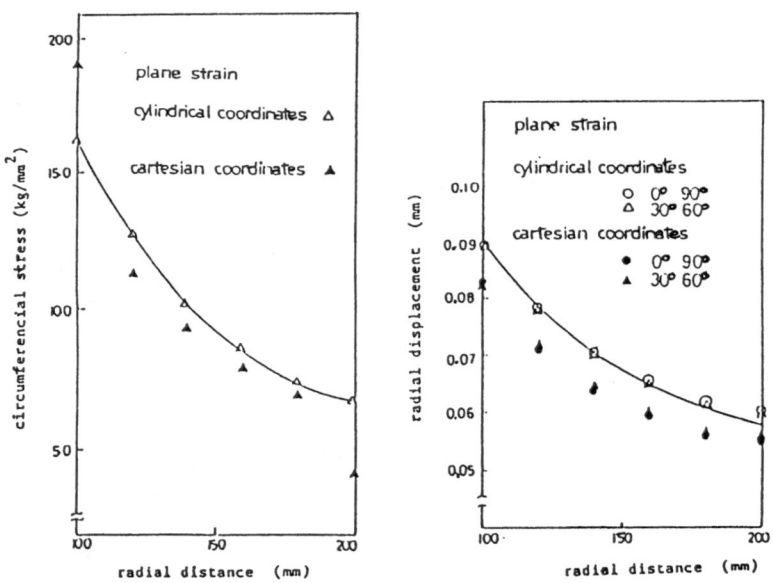

Fig. 2 A comparison of numerical results in
plane strain condition

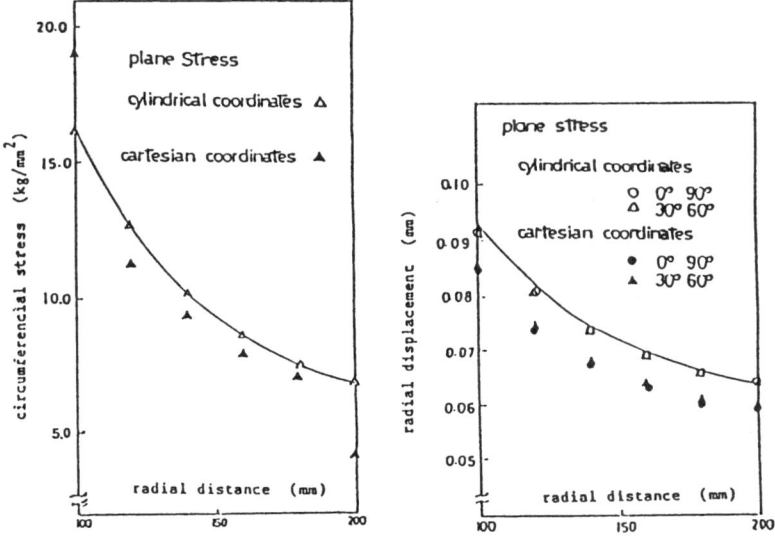

Fig. 3 A comparison of numerical results in
plane stress condition

systems are compared with each other:in Fig.2 for plane strain
and in Fig.3 for plane stress. In each figure, it is found
that the present approach based on the cylindrical system
provides more accurate results which are generally in good
agreement with the exact solutions.

4. APPLICATION FOR ROLLING

4.1 Calculation model
Fig.4 shows the model of four high cold mill system,including
the work rolls and the back-up rolls. For the purpose of the
numerical analysis, the chief parameters of the mill model are
defined as follows:

    Back-up roll diameter   1100 mm
    Work roll diameter   400 mm
    Barrel length of both roll   1800 mm
    Width of rolled sheet   1500 mm

At first, it is assumed that the boundary conditions of the
work roll through the present numerical analysis as follows:
    (1) The rolling pressure between the work roll and the sheet
is constant in the circumferential direction, but only varies the
values in the direction of width.
    (2) The work roll is simplified like a circular cylinder.
    (3) The effect of force which acts on the work roll neck is
evaluated by the way of applying the uniform shear stress and

bending moments on the side face.

(4) When one use the iterative procedure to evaluate the convergent of numerical solutions of the residual stress in the rolled sheet, the rolling pressure between the work roll and sheet is always influenced by the increments of the residual stress, at each calculations. But the pressure on the contact area between the work roll and the back-up roll surface on where the uniform pressure distribution was assumed before the simulation, will do not be vary in the simulation process .

For the cold rolling process is considered, and then the suitable dimensional parameters of the sheet in the preset report is determined as follows :

Width	1500 mm
Thickness	1.0 mm

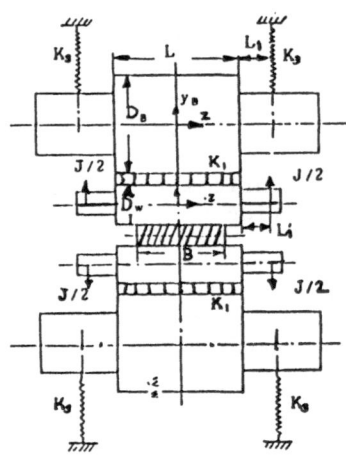

Fig. 4 Model of 4 high mill including work rolls
and back-up rolls

Before the iterative simulation for the residual stress, the deformation of the work roll is calculated with the condition of 20 % reduction of the sheet thickness. Then the work roll's profile is considered as the natural profile of the initial sheet. Namely this profile is considered as the standard profile for the calculation of the residual stress in the sheet.

4.2 Calculation of residual stress

After rolling ,some different distribution of the sheet thickness across the width from the initial one may occur because of the elastic deformation of the work roll.If the ratio of the initial sheet profile to the rolled sheet profile is not constant at any section of sheet width, so that the distribution of plastic elongation in the rolling direction would take place.

In the general case of the cold rolling, the proportion of the sheet width to thickness may be usually greater than one hundred, and therefore the widening of sheet in the cold rolling process can be negligible. So the deviation ratio of the reduction in the sheet thickness would correspond only to the deviation ratio of the plastic elongation in rolling direction, and the residual stress in sheet have to take place. As the cold rolling is operated, the work roll and the rolled materials should be always contact with each other, then the exit displacements of the work roll can be considered as the sheet compressive ratio to be used to calculate the deviation ratio of the plastic elongation strain in the rolled sheet.

The relation of deviation ratio between the compressive strain and the elongation strain can be written, when the rolled sheet has not been widened, as follows:

$$\Delta \varepsilon_H = \Delta \varepsilon_L$$

(11)

where $\Delta \varepsilon_L$ is the deviation ratio of the elongation strain, and $\Delta \varepsilon_H$ is the deviation ratio of compressive strain respectively. Here, write $\Delta \varepsilon_L$ and $\Delta \varepsilon_H$ in some detail as below:

$$\Delta \varepsilon_L = \varepsilon_L (Zi) - \varepsilon_L (Zj) \qquad \Delta \varepsilon_H = \varepsilon_H (Zi) - \varepsilon_H (Zj)$$

$$\varepsilon_L = \ln (L/L_c) \qquad \varepsilon_H = \ln (H/H_c) \qquad (12)$$

where Z is the coordinates along the direction of the sheet width, L_c and H_c is the standard point's length and thickness. If consider that the deviation of elongation strain in sheet may be absolved perfectly by that of elastic strain, the tensile stress in the rolling direction can be calculated by the Hook's low for using of the equation as follows:

$$\sigma_F (j) = E_M (\Delta \overline{\varepsilon}_H - \Delta \varepsilon_H(j)) + \sigma_{FM}$$

(13)

where E_m is the Young's modulus of the rolled materials, $\Delta \varepsilon_H$ is the mean value of the deviation of thickness strain, σ_F is the residual stress occurred in the rolled sheet, at the number j nodal point which is on the contact area between the work roll and the rolled materials, and σ_{FM} is the mean value of the tensile stress, in the present simulation, σ_{FM} is negligible. The integration of the residual stress through the arbitrary width of the sheet should be zero, because the sheet was not broken after rolled, this condition can be written as:

$$\sum_{j=1}^{K} \{\sigma_F (j) + \sigma_F (j+1)\} B_j h_j = 0$$

(14)

where Bj and Hj is the width and height of the area on where the σ_F acted.

It must be considered in the situation of cold rolling, that

the residual stress is a kind of tension which will make the feed
back action to the deformation of the work roll through the yield
condition of the materials. For instance, if the elongation ratio
of the plastic strain at the center is bigger than that at edge;
then the edge waves phenomena should occured. In this case , the
residual stress in the sheet is tension at the edge and
compressive stress at the center. The Tresca yield condition is
used for the rolled materials in this paper, so the increment of
the rolling pressure distribution is only regulated by the
.variation of the residual stress . For this reason, larger the
deviation of elongation plastic strain ratio is and greater   the
increment is given to the rolling presure . For the consideration
of the residual stress feed back effect, it is reasonable to use
the iterative calculation procedure for the calculation of
deformation   or residual stress. The yielding condition is   given
as follows:

$$P(j) - \sigma_F(j) = K \qquad\qquad (15)$$

where   P(j) is the rolling pressure acted on the point j , K is
the yield stress of the rolling materials. The flow chart of the
simulation program is shown in Fig.5

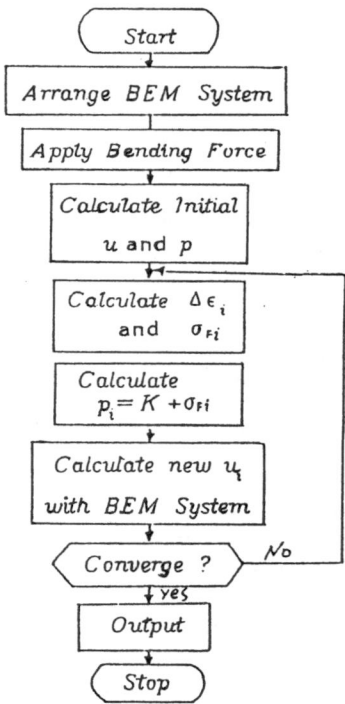

Fig. 5 Flow chart for residual stress calculation

Once the matrices of [H] and [G] are formed by BEM, it is very simple to operate the iterative calculation of residual stress without any change of those matrices. In addition, the discriminant standard of the convergent of numerical solutions is determind as the absolute difference between the numerical solution i'th iteration and that at i+1'iteration is equal or smaller than 10^{-4} . Furthermore when the reduction of the sheet is changed, so that the boundary element subdivision of the work roll should be redivited again because the geometric condition of the boundary of work roll change.

The residual stress distribution is shown in Fig.6. When the bending force acted on the work roll is changed in a reasonable range and the rolling reduction is constantly kept at 20% in this simulation. It is found that the residual stress varied directly as the variation of the bending forces.

When the positive bending force is applied on the side face of the work roll, the compressive stress and the tensile stress occured at the center and the edge of the sheet respectively. For this reason, the bending force control are usually used as a effective means of controlling the edge waves of the rolling sheet. The other means to control the shape of the rolled materials is to vary the profile of the work roll by the increase or the decrease of the rolling load. By this way the residual stress in the rolled sheet is also controlled too. In rolling process it is well known the rolling load at almost case is in linear relationship with the reduetion.

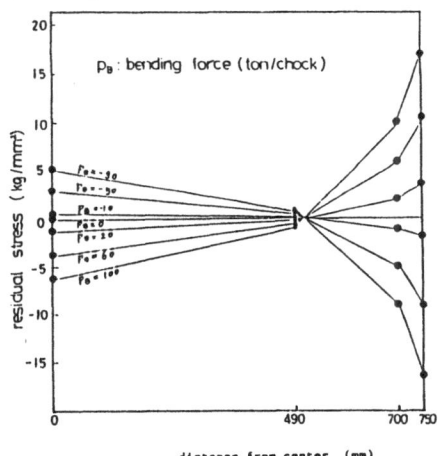

Fig. 6 Residual stress distribution when bending
force P_B is applied

The relationship between the reduction and the rolling load which is used in the present paper is shown in fig.7. If the rolling force and the reduction are known, the contact area width, between the work roll and sheet, the work roll and the back up roll could be determind by the classical method.

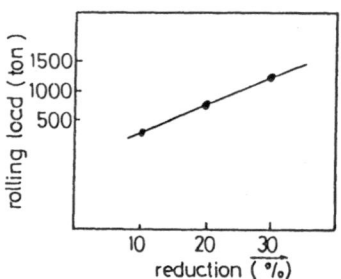

Fig. 7 Relationship between reduction and rolling load

In Fig.8 the residual stress distributions are shown when the reduction is changed from 20 % to 10,15,or 30 %.

Before the results shown in fig.8 are obtained, it was considerable that the rolling load will be increased as the reduction is increased, because larger rolling load should make larger deflective deformation of the work roll , so that the reduction of the sheet centre should be smaller, then the edge waves would take place ; inversely, if the reduction wae larger the center waves of the sheet should be found. In the Fig.8 it is found that when the reduction change from 20% to 30%, the result of the residual stress distribution is consistent with the expectations, but the other simulation result in the case of decrease of reduction are contrary to the expectations. There is,

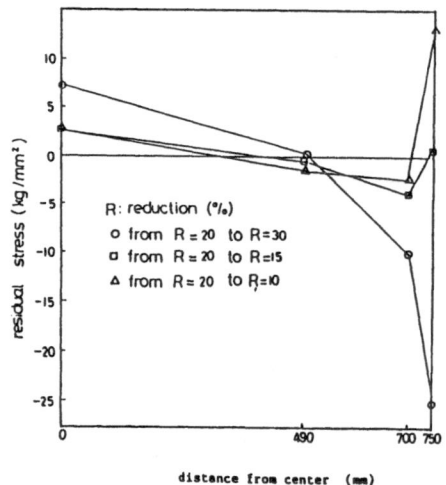

distance from center (mm)

Fig. 8 Residual stress distribution when reduction is changed from 20% to 10, 15 or 30 %

however, some tensile stress at the center of sheet and some compressive stress occurring at the 1/4 width from the center of sheet. It means that the complex deformations exist in the sheet.

The reduction changed from 20% to 30%, the initial distribution of the residual stress is obtained and then the bending force is used as the parameter for controlling the residual stress. The result of the residual stress distribution is shown in Fig.9. It is possible to find a conbination effect of using the reduction control and the bending force control,with the simulation result shown in Fig.9, and the residual stress can be considerably reduced, but not to be exact zero.

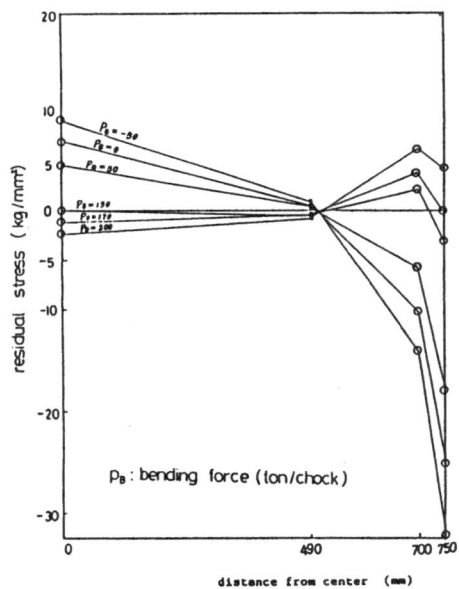

Fig. 9   Residual stress distribution when both bending forces and rection of thickness are controlled

5. CONCLUTION

In this paper, we presented the formulation of three dimentional elastic BEM analysis for rolling system in the cylindrical coordinates;some numerical results for the thick-wall cylinder under the uniform internal pressure have shown the effectiveness of using the cylindrical coordinates. The variation of work roll profile and the residual stress in sheet is discribed. By using the reduction control and the bending force control, the sheet profile can be improved .Through the simple examples in the paper, the BEM will be a useful tool for the materials working processes.

6. REFERENCE

1. C.A.Brebbia: Recent Advances in Boundary Element Method, Pentech Press, London(1978)

2. C.A.Brebbia: The Boundary Element Method for Engeneers, Pentech Press, London(1978)

3. Proceedings of the Second Japan National Symposium on Boundary Element Method. Sponsored by Japan Society for Computational Method in Engineering.(JASCOME) August 21.to.23,1985, Tokyo

4. Theory and Practice of Flat Rouling, Edited by Iron and Steel Institute of Japan(1984)

5. J.Kihara et al.: Experimented Evaluation of the Traction on the contact surface by the Aid of BEM, Journal of the Japan Society for Technology of Plasticity (vol.25, No.284, 1984)

SECTION XII COMPUTATIONAL ASPECTS

An Adaptive Indirect Boundary Element Method with
Applications
R.L. Johnston
*Department of Computer Science, University of Toronto, Toronto, Ontario,
Canada M5S 1A7*
G. Fairweather, A. Karageorghis
*Department of Mathematics, University of Kentucky, Lexington, KY 40506,
U.S.A.*

INTRODUCTION

The implementation of a boundary element method (BEM) is
complicated by the need to provide for evaluation of integrals
with singular integrand. This difficulty is normally handled
by devising specialized quadrature techniques for dealing with
the singularity. An alternative approach is to avoid the
problem altogether by employing an auxiliary bounadry ∂A to
formulate the BEM. However, this idea has not proven very
successful because accuracy of the numerical solution depends
very significantly on the precise shape and location of ∂A and
no viable algorithm for determining a suitable ∂A has, as yet,
been devised.

The adaptive boundary element method (AIBEM) is an indirect BEM
with auxiliary boundary. Its distinctive feature is that the
method itself determines a suitable location for ∂A. This is
done automatically via a nonlinear least squares procedure.
The resulting method is an accurate, highly versatile, easy-
to-use form of BEM.

In this paper, we give a brief description of the AIBEM and
demonstrate its usefulness in solving problems in (1) potential
flow, (2) the calculation of the stress intensity factor for a
crack in an inhomogeneous elastic body, and (3) plate-bending
problems.

---

[1]Present address: Department of Applied Mathematics, The University College of Wales, Aberystwyth, Dyfed SY23 3BZ, Wales.

DESCRIPTION OF THE METHOD

For simplicity, we describe the method in the context of the Dirichlet problem

$$(1) \qquad \nabla^2 u(P) = 0, \qquad P \in D,$$
$$(2) \qquad u(P) = f(P), \qquad P \in \partial D,$$

where D is a region in 2-space with boundary ∂D, and ∇^2 is the familiar Laplace differential operator. Let ∂D_A be an auxiliary boundary enclosing D (see Fig.1) and assume that the solution $u(P)$ of (1)-(2) has an analytic continuation defined at every point within the region enclosed by ∂D_A. Then it is well known that $u(P)$ can be represented as a simple layer potential

$$(3) \qquad u(P) = \int_{\partial D_A} \sigma_A(Q) \log r(P,Q) \, ds_Q, \qquad P \in D \cup \partial D,$$

where $r(P,Q) = |P-Q|$ is the distance between P and Q, and $\sigma_A(Q)$ is the (unknown) source density function. We remark that, in the case $\partial D_A = \partial D$, equation (3) is the form normally used in the derivation of the indirect BEM. The generalization to an auxiliary boundary was proposed by Oliveira[11] as a means of avoiding the need to evaluate improper integrals.

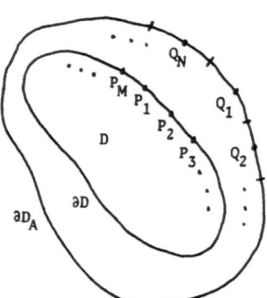

Figure 1.

A numerical procedure for determining σ_A can be formulated as follows. Let P_i, $1 \leq i \leq M$, be a set of "observation" points on ∂D. Evaluating equation (3) at each P_i and using the boundary data (2), we obtain the M conditions

$$\int_{\partial D_A} \sigma_A(Q) \log r(P_i,Q) \, ds_Q = f(P_i), \qquad 1 \leq i \leq M.$$

Next we subdivide ∂D_A into N ($\leq$ M) elements $\partial_j D_A$, $1 \leq j \leq N$, and let Q_j be the midpoint of $\partial_j D_A$. Then, assuming σ_A is piecewise constant, i.e.,

$$\sigma_A(Q) = \sigma_A(Q_j) = \sigma_j \qquad \text{for all } Q \in \partial_j D_A,$$

and applying the midpoint quadrature rule, we obtain the system of M equations in the N unknowns σ_j

$$(4) \qquad \sum_{j=1}^{N} w_j \sigma_j \log r(P_i, Q_j) - f(P_i) = 0, \qquad 1 \leq i \leq M,$$

where the w_j's are (known) quadrature weights.

It is usual to take M = N observation points, in which case the linear system (4) is N x N and can be solved via Gaussian elimination. However, one could also use M > N observation points. In this case, the system (4) is overdetermined but it can be solved in the least squares sense, i.e., minimize

$$(5) \qquad S_2(\underline{c}, \underline{Q}) = \sum_{i=1}^{M} \left[\sum_{j=1}^{N} c_j \log r(P_i, Q_j) - f(P_i) \right]^2,$$

with respect to the parameters $c_j = w_j \sigma_j$. This is also a linear problem and can easily be solved via a modified Gram-Schmidt procedure.

It has been observed by Heise[4] that accuracy of BEM's based on an auxiliary boundary is highly dependent on the location of the auxiliary boundary. Hence, the problem of improper integrals has essentially been replaced by that of choosing an appropriate ∂D_A. This can be resolved by letting the coordinates of the "quadrature" points Q_j, as well as the coefficients c_j, be parameters to be determined in the minimization of (5). In this way, the algorithm automatically adapts the size and location of the elements $\partial_j D_A$ to suit the problem. Hence, it can be viewed as an adaptive, indirect boundary element method (AIBEM). Of course the least squares minimization now becomes nonlinear (in the Q_j's) but one can easily use any one of several excellent nonlinear least squares packages currently available.

An alternative derivation of the foregoing method is given in Mathon and Johnston[7] where it is called the method of fundamental solutions (MFS). Let Q_j, $1 \leq j \leq N$, be a set of distinct points lying outside D and approximate the solution u(P) of (1)-(2) by a function u_N of the form

$$(6) \qquad u_N(\underline{c}, \underline{Q}; P) = \sum_{j=1}^{N} c_j K(P, Q_j), \qquad P \in D \cup \partial D,$$

where $K(P, Q_j) = \log r(P, Q_j)$ is a fundamental solution of the differential equation (1), $\underline{c} = [c_1, \ldots, c_N]^T$ and $\underline{Q}$ is a 2N-vector giving the coordinates of the singularities Q_j. Since u_N satisfies (1), the unknowns $\underline{c}$ and $\underline{Q}$ are chosen so as to approximate the boundary data (2) as well as possible. This is done by least squares. Selecting a set of points $P_i \in \partial D$,

$1 \leq i \leq M$, at which to carry out the least squares fit, it follows that this method leads to the minimization problem (5). Hence the two methods are the same. In the MFS context, it is easy to see how to extend the method to more complicated problems than (1)-(2). Also, analysis of the method is easier because it can be viewed simply as a problem in approximation of the boundary data.

From an implementation point-of-view, the MFS, or AIBEM, is quite straightforward. One need only specify the observation points P_i, the boundary data at each of them, initial guesses for the quadrature points Q_j and coefficients c_j, and then call a least squares routine. In the MFS algorithm that we have implemented, normalized fundamental solutions are used. For example, with Laplace's equation, we use

(7) $K(P,Q) = -\log[r(P,Q)/(\rho+r(Q))]$,

where ρ is the radius of the circle circumscribing D and $r(Q)$ is the distance of Q from the origin, which must lie inside D. Motivation for this choice lies in the fact that the argument of the log function will lie between 0 and 1, assuming $P \in D$, so that $K(P,Q)$ will be of one sign. We have observed that this normalization improves the rate of convergence of the least squares algorithm. To solve the least squares problem, we have used the subroutine LMDIF from the Minpack package.

It is sometimes difficult to decide a *priori* how many terms to use in (6) in order to achieve a desired accuracy. An idea of MacDonell[8] is to start with a certain number of singularities, say N_1. After a given number of iterations with the least squares algorithm, we increase the number by adding N_2 more singularities and iterate further with the larger set. Further singularities can be added until the required accuracy is reached. An additional benefit of this technique is efficiency because we do not have to work with the larger least squares problem all the time.

A MOVING BOUNDARY PROBLEM IN POTENTIAL FLOW ANALYSIS

The transient analysis of a cylindrical membrane type structure, inflated by internal pressure and subject to external wind pressure, was considered by Olson and Han[17]. By assuming the cylinder is sufficiently long, the problem can be treated as two-dimensional. Consider a cylindrical membrane inflated by internal pressure p_o and acted upon by external wind pressure p_w. Using polar coordinates, the differential equation for the deflection $w(\phi)$ for arbitrary wind pressure $p_w(\phi)$ is

(8) $$w'' + \left[\frac{(w+w')^2 - 2w'R'}{2R} \right] + w = g(\phi)$$

where

$$g(\phi) = R\left[1 + C(N_\phi - Rp_o) - \frac{R(p_o - p_w(\phi))}{N_\phi} \right],$$

N_ϕ is the membrane stress resultant, and R is the radius of curvature. For boundary conditions, the deflection will be zero at the base where the membrane is anchored.

The wind is modelled by potential flow. The boundary conditions are uniform velocity U_∞ far from the body and zero flow normal to the body and on the ground. Once the potential function u is known, the pressure is determined from the equation

(9) $$p_w(\phi) = p_\infty + \frac{1}{2}\rho\{U_\infty^2 - [(\frac{\partial u}{\partial x})^2 + \frac{1}{r^2}(\frac{\partial u}{\partial y})^2]\}.$$

Then, given the pressure, the deflections $w(\phi)$ are determined by solving equation (8). This produces a new profile for the body which, in turn, gives a new potential flow problem to be solved, and so on until steady state is reached.

We note that the velocity components $\partial u/\partial x$ and $\partial u/\partial y$ are required in the pressure calculation (9). Herein lies another advantage of the MFS derivation of the method. We simply differentiate (6) to obtain approximations for the velocity components. If the coordinates of Q_j are (a_j, b_j), we have

(10) $$\frac{\partial u}{\partial x} = \frac{\partial}{\partial x}u_N(\underline{c}, \underline{Q}; P) = \sum_{j=1}^{N} c_j \frac{(x-a_j)}{r(P, Q_j)},$$

and similarly for $\partial u/\partial y$. Hence, the derivative values in (9) can be obtained by a straightforward series evaluation.

In Johnston and Fairweather[6], the potential flow portion of the problem was solved using the AIBEM in cases where the body had circular and elliptical cross sections. The problem was modelled by a finite region with boundaries located sufficiently far away to assume the uniform velocity U_∞ at them. In all cases, good results (< 1% error) were achieved using 14 singularities, 4 of which were placed inside the body. It was observed that the accuracy achieved by using (10), and its counterpart for $\partial u/\partial y$, to compute derivatives was comparable to that achieved for the potential function u(P). Although not relevent to the membrane problem, bodies with wedge-shaped cross section were also considered. The difficulty here is that the sharp edge is a reentrant corner at

which the solution has derivative discontinuity. Again, good
results were obtained but it required 22 singularities, 10 of
which were placed inside the body.   We discuss this type of
problem in more detail in the next section.

Olson and Han[12] solved the foregoing transient problem, using
the AIBEM to solve the potential flow problem along with a
finite element method to solve the deflection equation (8).
Han[3] has extended this work to three-dimensional problems. In
all cases, the AIBEM gave excellent results.

CALCULATION OF STRESS INTENSITY FACTORS

We consider the problem of determining the stress intensity
factor (SIF) in an inhomogeneous isotropic elastic medium of
finite cross section, containing a line crack under shear. To
calculate the SIF, one must first solve a boundary value
problem for Laplace's equation, the solution of which exhibits
singular behaviour at the crack tip. A simple problem, that
possesses the type of difficulty encountered, is the so-called
Motz problem.   Referring to Fig.2, we want to find u(P) such
that

$$\nabla^2 u = 0 \qquad \text{in the rectangular region EFBOA,}$$

along with the boundary conditions

$$u = 1000 \qquad \text{on AE,}$$
$$u = 500 \qquad \text{on BO,}$$
$$\frac{\partial u}{\partial n} = 0 \qquad \text{on the remaining boundary.}$$

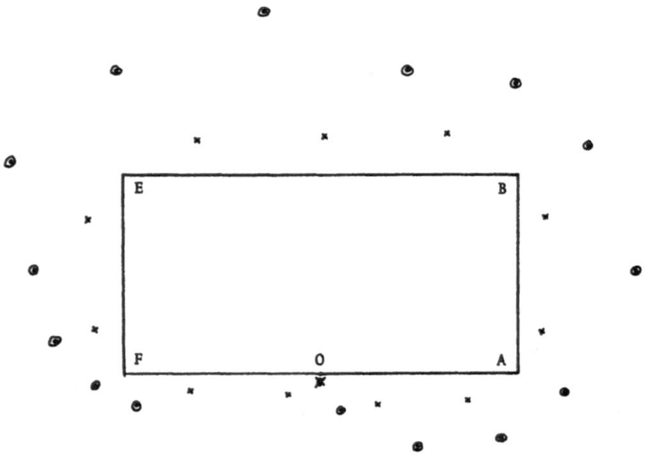

Figure 2

The segment AO corresponds to the crack with the point O being the crack tip where the normal derivative has a jump discontinuity. This problem was considered by Xanthis, Bernal and Atkinson[13]. They experimented with a variety of special procedures, such as using functions having the correct singular behaviour at the crack tip, in defining a direct BEM for the problem. Our purpose in considering this problem is to show that good accuracy can also be obtained with the AIBEM, but without having to resort to special procedures. One need only use a sufficient number of terms in (6) to obtain an adequate model of the singular behaviour.

Even without knowledge of the physical background of the problem, one can easily infer from the boundary data and geometry of the boundary ∂D that the solution is badly behaved at the crack tip. This information is useful in the initial placement of the singularities prior to calling the nonlinear least squares routine. For this problem, we used $N_1 = 18$ singularities distributed around the region at a distance 0.1 from the boundary. We placed 10 of them on a line parallel to the side AB. After 24 iterations of the least squares algorithm, we added $N_2 = 13$ singularities, distributed at a distance 0.075, with 6 of them on a line parallel to AB, and continued the iteration. Hence, a total of 31 terms in (6) were used. The number of observation points was M = 335, 165 of which were on the side AB. To avoid overflow, we normalized the boundary conditions by dividing by 1000.

Some typical results are summarized in Table 1. As measures of performance we have listed $S_2(\underline{c},\underline{Q})^{1/2}$, the root mean square error RMSE = $[S_2/M]^{1/2}$, and the values of u at the points $P_1 = (-0.5, -0.35)$ and $P_2 = (-0.5, -.107)$ on the crack. The numbers in brackets under the headings are the correct values. The first column indicates the number of iterations performed after the second set of singularities was added. In addition to

TABLE 1

# Iter.	$S_2(\underline{c},\underline{Q})^{1/2}$	RMSE	$u(P_1)$ (.57641)	$u(P_2)$ (.63445)	SIF (151.63)
11	.103	.0057	.5859	.6388	151.21
17	.062	.0034	.5795	.6382	151.39
23	.051	.0028	.5790	.6355	151.49
28	.046	.0025	.5784	.6352	151.49

these measures, we looked at the behaviour of the error along the side AB. After the iterations with N_1 singularities, the error had some rapid oscillations in the vicinity of the crack tip. But, after adding the second set of singularities, these oscillations decreased significantly in width as well as amplitude.

The final distribution of the singularities is shown in Fig.2. The circled dots represent the first N_1 of them, while the x's represent the additional ones. The rather large dot and x near the crack tip represent three singularities from the first group and two from the second. There was not sufficient resolution to depict them as distinct points. Since u(P) cannot be continued analytically through the crack tip, one should expect ∂D_A to touch ∂D at this point. The positioning of the singularities by the least squares algorithm bears this out.

There are a number of possible methods for computing the SIF. Basically, they can be classified as (i) limiting procedures that depend on values of the solution u(P) as P approaches the crack tip, and (ii) invariant integral procedures that involve integration of derivatives of u along AE, EF and FB. Due to the oscillatory behaviour of u_N in the immediate vicinity of the crack tip, we chose the latter approach. This involves evaluating the integral

$$F_2 = -\int_{AE}(u_x^2 - u_y^2)ds - \int_{EF}u_x u_y ds + \int_{FB}(u_x^2 - u_y^2)ds$$

and substituting into the formula

$$SIF = \left[\frac{4F_2}{7\pi}\right]^{1/2}$$

The factor 7 comes from the fact that we have normalized the dimensions of EFBA from 7 x 14 to 1 x 2.

Table 1 indicates convergence to the value 151.49 for the SIF, which is incorrect in the fourth digit. Given that the RMSE indicates, on average, a small error in the third digit of u (and its derivatives) plus the fact that integration is a smoothing operation, we should expect that there will be some error in at least the fourth digit of the SIF. Hence, the convergence to 151.49 is not surprising. If better accuracy is required, more singularities will have to be used (at the expense of a larger least squares problem to be solved).

BENDING OF A THIN PLATE

We now consider the problem of determining the transverse deflection of a thin plate under uniform load K per unit area. This problem requires the solution of the biharmonic equation

(11) $\nabla^4 u = 0$.

When the plate is clamped, the boundary conditions are

(12) $u = 0$ and $\frac{\partial u}{\partial n} = 0$,

whereas, when it is simply supported, they are

(13) $u = 0$ and $\nabla^2 u = 0$.

Given the solution u, the deflection at any point $P(x,y)$ is given by

$$w = (u + \frac{r^4}{64}) * \frac{K}{D} ,$$

where D is the bending rigidity and $r^2 = (x^2+y^2)$.

We consider an approximation for u in the form

(14) $u_N(\underline{c},\underline{d},\underline{Q};P) = \sum_{j=1}^{N} c_j K_1(P,Q_j) + \sum_{j=1}^{N} d_j K_2(P,Q_j)$,

where

$K_1(P,Q) = r^2(P,Q) * \log r(P,Q)$

$K_2(P,Q) = \log r(P,Q)$.

That is, we approximate the solution to the biharmonic problem by a linear combination of fundamental solutions of both the biharmonic equation and Laplace's equation.   This version of the MFS was suggested by Fairweather and Johnston[2] and by Bogomolny[1].   We call it   the biharmonic MFS (BMFS). Clearly, the approximation (14) satisfies the biharmonic equation (11). Therefore, we want to determine the coefficients c_j, d_j and locations of the singularities Q_j to satisfy either of the boundary conditions (12) or (13),   whichever applies at the time.   Once again, to improve the convergence rate of the least squares algorithm, we use normalized solutions. Specifically, we normalize the distance $r(P,Q)$ as in equation (7).

This problem, among others, was considered by Karageorghis and Fairweather[7]. Three specific problems were studied and some typical results are displayed in Table 2.   The columns under $|e|_{max}$ indicate   the maximum error   in approximating the indicated boundary condition.

(i) Clamped elliptic plate
The semi-major and semi-minor axes of the plate were a=1.0 and b=0.8333, respectively. This problem has symmetry with respect

to both the x- and y-axes. Hence, we need only work in the first quadrant. We used 4 singularities, with $N_1 = N_2 = 2$, so that, from symmetry, the total was 16. The number of iterations was 19 initially, and then 10.

(ii) Simply supported square plate
Again, this problem has symmetry with respect to both axes. The solution u has a mild discontinuity (in the second derivative) at each of the corners. Murashima, Nonaka and Nieda[10] also considered this problem using an auxiliary boundary method. They found that the accuracy of their results deteriorated near the corners. On the other hand, it is evident from Table 2 that the BMFS had no difficulty.

(iii) Simply supported elliptic plate with a circular hole
The elliptical plate is the same as before but with a circular hole of radius of r=0.2 in the centre. Exploiting symmetry, we placed 3 singularities outside the ellipse in the first quadrant and 2 inside the hole. Only one set of singularities was used. A similar problem was considered by Murashima, Nonada and Nieda[10] using an auxiliary boundary method. The BMFS produced comparable results.

TABLE 2

Problem	N	M	# Iter.	$|e|_{max}$		
				u	$\partial u/\partial n$	$\nabla^2 u$
(i)	4	48	19,10	.100E-5	.172E-5	
(ii)	4	48	13, 7	.557E-5		.541E-4
(iii)	3,2	24,24	25	.158E-4		.246E-4

CONCLUDING REMARKS

The foregoing examples demonstrate the power and versatility of the MFS, or AIBEM. One disadvantage is computational cost due to the nonlinear least squares algorithm. Indeed, one may well question the very idea of using a nonlinear algorithm to solve linear problems but this is the cost of automatic adaptivity. In this regard, we remark that we have done some preliminary work on a problem in heat exchanger design (see Ingham, Heggs and Manzoor[3]) where the boundary conditions are nonlinear. Application of the usual BEM requires some fundamental work whereas it is a natural extension for the MFS.

BIBLIOGRAPHY

1. Bogomolny A. (1985), Fundamental solutions methods for elliptic boundary value problems, SIAM J. Numer. Anal., Vol.22, pp.644-669.

2. Fairweather G. and Johnston R.L. (1982), The method of

fundamental solutions for problems in potential theory, in Treatment of Integral Equations by Numerical Methods, C.T.H. Baker and G.F. Miller (Eds.), Academic Press, London, pp.349-359.

3. Han P.S. (1986), Interactive nonlinear analysis of wind-loaded pneumatic membrane structures, Ph.D. Thesis, Dept. of Civil Engineering, Univ. of British Columbia.

4. Heise U. (1978), Numerical properties of integral equations in which the given boundary values and the sought solutions are defined on different curves, Comp. Structures, Vol.8, pp.199-205.

5. Ingham D.B., Heggs P.G. and Manzoor M. (1981), Boundary integral equation solution of non-linear plane potential problems, IMA Jour. Numer. Anal., Vol.1, pp.415-426.

6. Johnston R.L. and Fairweather G. (1984), The method of fundamental solutions for problems in potential flow, Appl. Math. Modelling, Vol.8, pp.265-270.

7. Karageorghis A. and Fairweather G. (1986), The method of fundamental solutions for the numerical solution of the biharmonic equation, submitted to Jour. Comp. Phys.

8. MacDonell M. (1985), A boundary method applied to the modified Helmholtz equation in three dimensions, and its application to a waste disposal problem in the deep ocean, M.Sc. Thesis, Department of Computer Science, University of Toronto.

9. Mathon R. and Johnston R.L. (1977), The approximate solution of elliptic boundary value problems by fundamental solutions, SIAM J. Numer. Anal., Vol.14, pp.638-650.

10. Murashima S., Nonaka Y. and Nieda H. (1983), The charge simulation method and its applications for the two-dimensional elasticity, Proc. Fifth Int. Conf. on Boundary Elements, C.A.Brebbia, T.Gutagami and M.Tanaka, (Eds.), Hiroshima, Japan.

11. Oliveira E.R. (1968), Plane stress analysis by a general integral method, J.Engrg.Mech.Div. ASCE, Vol.94, pp.79-101.

12. Olson M. and Han P.S. (1984), Interactive analysis of wind loaded-membrane structures, Proc. 2nd Int. Conf. on Numerical Methods for Nonlinear Problems, Barcelona.

13. Xanthis L., Bernal M.J.M. and Atkinson C. (1981), The treatment of singularities in the calculation of stress intensity factors using the boundary integral equation method, Comp. Meth. Appl. Mehc. Engrg., Vol.26, pp.285-304.

New Developments in the BEASY Boundary Element Analysis System

A.C. Mercy, S. Nageswaran and J. Trevelyan
Computational Mechanics Institute, Southampton, U.K.

ABSTRACT

Since its first release onto the market in 1981, the BEASY analysis package has become firmly established as the major boundary element analysis system available. The advanced mathematics involved in boundary methods of analysis are exploited by BEASY to offer exciting advantages over conventional finite element programs. BEASY offers engineers simple and easily learnt data preparation, significant reductions in mesh generation time and improved accuracy of results.

In spite of these important advantages, there is surprisingly a widespread ignorance of boundary methods of analysis and a reluctance to part with the familiar finite element methods. This stems from the fact that many engineers have become accustomed to their (often outdated) programs and are reluctant to adopt new methods.

This paper introduces the basic concepts behind the BEASY program by the use of several examples. New developments, which are included in a recent release of the package, are presented and some applications discussed.

INTRODUCTION

During the last five years, the boundary element method (BEM) has developed sufficiently to become a real alternative to the finite element method (FEM) in engineering analysis. Indeed, for many types of problem commonly encountered by engineers, BEM offers significant advantages over FEM in terms of both speed and accuracy[1]. BEASY uses the BEM to analyse 2D, 3D and axisymmetric geometries in problems of potential, stress and thermal stress[2].

With the continuing advances in the power of computers, by far
the major proportion of the cost of performing a numerical
analysis of an engineering component is now the preparation of
the data describing the problem. Numerous solid modellers and
CAD packages have arrived on the market in an attempt to speed
up the creation of finite element models. These do indeed
reduce the time and effort required to generate the input
data, but the results of these pre processors (i.e. the finite
element mesh) require extensive checking - especially for
three-dimensional models.

Using boundary elements, the only part of the component that
needs defining in terms of elements is the boundary, or
surface. So a boundary element mesh for a 3D problem consists
of two-dimensional elements modelling the surface only
(Figure 1). Similarly, the elements used to analyse a 2D or
axisymmetric problem are line elements which completely
enclose the problem 'domain' (Figure 2). Thus boundary
elements are always one dimension less than the problems they
are used to solve. This means that, whatever developments are
made in FEM pre processors, data preparation for boundary
elements will always be an order of magnitude faster.

This simplicity of mesh creation is emphasised when an
engineer wishes to change his model in some way. (As with any
engineering design, the first approach is never the final
solution - changes always occur). Figures 3a and 3b show
a boundary element mesh and a finite element mesh for the 2D
analysis of a plate under uniform, in-plane tension. If this
model is to be modified by the inclusion of a small hole in
the plate, then the purely local change can be made simply and
quickly in the boundary element mesh (Figure 4a). However,
the refinement necessary in the finite elements around the
hole changes the entire mesh (Figure 4b), usually requiring a
new mesh to be created.

A further advantage of BEASY over the various finite element
packages is its accuracy, particularly in stress analysis.
This superior accuracy arises from three major reasons:

1. In the FEM, every element is assumed to displace in some
    form, usually quadratic (Figure 5). This applies an
    artificial 'stiffness' which in reality does not exist.
    The BEM makes no such assumptions, except on the boundary.

2. Using FEM, the displacements of all nodes are calculated
    first. Stresses are then computed from these displace-
    ments; a step which involves certain approximations. BEM
    forms its equations in terms of both displacements and
    tractions (surface stresses). Both are therefore
    computed at the same stage and to the same degree of
    accuracy.

3. BEASY uses discontinuous elements (Figure 6). Here the
    nodes are not placed at the extremities of the element, but
    are positioned within the element. There are no nodes
    shared with adjacent elements. Stresses are therefore not
    forced to be continuous where natural discontinuities will
    occur. This is particularly important in problems
    involving stress concentrations.

Another important feature of BEASY is the simplicity and
economy with which 'infinite' and semi-infinite' problems may
be solved. These are problems in which the material to be
analysed extends to a large distance away from the region of
interest. This type of problem is commonly found in the
fields of electromagnetics, geotechnics and corrosion
protection, and may well be encountered in many other areas.
In this type of problem, there is no 'outer' boundary, and the
boundary element mesh is simply the surface of the local area.
For example, the analysis of the area around a small hole in a
large plate (Figure 7) could be performed using the boundary
element mesh shown in Figure 8.

There are disadvantages associated with using boundary elements,
and these fall into two categories:

1. Those engineering fields which cannot yet be solved by a
    general boundary element program (although possibly by a
    specialised program). Such fields are:

            * Dynamic response analysis

            * Non-linear material analysis

            * Plate bending analysis

2. Those disadvantages found when using existing boundary
    element programs to analyse 'solvable' types of problems.
    The particular drawbacks are:-

    * The underlying mathematics is more complicated than for
      FEM. Although this is a problem for program developers
      rather than for users it does tend to increase the
      computer time, element for element, over FEM.

    * The fully populated matrices involved in BEM cause the
      disc space taken per element to be higher than in FEM.
      This is an unfair comparison, however, since there are
      far fewer elements using BEM, and the superior accuracy
      of BEM means that remarkably good results can be
      obtained using extremely coarse meshes[3].

THE BEASY SUITE

BEASY is a general purpose boundary element analysis package
for the solution of 2D, 3D and axisymmetric problems in:

* Potential analysis: Heat transfer

                        Electrostatics

                        Corrosion protection

                        Any other phenomenon governed by the
                        Laplace or Poisson equation

* Elasticity: Linear stress analysis (including
                        thermal stress)

                        Newtonian fluid flow

Pre and post processors are available to simplify the
modelling process and to aid the interpretation of results.
The general structure of the programs is illustrated in
Figure 9, which shows the path taken through pre processing,
analysis and post processing. It should be noted that for a
thermal stress analysis the BEASY analysis program should be
run twice. The first run performs a steady-state heat
transfer analysis of the problem subject to the thermal
boundary conditions provided. The analysis produces a
solution which completely satisfies the Laplace equation.
This solution is taken as part of the boundary conditions (as
a body force term) in the second run of the BEASY analysis
program, which solves the linear stress equations.

Pre and post processing may be performed in two ways: either
using the BEASY pre and post processors, or using some other
CAD package but using an interface to create BEASY input data
from their output files and to allow the BEASY results to be
read by the external program. The use of these CAD packages
requires some special action when creating boundary element
models, since they are usually geared to creation of finite
element meshes. In general, the programs are used to build
up meshes of 'shell' elements when 3D boundary elements are
required, or meshes of 'bar' or 'beam' elements when 2D or
axisymmetric boundary elements are required.

Element types
The elements used by BEASY are shown in Figure 6. Constant,
linear and quadratic elements are available for all types of
analysis, referring to the order of the interpolation
functions over the elements. Discontinuous elements are
used - these are characterised by nodes appearing within the
body of the element and not on the edges.

Discontinuous elements exhibit various advantages which make them preferable to continuous elements (those having nodes shared with adjacent elements). The major advantages are:

1. Improved accuracy, particularly in areas of stress concentration (as described above).

2. Ease of refining meshes in areas of interest. Like finite elements, boundary elements should become smaller where conditions are changing rapidly. This refinement is greatly simplified by the fact that adjacent elements need not share a common edge of the same length (i.e. "fanning" of meshes is not required as simple transitions may be made from small elements to large elements). It is sufficient for the entire object to be completely covered by non-overlapping elements. Figure 10 shows an example of the type of mesh refinement made possible in this way by the use of discontinuous elements.

3. Similarly, since the nodes are not shared and therefore no continuity of displacements needs to be enforced at element edges, it is permissible for elements of different orders to be placed adjacent to each other. This reduces the solution time for those problems having areas of both intense activity and low activity. Constant, linear and quadratic elements may be mixed freely.

In addition to the line elements (for 2D and axisymmetric analysis) and quadrilateral elements (for 3D analysis), a third family is available for 3D potential analysis. This is the family of tube elements (Figure 11) which are popular for the corrosion protection analysis of offshore and marine structures. It is uneconomic to perform a global analysis of such a structure using the conventional quadrilateral elements as shown in Figure 6. To obtain an overall view of the electrostatic behaviour of the sea water, the tube elements are used to model the boundary to the sea water (i.e. the surface of the metal legs of the platform). These are by no means as accurate, since they assume negligible variation around the circumference of each section, but are often used to obtain a set of boundary conditions for a more rigorous local analysis.

Zones
A useful facility of the BEASY analysis is that of 'zoning' or splitting of a problem into two or more distinct regions. This may be done for various reasons[4], but the major reasons are:-

* to enable solution of problems involving more than one set of material properties

* to eliminate numerical problems arising from geometric considerations (e.g. large aspect ratios, etc.).

In this way, zones are 2D regions for 2D (and axisymmetric problems) or 3D regions for 3D problems. The use of zones requires elements to be placed on the interface between the two regions. Equations are formed for each individual zone in the problem, and these sets of equations are combined by considering the coupling between zones at these 'interface elements'.

Two new developments in BEASY are concerned with the treatment of the zones and interface elements defined for the problem:

1. BEASY is now totally zone-based, in that the information for different zones is contained in separate working files. In this way, the analysis may be performed for part of the model described in the input data file. Consider, as an example, a thermal stress analysis of an insulated cylinder (Figure 12). It is decided to analyse this using the axisymmetric analysis facility of BEASY, giving rise to the mesh illustrated in Figure 13. Note that no elements are placed on the <u>axis</u> of symmetry (the vertical axis) nor on a <u>plane</u> of symmetry (the horizontal plane at the bottom of the mesh). Note also the zoning of the problem into the cylinder itself and the insulating material.

   This analysis is interesting in as much as the insulating material, while of great importance in the thermal part of the analysis, plays no part at all in the stress analysis. To avoid unnecessary computation in the stress part of the analysis, the cylinder alone may be considered here (with a minor alteration to the data file).

2. Loading may, in Version 3 of BEASY, be applied to inter-face elements. It is often found that at junctions between different parts of a problem (as modelled by different zones) the default condition of continuity of all variables is not applicable. Here it is necessary to apply some form of 'interface condition' to model the behaviour.

   This new development allows the solution of a wide variety of problems which have hitherto been uneconomical, if not impossible, to solve. A good example is in the analysis of adhesive joints. Here the extreme aspect ratios give rise to difficulties however the problem is analysed. If finite elements are used, extensive data preparation and pre processing is required in the transition between small elements (around the adhesive layer) and larger elements away from the area of interest. If conventional boundary element analysis is used, then numerical problems arise from the aspect ratios encountered in the glue layer.

A good approximation to the behaviour of such problems may be found using BEASY. The adhesive layer is modelled as an interface between discrete zones (on either side of the joint). The behaviour of the adhesive may be treated as a spring of given stiffness tangentially, while the behaviour normal to the interface is often suitably modelled by the default (continuity) condition.

This type of approximation, while not giving any solutions for stresses in the vulnerable glue material itself, gives a set of boundary conditions (on either side of the adhesive) for a second, local, analysis (Figure 14).

Internal solution
Although BEASY allows engineers to find problem solutions on the boundary, it is often desirable to find solutions within the body of the material. This may be done in a simple and economic way by the use of 'internal points'. These are points, specified by the engineer, at which results are computed following the completion of the 'boundary solution'. The points may be generated simply by the pre processor in such a way as to assist greatly in the interpretation of results using a post processor.

Because of the step-by-step analysis adopted in the BEASY suite[2,4], it is very economical to calculate internal solutions again and again since this is the last stage in the analysis. The previous stages need not be performed repeatedly since these will be identical every time if only internal point results are required.

Loading
A major feature of the latest developments in BEASY is the increased flexibility in boundary condition types. This has already been mentioned above (with regard to zone interface elements) but a more complete discussion follows here.

The loading applied to the boundary element model may be categorised into four different types (for both potential and stress analysis). See Tables 1 and 2 for details.

1. Element boundary conditions The most popular and common type of loading is applied over the length of an element on the problem boundary. A new development in the program allows higher order variation across each element. Here the engineer can define the value of a particular boundary condition at a discrete number of points along a line, and

Loading Type	Boundary	Interface	Node
POTENTIAL	*	*	*
NORMAL FLUX DENSITY	*		
AMBIENT POTENTIAL	*	*	
HEAT TRANSFER COEFFICIENTS	*	*	
X-DISPLACEMENT	*	*	*
Y-DISPLACEMENT	*	*	*
Z-DISPLACEMENT	*	*	*
NORMAL DISPLACMENT	*	*	*
TANGENTIAL DISPLACEMENT 1	*	*	*
TANGENTIAL DISPLACEMENT 2	*	*	*
X-TRACTION	*		
Y-TRACTION	*		
Z-TRACTION	*		
NORMAL TRACTION	*		
TANGENTIAL TRACTION 1	*		
TANGENTIAL TRACTION 2	*		
X-ADDED TRACTION		*	
Y-ADDED TRACTION		*	
Z-ADDED TRACTION		*	
NORMAL ADDED TRACTION		*	
X-X SPRING	*	*	
Y-Y SPRING	*	*	
Z-Z SPRING	*	*	
X-X INTERFACE SPRING		*	
Y-Y INTERFACE SPRING		*	
Z-Z INTERFACE SPRING		*	
NORMAL SPRING	*	*	
NORMAL INTERFACE SPRING		*	
SLIDING INTERFACE		*	
ADDED FLUX DENSITY		*	
MEMBRANE		*	

Table 1 Availability of boundary and interface conditions in
current version of BEASY

a smooth curve (or spline) is fitted such that it passes through all of these points. These 'spline boundary conditions' are a new feature of the BEASY pre processor.

2. Interface conditions Loading may, in the latest version, be applied at an interface between zones. This is often useful in problems such as joints between materials where the behaviour of the joint may be approximated in some way. Typical interface conditions for a thermal analysis may be an added flux density or a membrane dividing adjacent zones. Typical interface conditions for a stress analysis may be a spring in some local or global direction, or a sliding interface. Table 1 contains a complete list of boundary conditions available in the new version of BEASY.

3. Nodal boundary conditions A further new development in the BEASY package is the introduction of nodally applied boundary conditions. The types of loading which may be applied in this way are indicated in Table 1.

4. Body forces These are types of loading which cannot be assumed to act only on the boundary, and cannot therefore be applied to boundary elements. In the current version, the available types of body force are as shown in Table 2.

Analysis Type	Body Forces Available
Thermal (potential)	Point sources of flux Line sources of flux Distributed sources of flux Point, line and distributed sinks
Stress	Thermal loading Rotational loading about up to three axes Acceleration in up to three directions

Table 2 Availability of body forces in current version of BEASY

Pre and post processing

The general architecture of the BEASY suite (Figure 9) shows the programs BEASYG and BEASYP. These are pre and post processors which are tailored to the BEASY boundary element analysis system. Alternatively, interfaces are available to commerical CAD packages for users who prefer to continue using their existing software to generate their boundary element models. Interfaces currently available include:

* BEASY/SUPERTAB interface for pre and post processing

* BEASY/PATRAN-G interface for pre and post processing

* BEASY/FEMVIEW interface for post processing

Both the pre and post processors (BEASYG and BEASYP) are fully interactive programs. Geometry and mesh generation is carried out using simple and easily learnt commands. Model alteration (of geometry, mesh and applied boundary conditions) is aided by interactive plotting using a variety of graphics devices.

The post processor provides complete colour graphical presentation of the results. The forms of plot currently available are:-

Geometry and mesh plots

Boundary condition plot (symbolically superimposed on the mesh)

XY plot of boundary conditions

Deformed shape plot (superimposed on the original geometry)

Contour plot of results through interior of 2D body

Contour plot of results on surface of 3D body

Contour plot of results on a section through a 3D body

XY plot of results.

Facilities are present for full rotation and zoom of views. Areas or groups of elements can also be chosen if examination of results is required for particular regions.

Graphical devices currently supported are:-

Calcomp	Sun
Tektronix 4010/4100	Hewlett Packard
Apollo	Ramtek

APPLICATIONS OF BEASY

Analysis of adhesive joints
The introduction of interface type 'boundary' conditions
enables BEASY to be used to analyse a wider range of
problems. A typical example is the stress analysis of bonded
joints. Here, the extreme aspect ratios encountered in glue
layers, which have hitherto presented serious difficulties in
all forms of numerical analysis, are no longer a problem since
the adhesive layer need not be modelled as a material. An
approximation to the overall behaviour of the joint may be
found using an "interface spring" condition to model the
adhesive. The stresses computed near the interface may
subsequently be used as boundary conditions in a local analysis
of the glue itself (which is likely to be the material to
fracture) upon failure of the joint.

The behaviour of the adhesive layer may be approximated
mathematically as shown in Figure 15. In a direction normal
to the joint, continuity of all variables is assumed, i.e.
tensile and compressive stresses are transmitted directly
across the joint. In the tangential direction, a spring of
some stiffness is applied. The stiffness is usually
calculated from the shear properties of the adhesive material.
This type of approximation at the interface is clearly a
fairly crude representation of an extremely complex phenomenon,
elastic bahaviour being assumed throughout. However, it is
intended as a preliminary analysis, and the typical boundary
element mesh (Figure 16) for this type of problem emphasises
the substantial advantages at the data preparation stage.
Moreover, the results which have been obtained have been found
to compare well with those from more sophisticated and time-
consuming approaches.

Figure 17 shows the variation of direct stress over the interface elements of the mesh shown in Figure 16, when the lap
joint is subjected to a direct tensile load as shown in the
figure. It should be noted that the stresses plotted are
those occurring in the metal, and not in the glue. (No
stresses in the glue may be calculated using this interface
spring approximation). Fillets in the adhesive material which
occur at the ends of such joints may be modelled in a similar
way, by increasing the 'spring' stiffness as required in these
regions.

Centrifugal pump
This example refers to the axisymmetric stress analysis of a
centrifugal pump casting. Although such components are not
truly axisymmetric, each individual section may be assumed
to behave in this way. The BEASY mesh used for the analysis
is shown in Figure 18, and may be seen to be divided into five
zones with a line of internal points around a critical area.

Zoning of this problem was used only because of the aspect ratios found.

The outlet was subjected to a pressure of 255 psi, and the inlet to a pressure of 50 psi. Between these two regions, a parabolic variation was assumed which was specified using the new spline boundary condition facility. The extreme edge at the outlet end was assumed to be rigidly fixed to a plate as shown in Figure 18.

The computed deflected shape is presented in Figure 19, and shows the majority of the activity taking place at the outlet section. The inlet merely undergoes a uniform displacement in the axial direction. A point of inflection may be observed (as indicated in the figure) around the outlet.

Two types of stress plot may be used in the new version of BEASY - xy plots and contour plots. Figure 20 shows a zoomed view of the highly stressed area around the outlet, upon which is superimposed a 'domain' contour plot of the maximum principal stresses. This plot clearly shows the positions of two stress concentrations, being separated by the point of inflection as noted from the displaced shape plot. A typical xy stress plot is given in Figure 21, in which the three principle stress components around the outlet are plotted in graphical form. This plot verifies the value of maximum principal stress at the stress concentration.

This application clearly shows the benefits involved with using BEASY in both data preparation and interpretation of results (particularly in stress contour form) for this type of problem.

3D Aerospace problem
The third and final example problem to be presented in this paper refers to a large 3D problem involving the linear stress analysis of an aerospace component. The boundary element mesh (Figure 22) consists of 120 quadratic quadrilateral elements, and symmetry is not applied on account of the eccentricity of the boundary conditions. Readers may like to note the simplicity with which mesh refinement can be carried out using the discontinuous element types in BEASY (Figure 6).

The displaced shape presented in Figure 23 shows the twisting of the component under the eccentric loading, and also the S-shaped deformation arising from the effect of the two stiffeners.

A surface contour plot of maximum principal stress on the top of a stiffener shows the severe stress concentration which builds up in this region (Figure 24).

CONCLUSIONS

The Boundary Element Method is a powerful technique for analysing engineering problems. BEASY is a system for the solution of general problems and is at the leading edge of this technology. Work is continually being carried out to maintain this position:-

BEASY has a number of advantages for the engineer:-

* Models are far easier to create than for traditional domain methods such as Finite Elements. This results in dramatic savings in man-time.

* Once created, models are extremely easy to change. Localised changes to a model only affect that area of the mesh. With domain methods local changes can affect the entire global mesh.

* BEASY is extremely accurate especially for problems involving stress concentrations.

* Infinite or semi-infinite domain problems can be analysed as simply as solid models.

* Results are far easier to interpret than for domain methods. Results are given directly only on the surface of the problem which for most analyses is where they are required. Internal point results are only given at those points specifically requested. For this reason it is very easy to interface BEASY to commercial modelling packages such as PATRAN or SUPERTAB.

Disadvantages of the BEM are:-

* It is difficult to produce general programs for non-linear analysis, although research into these areas is continually being conducted.

* BEASY is unsuitable for plate bending problems. FE plate and shell elements exist which efficiently model these problems.

BEASY is thus suitable for two types of engineering company:-

Firstly, it provides a powerful tool to complement the use of Finite Elements. The advantages of BEASY can be used to best advantage for initial design stages of a project. These first analyses are often iterative in nature requiring a series of changes to the models followed by further analysis runs. This is commonly the most time consuming period of a project. Any method which reduces turn around times and man power costs must be of value.

Secondly, BEASY can provide to those companies that do not yet perform any analysis, an accurate yet easy to use capability at a realistic cost. The system can be used by engineers without the need for expensive specialist training.

REFERENCES

1. Brebbia C. A., Umetani S., Trevelyan J., "Critical comparison of Boundary Element and Finite Element methods for stress analysis" Boundary Element Technology (BETECH) Conference, Adelaide, Australia, Nov. 1985.

2. BEASY Users' Manual, Version 3.00. Computational Mechanics BEASY Ltd, 52 Henstead Road, Southampton, UK.

3. Brebbia C. A., Trevelyan J., "On the accuracy and convergence of boundary element results for the Floyd pressure vessel problem". To be published, Int. J. Computers and Structures.

4. Brebbia C. A., Mercy A. C., "Use of boundary elements as a computer aided design tool". Computational Methods and Experimental Measurements, 1986, Proceedings, Springer-Verlag.

5. Brebbia C. A., Telles J. C. F., Wrobel L. C., "Boundary Element Techniques", Springer-Verlag, 1984.

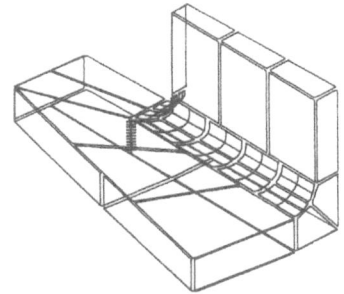

Figure 1   Typical 3D boundary
element mesh

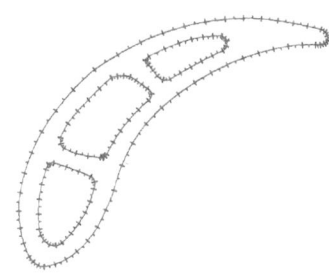

Figure 2   Typical 2D boundary
element mesh

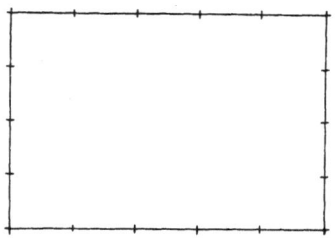

Figure 3(a)   Boundary element
mesh for plate
problem

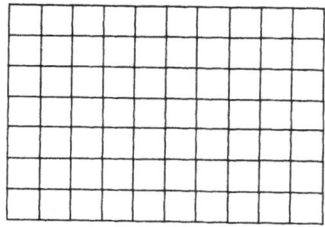

Figure 3(b)   Finite element
mesh for plate
problem

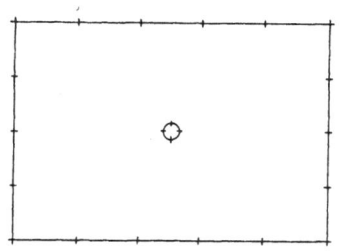

Figure 4(a)   Introduction
of a hole in
plate (BEM)

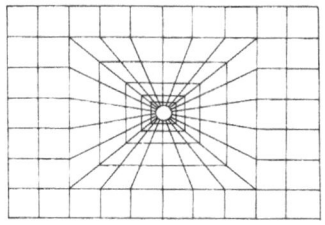

Figure 4(b)   Introduction
of a hole in
plate (FEM)

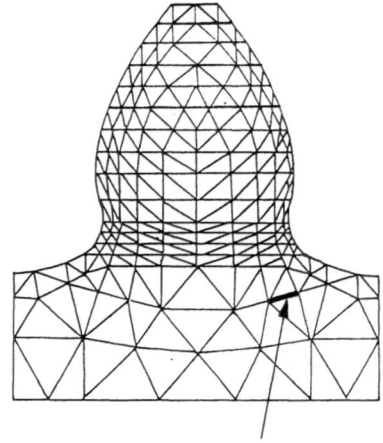

Displacement assumed to be
quadratic over this line

Figure 5   Displacement
constraint in f.e.
domain

Constant Line Element	
Linear Line Element	
Quadratic Line Element	
Constant Quadrilateral	
Linear Quadrilateral	
Quadratic Quadrilateral	

× Mesh Point (defines geometry)
● Node Point

Figure 6   BEASY Element types

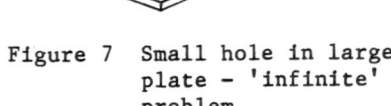

Figure 7   Small hole in large
plate - 'infinite'
problem

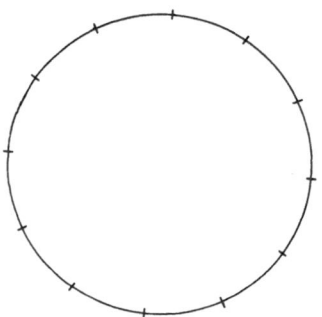

Figure 8   BEASY mesh for hole
in plate problem

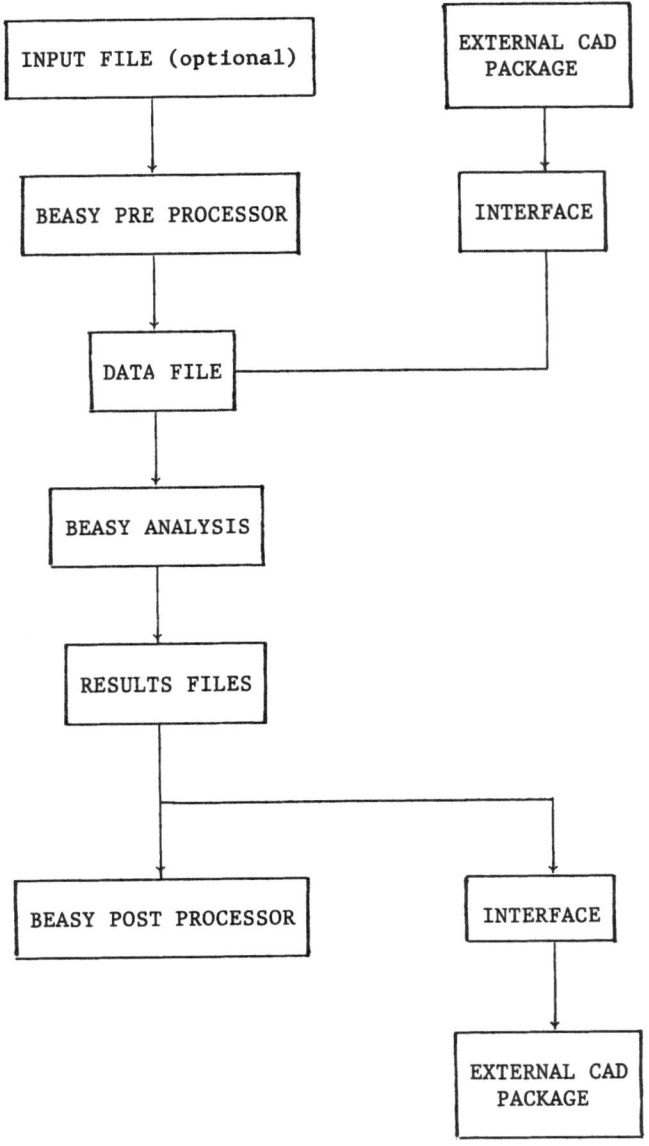

Figure 9   General architecture of BEASY, Version 3.01

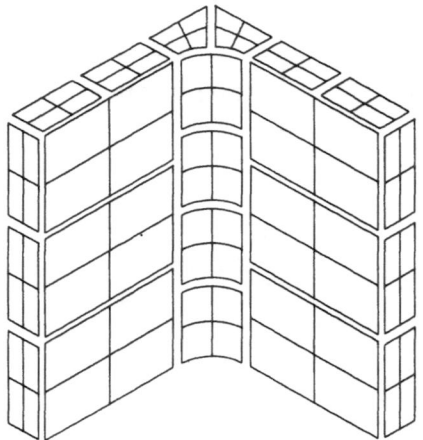

Figure 10   Mesh refinement using
discontinuous elements
(elements shrunk for
clarity)

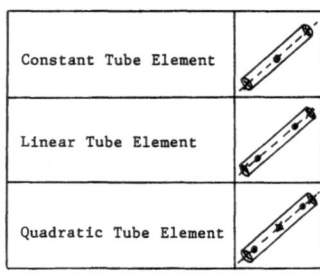

X Mesh Point (defines geometry)
● Node Point

Figure 11   BEASY tube
elements

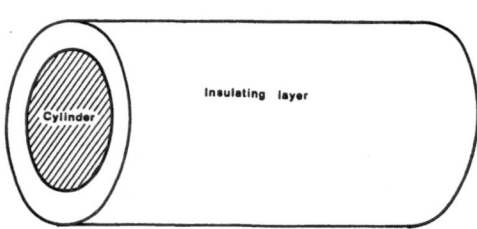

Figure 12   Insulated cylinder

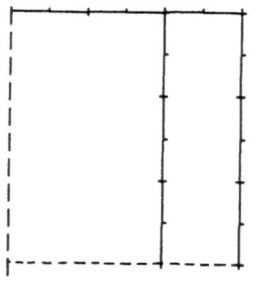

Figure 13   Two zone BEASY
mesh for cylinder
problem

First analysis results form boundary
conditions for more refined local
analysis

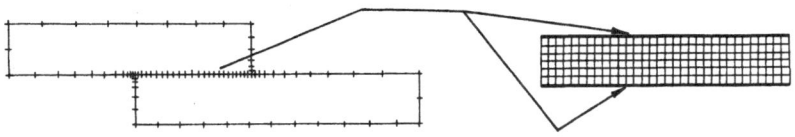

(a) Preliminary BEASY analysis   (b) Local analysis

Figure 14   Two stage analysis of adhesive joint

Spring interface   Normal tractions
condition in the   transmitted
tangential      directly across
direction   interface

Figure 15   Mathematical model of adhesive layer

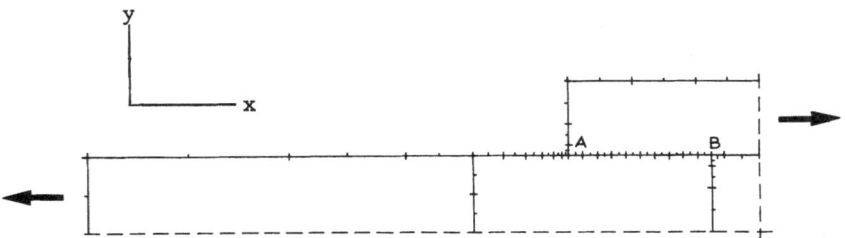

Figure 16   BEASY mesh for typical joint

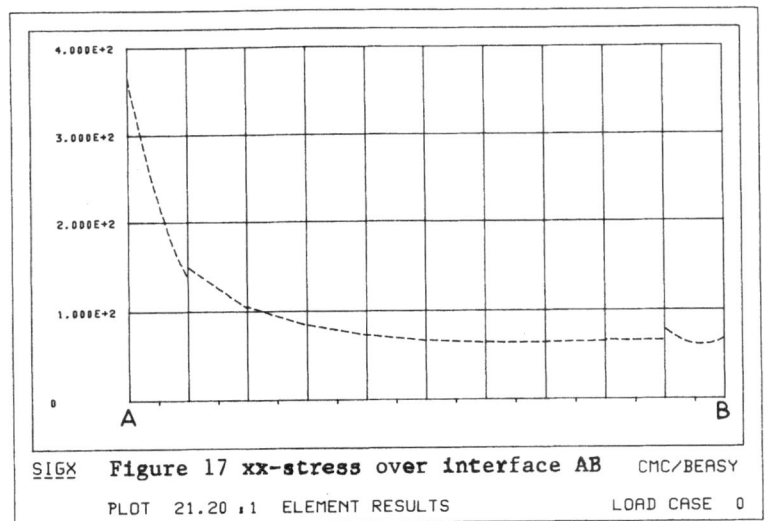

SIGX   Figure 17 xx-stress over interface AB   CMC/BEASY

PLOT 21.20 ,1   ELEMENT RESULTS   LOAD CASE 0

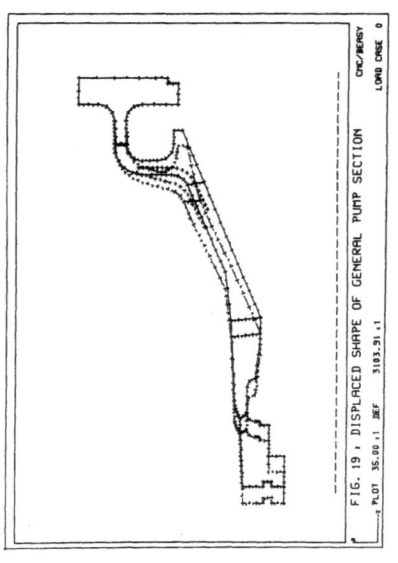

FIG. 19 , DISPLACED SHAPE OF GENERAL PUMP SECTION

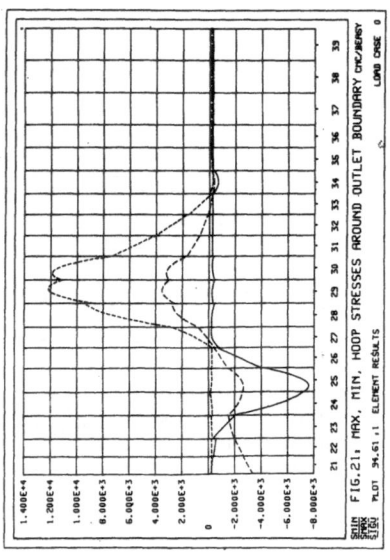

FIG. 21, MAX. MIN. HOOP STRESSES AROUND OUTLET BOUNDARY

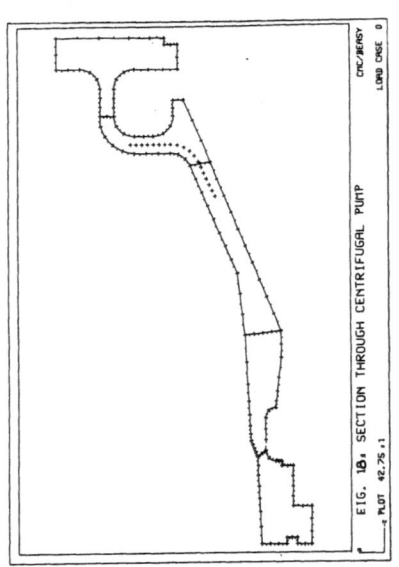

FIG. 18 , SECTION THROUGH CENTRIFUGAL PUMP

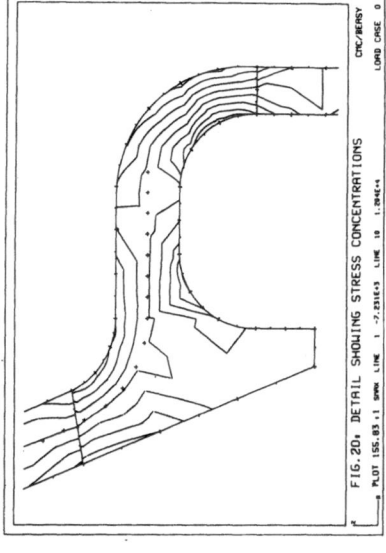

FIG. 20, DETAIL SHOWING STRESS CONCENTRATIONS

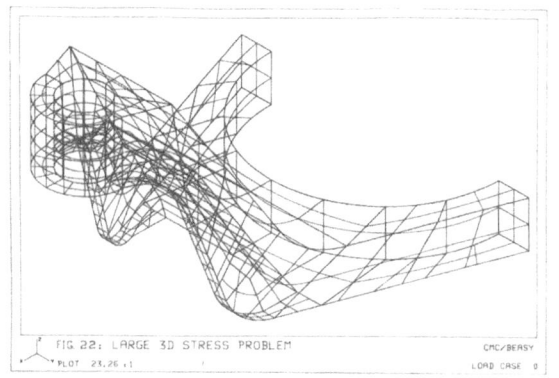

FIG. 22: LARGE 3D STRESS PROBLEM

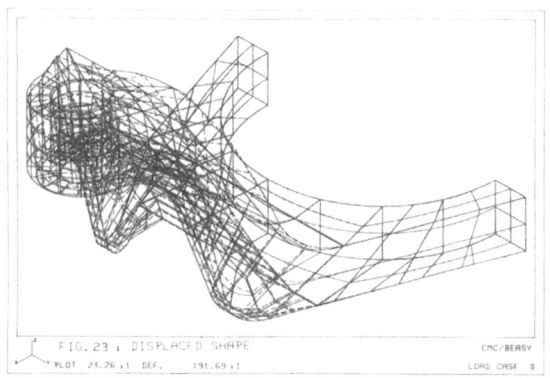

FIG. 23 : DISPLACED SHAPE

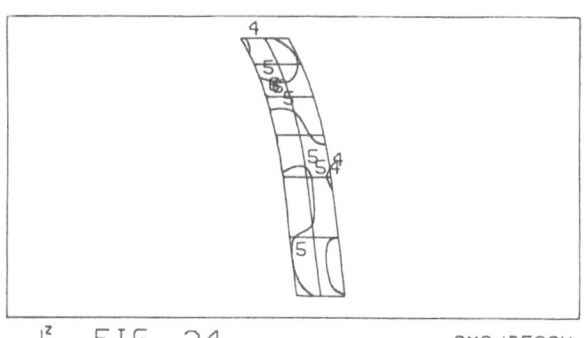

FIG. 24.

CMC/BEASY

PLOT 34.99: 1 SMAX LINE 1 $-7.055E + 5$ Load Case 0
LINE 10 $9.062E + 5$

Interactive Laplace's Equation Analyzing System ILAS

S. Murashima
Department of Electronics, Faculty of Engineering, Kagoshima University, Koorimoto 1-21-40, Kagoshima, Japan
Y. Nonaka
Department of Geography, Faculty of Education, Miyazaki University, Funatsuka 1-1, Miyazaki, Japan

INTRODUCTION

The Charge Simulation Method[1,2,3] (we will call this CSM hereafter) is a simple and effective method for solving Laplace's equation. CSM is a super-position method of Green's function. It corresponds to the boundary type method and collocation method in terms of weighted Residuals. When CSM is compared with FEM, it has the same features as BEM because CSM and BEM belong to the boundary type method. However it has several advantages over BEM as follows:
1) the number of unknown constants in CSM is smaller than BEM,
2) the numerical integration is not necessary,
3) the computing time is very short,
4) the error estimation is easy,
5) the solution and its derivative are given analytically.
6) CSM is suitable for implementing an analyzing system on small computer because that the data we have to handle is very small.
The disadvantages of CSM are as follows:
1) the CSM is not straightfoward because the charge points and contours points must be determined by trial and error. It is dificult especially for biginner to determine a good arrangement of charge points and contour points,
2) the property of the simultaneous equation we have to solve is usually worse than that for BEM,
3) there is no assurance for decreasing of accuracy when we increase the number of unknown constant. It depends on the arrangement of charge points and contour points. Implementing CSM as an computer

aided analyzing system of Laplace's equation, we can
reduce above disadvantages and increase advantages.
The interactive Laplace's equation analyzing system
ILAS is a menu-driven small turn key command system
of CSM performing the numerical solution of two
dimensional and axi-symmetrical potential problems.
This system is coded in BASIC and especially dedi-
cated for the interactive use on small personal
computer. ILAS has some utility programs such as a
contour map editor , a charge pattern generator and
a file manager. These utility programs are implemented so that they reduce the disadvantages of CSM
and increase the merit of CSM. There are some
problems in present ILAS. We have to replace BASIC
by Fortran in order to improve the computing time.
Because of the peripheral utilities , however , the
total performance of ILAS is satisfactory.

CHARGE SIMULATION METHOD

<u>Principle of the CSM</u>
Here we consider the Dirichlet problem of Laplace's
equation in a Domain D, which is shown in Fig.1. For
simplicity, we consider a problem in two dimensions:

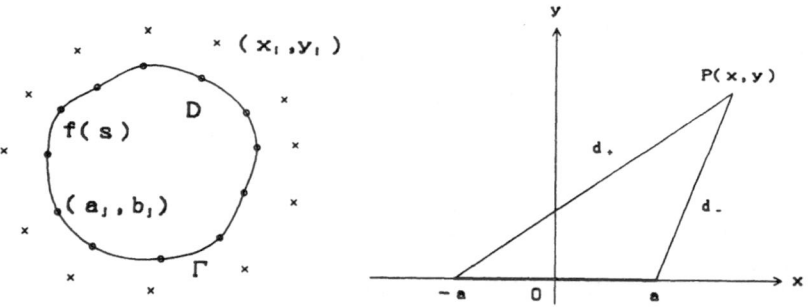

Fig.1 Dirichlet Problem   Fig.2 Strip line charge

$$\nabla^2 \Psi(x,y) = 0 \qquad (x,y) \text{ in } D \qquad (1)$$

$$\Psi(x,y)_\Gamma = f(s) \qquad (x,y) \text{ on } \Gamma \qquad (2)$$

where Γ is the boundary, and $f(s)$ is the boundary
value at any boundary points. In the ·CSM, the
general solution $\Phi(x,y)$ of the proceeding
Laplace's equation is expressed by means of super-
position of potential function due to several
charges at charge points (x_i,y_i) external to D. That

1S

$$\Phi(x,y) = \sum_{i=1}^{N} Q_i \ G(x,y;x_i,y_i) \tag{3}$$

where N is the number of charges. Q_i is unknown constant. called the weighted facter or the magnitude of i-th charge. and $G(x,y;x_i,y_i)$ is Green function in the unbounded region due to unit line charge at (x_i,y_i) . In two dimensional problem, the function $G(x,y;x_i,y_i)$ is as follows:

$$G(x,y;x_i,y_i) = \frac{1}{2\pi}\log\sqrt{(x-x_i)^2+(y-y_i)^2} \tag{4}$$

The expression (4) is a harmonic on the domain D. because the charge points (x_i,y_i) are not inside the domain D. Hence the general solution (3) satisfies eq.(1). To determine the solution, we impose on eq.(3). the boundary condition at suitably chosen contour points (a_j,b_j) of the same number as that of the charge points. Then

$$\Phi(a_j,b_j) = \sum_{i=1}^{N} Q_i G(a_j,b_j;x_i,y_i) = f(s_j) \tag{5}$$

where (a_j,b_j) is the j-th contour point. and s_j also denotes the contour point. We here assume that (a_j,b_j) is on the boundary and (x_i,y_i) is outside the boundary.

Solving this system of N linear eq.(5), we can determine the N charge Q_i. Substituting Q_i into eq.(5) we get an approximate solution of CSM. Of course. whether or not the calculated set of charges fits the boundary conditions must be checked. In 2-dimensional case. we can use another type of fundamental solution. Typical fundmental solution often used besides the expression (4) is the potential function due to a unit strip line charge located on (a.0) and (-a.0) as shown in Fig.2.

$$G(x,y)= \frac{1}{2\pi}\log(X +\sqrt{X^2-1}) \tag{6}$$

where

$$X = \frac{d_+ + d_-}{2a} , \quad d_+ = \sqrt{(x+a)^2 + y^2} , \quad d_- = \sqrt{(x-a)^2 + y^2} \quad (7)$$

In 1969 , H.Steinbigler first used CSM in his dissertation to compute the electrical field around a high voltage apparatus. He used the potential functions due to Ring charge , straight line charge of finite length and point charge. His CSM achieved a very high accuracy especially for axi-symmetric problems.

a) Ring Charge[1,2](Fig.3)

$$G(r,z;e,d) = \frac{K(k)}{2\pi^2 \sqrt{(r+e)^2 + (z-d)^2}} \quad (8)$$

where

$$k = \sqrt{\frac{4re}{(r+e)^2 + (z-d)^2}}$$

$$G(r,z;e,d) = \frac{1}{4\pi \sqrt{e^2 + (z-d)^2}} \quad r \longrightarrow 0 \quad (9)$$

$$G_r(r,z;e,d) = -\frac{\{e^2 - z^2 + (z-d)^2\}E(k) - \{(r-e)^2 + (z-d)^2\}K(k)}{4\pi^2 r \sqrt{(r+e)^2 + (z-d)^2} \{(r-e)^2 + (z-d)^2\}} \quad (10)$$

$$G_z(r,z;e,z) = -\frac{(z-d)E(k)}{2\pi^2 \sqrt{(r+e)^2 + (z-d)^2} \{(r-e)^2 + (z-d)^2\}} \quad (11)$$

where K(k) and E(k) are Elliptic Integrals of first and second kind, respectively.

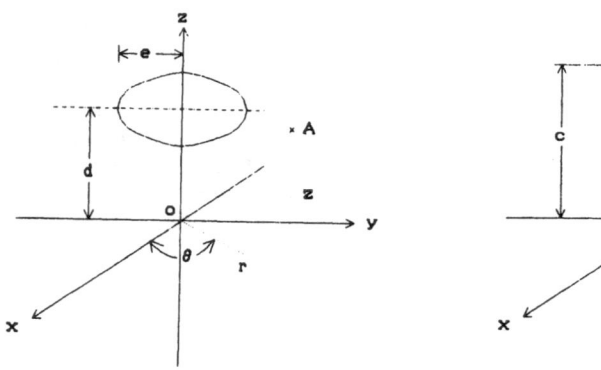 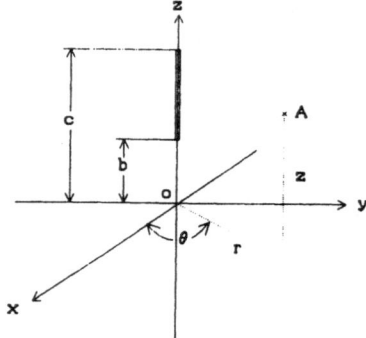

Fig.3 Ring charge   Fig.4 Line charge

b) Line Charge of Finite Length(Fig.4)

$$G(r,z;c,b) = \frac{1}{4 \pi} \ln \left\{ \frac{c-z+ \sqrt{r^2+(c-z)^2}}{b-z+ \sqrt{r^2+(b-z)^2}} \right\} \qquad (12)$$

$$G(r,z;c,b) = \frac{1}{4 \pi} \ln \frac{c-z}{b-z} \qquad r \text{---}>0 \quad (13)$$

$$G_r(r,z;c,b) = \frac{1}{4 \pi r} \left\{ \frac{c-z}{\sqrt{r^2+(c-z)^2}} - \frac{b-z}{\sqrt{r^2+(b-z)^2}} \right\} \qquad (14)$$

$$G_z(r,z;c,b) = \frac{1}{4 \pi} \left\{ \frac{1}{\sqrt{r^2+(c-z)^2}} - \frac{1}{\sqrt{r^2+(b-z)^2}} \right\} \qquad (15)$$

If we use Green's function in 3-dimension ,i.e.. $1/(4 \pi r)$, it is impossible to get such high accuracy. In the 1970s. several reports related to the application of CSM were published especially in Japan and West Germany. Besides the ring charge and line charge. the circular disk charge[4] is reported to be useful for flat electrode.
In the case of general 3-dimensional problem. the triangular plate charge[5] is very useful.

The Properties of the error in CSM[6]
Here we define the error $e(x,y)$ as the difference between calculated potential $\Phi(x,y)$ and the exact potential $\Psi(x,y)$. as follows:

$$e(x.y) = \Phi(x.y) - \Psi(x,y). \qquad (16)$$

Since $\Psi(x,y)$ satisfies eq.(1) and (2). and $\Phi(x,y)$ satisfies eq.(1) , we get:

$$\nabla^2 e(x,y) = 0, \quad (x,y) \quad \text{in} \quad D, \qquad (17)$$

$$e(x,y) = f(s) - \Phi(x,y), \quad (x,y) \text{ on } \quad \Gamma \qquad (18)$$

The second term of the right hand side of eq.(18) is already known. Hence, the error $e(x,y)$ is the solution of the first kind of boundary value problem. with boundary value $f(s) - \Phi(x,y)$. In other words, the error has the same properties as the potential. This means that in CSM. the maximum of the absolute error appears on the boundary. This point is very important because we can know the maximum error of the analysis without comparing with other solution.

FUNCTIONS OF ILAS

Concept of ILAS
Thinking about the case we solve a certain boundary value problem. The first thing we have to do is to determine the method suitable for the problem. Supposed that we can find a good method for the problem , is it all finished? It is not the case. Talking about a boundary value problem, not only to solve the problem but also we have many things to do. For examples we have to input a fairly large data set and estimate error of the analysis. After solving the problem in question we have to display the graphical results such as 3-dimensional shape or cotour map and so on. So far what is called the good numerical method is mentioned about on the view point of computing time and accuracy. This view point is not what we are standing on. We consider that numerical method should be estimated totally. This means that not only computing time and accuracy but also peripheral utilities should be estimated.
In order to show that CSM is such method. we are trying to install a computer aided CSM system called ILAS which means Interactive Laplace's equation Analyzing System.

Environment of ILAS
ILAS needs following hardware:1) NEC-9801 N88-Basic System 2) 256kbyte main memory 3) auxiliary memory more than 640kBYTE.
If N88-Basic Interpreter on MS-DOS is available, ILAS can be run on IBM-5550. If N88-Basic compiler is available. 3-times high speed is obtained.

ILAS Family
In this research, following 5 programs are planed to develope.

ILAS2 ILAS2 is a basic Ilas program to analyze two dimensional Laplace's eauation in homogenious region under Dirichlet .Neumann or Mixed boundary value condition. ILAS2 also supports the problem consisting of two regions of different constant. As fundamental solution . ILAS2 uses following charges: 1) line charge given by eq.(4), 2) strip line charge given by eq.(5) 3) etc

ILASR This program analyzes 3-dimensional and

rotational symmetric Laplace's equation. Except for
the fundamental solution, the ILASR is almost same
as the ILAS2. As the fundamental solution. ILASR
uses 1) ring charge 2) line charge of finite length
3) disk charge and 4) point charge.

ILASG This program solves 3-dimensional problem by
superposing 1)triangular plate charges, 2) ring
charge, 3)line charge of finite length. 4) disk
charges 5) point charges. The main part of this
program consists of several 3-dimensional graphic
tools. The fundamental solution mainly used in ILASG
is the function due to the triangular plate charge.

ILAS2F . ILASRF These two programs are for
solving free boundary problem in 2-dimension
(ILAS2F) and in 3-dimension with rotational
symmetry(ILASRF).

UTILITY PROGRAMS OF ILAS2

Input program of Boundary Shape
The function of this program is to input , edit and
display the boundary shape. The boundary shape
consists of sevral segments like a line , eliptic
arc or spline curve. In this program , the type of
boundary condition are given on each segment. The
boundary data made by this program is saved into an
auxiliary memory. This data is read by the pattern
generator in next step.

Generation of Charge Pattern
The function of this program is to generate a charge
pattern reasonable for the boundary shape or
boundary condition. For biginner in CSM it is very
difficult to avoid the bad and unsuitable charge
pattern in order to obtain reasonable accuracy. In
this program , several typical charge patterns which
yield reasonable accuracy are provided for each
boundary segment ,i.e., straight line , curved line
or spline curve. The generated charge pattern are
displayed on CRT with the boundary shape which are
saved in previous step. Confirmed charge pattern are
saved as Data file with extension DTn. An example of
pattern generator screen are shown in Fig.5.

ANALYSIS by CSM and Error Estimation
After reading the data saved in previous step, this
program detemines the charge amplitude and the
results are added to data file. If the accuracy is
not enough , we have to modify the charge pattern
and calculate the charge amplitudes again.

Contour Map Editor

This program draws contour map of CSM solution and edit the contour map. The algorithm used for making contour lines is not ordinary one. In this program we need not calculate the function value on grid points. The program finds the point on contour line by using function values and its derivatives. Because that the CSM is analytical, the derivatives of CSM solution can be derived easily and accurately. This program also draws 3-dimensional shape with hidden-line elimination. These graphes are saved into mini-floppy disk and are reused in the edit mode. Examples of contour map editor screen are shown in Fig.6,7,8 and 9.

File Manager

Ilas use several type of files ,i.e., data file or graphic file. This file manager handle the hierarchical tree structure the root of which corresponds to the boundary file. Several solutions are obtained for each boundary file. Several contour maps or 3-D figures are drawn for each solution.

CONCLUSION

Interractive Laplace's eqation analyzing system ILAS is described. It is based on the Charge Simulation Method which is the non-direct Boundary Element Method. ILAS has several utility programs which increase the advatages and reduce the defects of CSM. This system is developed on the stand point that the numerical method should be estimated total work which includes preparations , analysis of boundary value problem and post-processings. Because of the simplicity of CSM ,we can conclude that ILAS has sufficiently high performance . Remarkable point of ILAS is that this is installed on a small personal computer like IBM 5550 and NEC-9801.

REFERENCES

1) H.Steinbigler:Dissertation, Tech. Univ. Munchen (1969)
2) H.Singer,H.Steinbigler and P. Weiss:IEEE, **PAS-93** (3), 1660, 1974
3) Sadayuki Murashima: Charge Simulation Method and Its Applications , Morikita Shuppan Co.LTD, Tokyo, 1983
4) T.Takuma: Charge Simulation Method including Disk charge, Journal of IEEJ **97-A** , 411 ,(1978).
5) T.Kouno and T.Takuma:Numerical Calculation of Electric Field, Ohmsha Co.lTd, Tokyo, 1980
6) S.Murashima, M.Kato and E.Miyachica : On the

Properties of the Error in the CSM.   Journal of IEEJ
98-A.P.39, 1978.
7) S.Murashima and H.Kuhara:An Approximate Method to
Solve Two-Dimensional Laplace's Equations on a
Riemann Surface, Journal of Information Processing,
Vol.3. No.3. P.127 139, 1980.

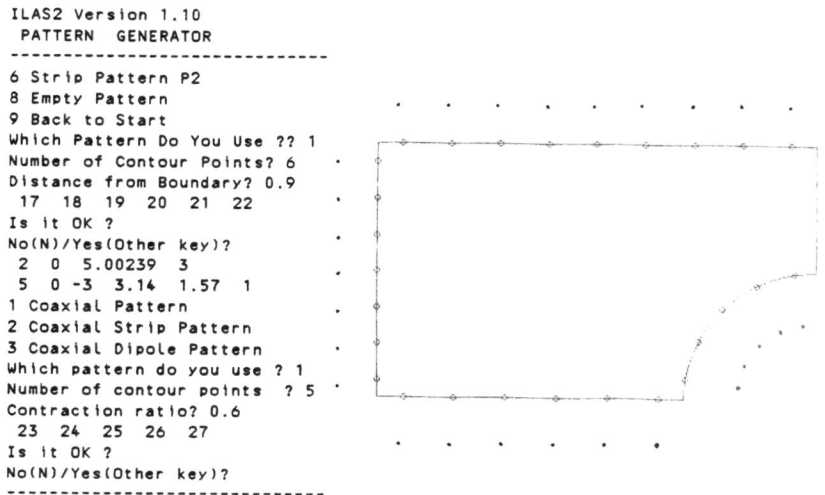

```
ILAS2 Version 1.10
 PATTERN  GENERATOR
-------------------------------
6 Strip Pattern P2
8 Empty Pattern
9 Back to Start
Which Pattern Do You Use ?? 1
Number of Contour Points? 6
Distance from Boundary? 0.9
 17  18  19  20  21  22
Is it OK ?
No(N)/Yes(Other key)?
 2   0  5.00239  3
 5   0  -3  3.14  1.57  1
1 Coaxial Pattern
2 Coaxial Strip Pattern
3 Coaxial Dipole Pattern
Which pattern do you use ? 1
Number of contour points  ? 5
Contraction ratio? 0.6
 23  24  25  26  27
Is it OK ?
No(N)/Yes(Other key)?
-------------------------------
```

Fig.5   An example of Pattern Generator screen

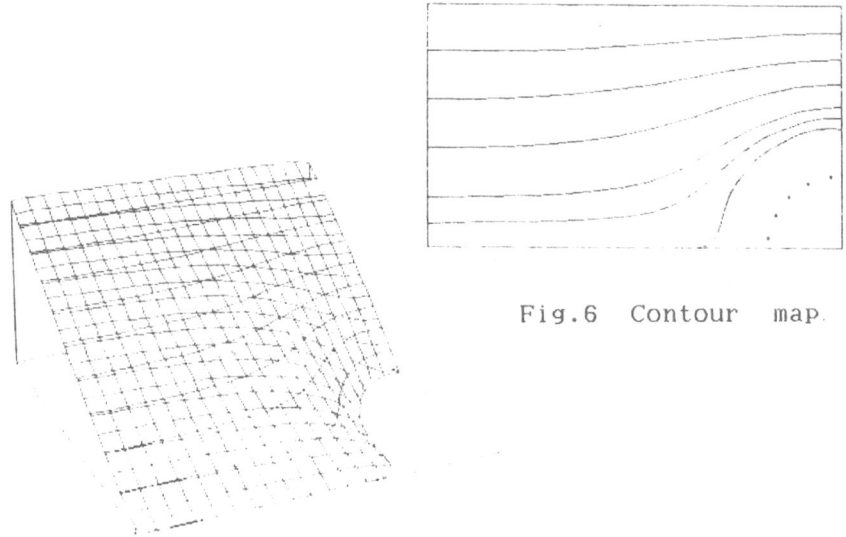

Fig.6   Contour   map

Fig.7 3-dimensional figure

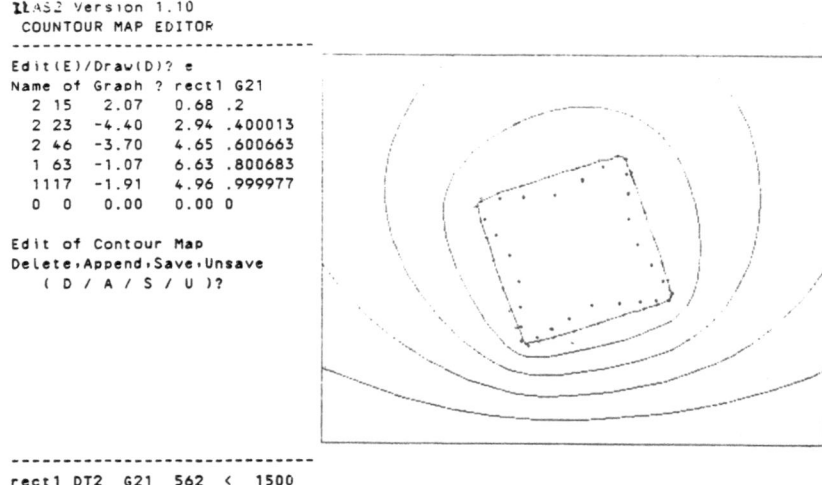

Fig.8 The contour map of a CSM solution by 24 line charges for a Dirichlet problem.

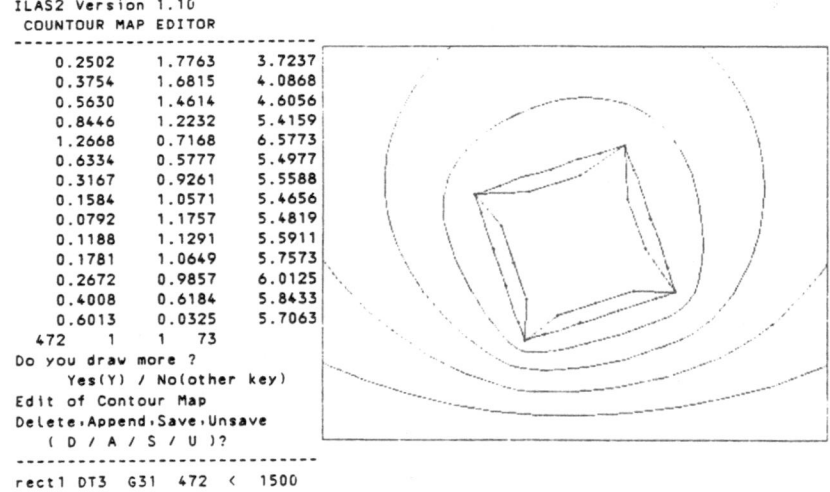

Fig.9 The contour map of another CSM solution by 14 strip line charges for the same Dirichlet problem as Fig.8.

A Large System BEAP-II of Boundary Element Analysis with Sparse Algorithm

B. Hua, Z. Mao, X. Ye
Shanghai Institute of Computer Technology, Shanghai, China
Y. Zhang and G. Luo
Jiao Tong University, Shanghai, China
H. Xiong
Second Industry University, Shanghai, China

ABSTRACT

In this paper a large system BEAP-II of boundary element analysis developed by the authors has been presented. This system can solve the stress analysis problems of two and three dimensional continuum and the stress strength factor of crack. On numerical solution due to having used the subregion and the unsymmetric sparse algorithm[5] to boundary element method,then one or two matrix sections are reserved in internal storage. Therefore this program can solve large engineering problems. BEAP-II possesses unified pre and post processors, solver and element library. Some applied examples in engineering can be found in author's other papers [6,7].

INTRODUCTION

The boundary element method (BEM) being developed quickly in recent years is an effective numerical analysis method.Its kernel idea is that the conception on the element coupled with the boundary integral equation.Therefore many successful applications occur in many fields, in particular, the solid mechanics. The research works have been collected in all previous proceedings of the international conferences on boundary element method edited by Brebbia .

The research and application of boundary element method have deeply entered into the engineering practices. Some kinds of program occur in succession, for example, France's CASTOR, Britain's BEASY, Japan's BEM, etc. In China the developments of program packages on boundary element method are proceeding. A large program BEAP-II that possesses the consideration on both generality and standardization has been developed by authors. This program can solve large engineering problems and possesses unified pre and post processor,solver and element library. The numerical solution is made by using the subregion method coupled with the unsymmetric sparse algorithm, so that the coefficient matrix will be banded. It can be used to solve large engineering problems on different kinds of computer.

The computational results show that the boundary element method is very effective for the stress analysis and computing stress strength factor of crack. And by using rather rough meshes the fine results can be also given.

FUNDAMENTAL PRINCIPLE

Fundamental boundary integral equation of linear elastic problem is based on the principle of virtual displacement[1].If we extend this principle to that case where virtual displacement δu_i does not satisfy boundary condition of displacement, the equilibrium equation of the system can be shown as:

$$\int_{\Omega} (\sigma_{ij,j} + f_i) U_i \, dv = \int_{\Gamma_2} (t_i - \bar{t}_i) U_i \, ds - \int_{\Gamma_1} (u_i - \bar{u}_i) T_i \, ds$$

Utilizing the properties of the fundamental solution above formula can be simplified to following boundary integral equation:

$$c_{ij} u_j(P) + \int_{\Gamma} T_{ij}(P,Q) U_j(Q) ds = \int_{\Gamma} U_{ij}(P,Q) t_j(Q) dS(Q)$$
$$+ \int_{\Omega} U_{ij}(P,Q) f_j(Q) dV(Q) \qquad (2.1)$$

Where u_j and t_j denote displacements and tractions respectively, U_{ij} and T_{ij} denote the fundamental solution of the displacement and traction respectively, f_j denote body forces(three dimension)or surface forces (two dimension).

The program includes boundary elements of two and three dimensional problems, in which boundary elements are constant, linear and quadratic elements for two dimensional problem and 8-node isoparametric element for three dimensional problems. Let us consider three dimensional body not losing the generality. Assuming Γ be boundary of a three dimensional body Ω ,and be discretized into Ne elements and Np nodes. All of the nodes are unifiedly numbered. In identical subregion order of the node numbers is succesive. Hereby substituting the expressions of the displacement and coordinate transformation of 8-node isoparametric element into (2.1), then we obtain :

$$c_{ij}u_j(P) + \sum_{n=1}^{Ne} \int_{S_n} T_{ij}(P,Q) \sum_{K=1}^{8} N_K(\xi,\eta) U_j^K G_n d\xi d\eta$$
$$= \sum_{n=1}^{Ne} \int_{S_n} U_{ij}(P,Q) \sum_{K=1}^{8} N_K(\xi,\eta) t_j^K G_n d\xi d\eta$$
$$+ \sum_{m=1}^{Me} \int_{\Omega_m} U_{ij}(P,Q) f_j(Q) J_m dV(Q) \qquad (2.2)$$

Using numerical integration the equation system (2.2) becomes

$$c_{ij}u_j(P) + \sum_{n=1}^{Ne} \sum_{r=1}^{P_r} \sum_{s=1}^{P_s} W_{rs} T_{ij} \sum_{K=1}^{8} N_K(\mathfrak{Z}_r, \eta_s) u_j^K G_n$$
$$= \sum_{n=1}^{Ne} \sum_{r=1}^{P_r} \sum_{s=1}^{P_s} W_{rs} U_{ij} \sum_{K=1}^{8} N_K(\mathfrak{Z}_r, \eta_s) t_j^K G_n$$
$$+ \sum_{m=1}^{Me} \sum_{r=1}^{P_r} \sum_{s=1}^{P_s} \sum_{t=1}^{P_t} W_{rst} U_{ij} f_j J_m \qquad (2.3)$$

in which P_r, P_s and P_t are the numbers of Gaussian points, $\mathfrak{Z}_r$, η_s and $\mathfrak{Z}_t$ are Gaussian point coordinates, W_{rs}, W_{rst} are weighting factors.

Introducing several notations into the last equation system it can be written as:

$$c_{ij}u_j(P) + \sum_{n=1}^{Ne} H_i^{(n)} u^{(n)} = \sum_{n=1}^{Ne} G_i^{(n)} t^{(n)} + F_o$$

in which $H_i^{(n)}$ and $G_i^{(n)}$ are called the element matrices of point i under consideration, $u^{(n)}$ are the displacements of element nodes, $t^{(n)}$ are the tractions of the element nodes, and they have the forms respectively as follows:

$$H_i^{(n)} = \begin{bmatrix} t_{1,1}, \cdots, t_{1,24} \\ t_{2,1}, \cdots, t_{2,24} \\ t_{3,1}, \cdots, t_{3,24} \end{bmatrix}, \qquad G_i^{(n)} = \begin{bmatrix} u_{1,1}, \cdots, u_{1,24} \\ u_{2,1}, \cdots, u_{2,24} \\ u_{3,1}, \cdots, u_{3,24} \end{bmatrix}$$

$$u^{(n)} = \{u_1^{(n)}, v_1^{(n)}, w_1^{(n)}, \cdots, u_8^{(n)}, v_8^{(n)}, w_8^{(n)}\}^T$$

$$t^{(n)} = \{t_{x_1}^{(n)}, t_{y_1}^{(n)}, t_{z_1}^{(n)}, \cdots, t_{x_8}^{(n)}, t_{y_8}^{(n)}, t_{z_8}^{(n)}\}^T$$

Furthermore, the last equation system can be simplified as:

$$Tu = Ut + F_o \qquad (2.4)$$

Finally, the equation system (2.4) may be written as:

$$AX = F \qquad (2.5)$$

where X is a unknown matrix of the boundary traction and the boundary displacement, F is the sum of the matrix produced from some columns in U and T multiplying given boundary tractions and boundary dispacements and the matrix F_o. A is a dense unsymmetric matrix. If the region is divided into subregions, then A may be a banded matrix.

The general procedure by which the global matrix A and right term F are formed is as follows:

$$\sum_n H_i^{(n)} \Rightarrow A \quad , \qquad \sum_n G_i^{(n)} t^{(n)} + F_o \Rightarrow F$$

when some displacements of boundary nodes are known we must

interchange some columns in $\sum_n H_i^{(m)}$ with corresponding columns in $\sum_n G_i^{(m)}$. The submatrix on the diagonal line of matrix A can be determined as later paragraph. If Ω be divided into subregions, then assembling A is controlled by using the numbers of starting submatrices and ending submatrices per line, and automatically performed by a program module.

NUMERICAL TECHNIQUES

1. Sparse algorithm

In general, a large region Ω can artificially be divided into some subregions, for instance, two subregions Ω_1 and Ω_2, as Fig.3.1. We can form the systems of linear equations on the boundary element for each subregion as follows:

$$[T' \ T_I'] \begin{bmatrix} U' \\ U_I' \end{bmatrix} = [U' \ U_I'] \begin{bmatrix} t' \\ t_I' \end{bmatrix}$$

$$[T^2 \ T_I^2] \begin{bmatrix} U^2 \\ U_I^2 \end{bmatrix} = [U^2 \ U_I^2] \begin{bmatrix} t^2 \\ t_I^2 \end{bmatrix} \tag{3.1}$$

If the compatible condition $u_I = u_I' = u_I^2$ and equilibrium condition $t_I = t_I' = -t_I^2$ are introduced into (3.1), then the system of linear algebraic equations on the boundary element for total region can be obtained as follows:

$$\begin{bmatrix} T' & T_I' & -U_I' & 0 \\ 0 & T_I^2 & U_I & T^2 \end{bmatrix} \begin{bmatrix} u' \\ u_I \\ t_I \\ u^2 \end{bmatrix} = \begin{bmatrix} U' & 0 \\ 0 & U^2 \end{bmatrix} \begin{bmatrix} t' \\ t^2 \end{bmatrix} \tag{3.2}$$

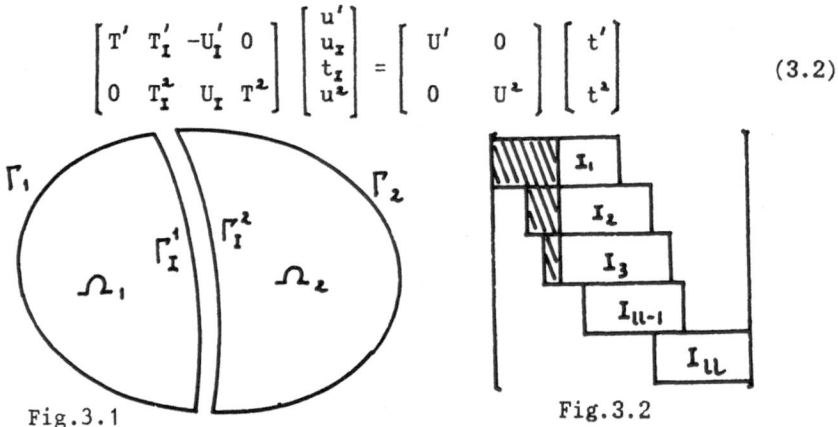

Fig.3.1   Fig.3.2

Taking more subregions, then the coefficient matrix A in equation system (2.4) is an unsymmetric banded matrix, as Fig.3.2.

A new banded algorithm of partitioning section-block for solving

large system of unsymmetric linear equations is given by authors
in reference 5. This algorithm in fact has attended three
objects: the storage capacity should be less, the operation
speed should be quick and the bandwidth should not be limited.

The unsymmetric matrix is stored in the scheme of variable
bandwidth storage by rows by partitioning sections. According
to the integrated number of elements in band region, we should
divide the matrix A into 11 sections, as Fig.3.2.

The matrix A section by section is written in the disk by the
scheme of variable bandwidth. During the computations one or two
matrix sections are reserved in the internal storage. The i--th
matrix section contains $mm(i,1)$ rows and $mm(i,2)$ columns
submatrices, which are stored in array $a(1:nl,1:mb)$, a being an
operation region of the internal storage.
Using the scheme of storage as above mentioned, we form every
section and let them be written in the disk in turn, then use
the decomposition of block LU by rows, the computation being
performed by the variable bandwidth scheme in a in the form of
stripe. During the process the submatrix is a computation unit.

2. Numerical treatment

(1) Singularity on 8-node isoparametric element in three
dimensional problem.
(a) Treatment on $1/r$ singularity. Let Xa is a point under
consideration of the discrete equation system (2.2), and Sn is
a boundary element. For Xa $\notin$ Sn, by computing for the
integrated terms in (2.2) the singularity does not occur. For
Xa $\in$ Sn, the element can be divided into triangles each with a
vertex at Xa [2,3,6]. For each of triangles coordinates using
linear shape function of 4-node isoparametric element are
defined. From this we conclude $U_{ij} \in O(1/r)$ when $p \to Xa$. And the
integrand $U_{ij} N_K GJ$ is bounded quantity.

(b) Treatment on $1/r^2$ singularity. Let p is a node on the
boundary element Sn, its order number of numbers of nodes in
the element is l-th, and $f_j \equiv 0$, then integral equation (2.1)
and (2.2) can rewrite as follows:

$$c_{ij}u_j(P) + \sum_n \int_{S_n \ni p} N_\ell(\mathfrak{f},\eta) T_{ij}(P,Q) dS(Q) u_j^{(\ell)}$$

$$+ \sum_n \int_{S_n \ni p} \sum_{\substack{k=1 \\ k \neq \ell}}^{8} N_K(\mathfrak{f},\eta) T_{ij}(P,Q) u_j^{(K)} dS(Q)$$

$$+ \sum_m \int_{S_m \not\ni p} \sum_{k=1}^{8} N_K(\mathfrak{f},\eta) T_{ij}(P,Q) u_j^{(K)} dS(Q)$$

$$= \int_S U_{ij}(P,Q) t_j(Q) dS(Q) \qquad\qquad (3.3)$$

Take $t_j = 0$, and u_j being identity displacements, then we obtain:

$$c_{ij} + \sum_n \int_{S_{nap}} N_\ell(\zeta,\eta) T_{ij}(P,Q) dS(Q)$$

$$= -\sum_n \int_{S_n,\ni p} \sum_{\substack{K=1 \\ K \neq \ell}}^{8} N_K(\zeta,\eta) T_{ij}(P,Q) dS(Q)$$

$$-\sum_m \int_{S_{m\ni p}} \sum_{K=1}^{8} N_K(\zeta,\eta) T_{ij}(P,Q) dS(Q) \tag{3.4}$$

In the formula (3.4), the singular point is not contained in its right second term, and so $N_\ell(\zeta,\eta)$ is not contained in its right first term, that the singular point is not contained also in this term[6].

(2) On the discontinuity of the traction. Let us consider a point on the surface, where the traction is discontinuous, according to a discontinuity of outward normal, as Fig.3.3. We can use 'multiple point' concept. Let x, x' and x'' are multiple points, and they satisfy displacement continuity condition at the multiple point:

$$u_j(x) = u_j(x') = u_j(x'') \tag{3.5}$$

however the tractions at these points may be different values.

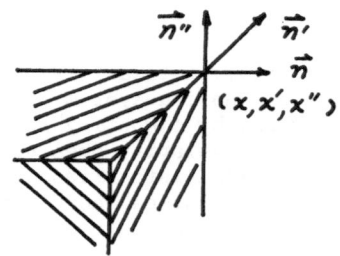

Fig.3.3

For the other cases where the formula (3.5) is not suitable, the conditions of stress tensor symmetric and strain-displacement compatibility can be used[4].

(3) On precision at internal point. The displacement and stress at the internal point can be computed by using the displacements and stresses at the boundary points, D_{Kij}, S_{Kij} and the integral formulas of the displacement and stress at internal point', in which S_{Kij} (for instance three dimension) contains the factor $1/r^3$, therefore the size of the boundary elements and the number of Gaussion integral points in elements must be arranged[6] according to the interval of value r. This program possesses the function automatically selecting Gaussian integral point.

3. The structure of the program

Total structure of the program BEAP-II consists of some parts as follows:

COMPUTING EXAMPLE

The more than ten problems of the stress analysis and stress strength factor of crack have been solved by using program BEAP--II, in detail, it may be seen in references 7 and 6.

CONCLUSION

(1) BEAP--II program can be applied to the stress analysis and computing stress strength factor of crack of the two and three dimensional continuum problems.

(2) Based on the subregion method coupled with the variable bandwidth algorithm of section-block, large engineering problems can be effectively solved by using BEAP-II.

(3)The numerical treatment on the singularity and discontinuity of traction has been taken account in this program, therefore the examples being computed show that the numerical precision is excellent.

(4) Due to the modularity and generality of the program, the element library may be arbitrarily extended.

ACKNOWLEDGEMENT

The authors would like to thank associate professor Jiang Shanbiao, for whose help in the check of this paper has made the publication possible.

REFERENCES

(1) Brebbia C.A., The Boundary Element Method for Engineers, 1978.
(2) Banerjee P.K. and Butterfield R., Developments in Boundary Element Method--1, 1979.

(3) Lachat J.C. and Watson J.O., Effective Numerical Treatment
of Boundary Integral Equation: A Formulation for Three
Dimensional Elastostatics, Int.J.Num.Mech.Eng.10, P991~1005,
1976.
(4) Brebbia C.A., Recent Advances in Boundary Element Methods,
1978.
(5) Hua Bohao and Ye Xiangrong, An Algorithm for Solving Large
System of Unsymmetric Linear Equations in Boundary Element
Method,Proceedings of the International Conference on Boundary
Element Method in Engineering, Beijing, China, 1986,10.
(6) Hua Bohao and etc., The Applications of Boundary Element
Method in Stress Analysis, First National Conference on
Boundary Element Method, China, 1985.12.
(7) Hua Bohao and etc., The Applications of the Program BEAP-II
in Several Typical Engineering,(to be published).

SECTION XIII NUMERICAL AND MATHEMATICAL ASPECTS

Mixed Methods in BEM for Elasto-Plastic Problems

E. Schnack
Institute of Solid Mechanics, The University of Karlsruhe, 7500 Karlsruhe 1, F.R. Germany

SUMMARY

The BEM is much more suitable than the FEM to solve problems with high field gradients (Brebbia [1,2]). However, the FEM assures, contrarily to BEM, symmetrical, positive definite stiffness matrices. This is very advantageous, especially for problems with a high number of degrees of freedom. The question arises now, in which form a coupling of BEM and FEM is possible in order to take profit of these great advantages of BEM against FEM (Zienkiewicz[3,4], Schnack[5], Wendland[6]). For this purpose, the body is divided in subdomains. One can use, then, BEM elements for domains of high stress concentrations. For these domains, a stress field is defined at first, which satisfies the equilibrium and boundary conditions exactly. The stress field will result in a displacement field, which defines the displacement vector of the boundary for the mentioned subdomains. Independently, for the same boundary, a displacement initial formulation will be formulated, which contains the trial functions of the adjacent finite elements. In the following, an optimal approximation of both of the displacement fields on the boundary will be achieved by a penalty term.

The result is the so-called 'generalized compatibility condition' within mixed methods. Additionally, the special element, formulated by the BEM, must be in equilibrium with the adjacent finite elements. As a consequence, the 'generalized equilibrium equation' will be derived from the principle of virtual work, too.

In case, if no formulation of the stress field and accompanying displacement field is possible in a theoretical way (e.g. Airy's formulation or Boussinesq-Neuber-Papkovitch formulation), the displacement field can be computed by the collocation method from a formulated traction field by a further integral equation of quasi Fredholm-type.

Results of notch and fracture mechanics show the high rate of convergence by use of this mixed method. Therefore, the method prescribed above is extended to problems with elasto-plastic material law.

INTRODUCTION

In the following prescription, initial and convective effects are out of interest for the elasto-plastic material behaviour in this paper. So it is defined for a typical variable ζ:

$$\dot{\zeta} = \frac{\partial \zeta}{\partial t} = \frac{d\zeta}{dt} \,. \tag{1}$$

The domain is given as two or three dimensional one:

$$V \subset \mathbb{R}^n, \quad n = 2,3. \tag{2}$$

For the continuum, only small deformations are allowed, so with the rate formulation of Equation (1) follows:

$$\dot{e}_{km} = \frac{1}{2} \, (\dot{u}_{k,m} + \dot{u}_{m,k}). \tag{3}$$

The strain rates are split up in elastic and plastic terms:

$$\dot{e}_{km} = \dot{e}_{km}^{(e)} + \dot{e}_{km}^{(p)}. \tag{4}$$

In the case of anisotropic elasticity, the elastic strain rates can be formulated as a function of stress rates by means of an elastic compliance tensor $E_{km\ell p}^{-1}$ with up to 21 independent constants:

$$\dot{e}_{km}^{(e)} = E_{km\ell p}^{-1} \, \dot{\tau}_{\ell p}. \tag{5}$$

In the following, a loading function F becomes necessary for definition of plastic strain rates:

$$F \, (\tau_{km}, \, e_{km}^{(p)}, \, w^{(p)}) \leq 0. \tag{6}$$

The plastic strain energy density is commonly defined:

$$w^{(p)} = w^{(p)} \, (x_i,t) = \int_{-\infty}^{t} \tau_{km} \, (x_i,t') \, \dot{e}_{km}^{(p)}(x_i,t') \, dt'. \tag{7}$$

Now, Equation (6) is written explicitly in the sense of the von-Mises idea with an equivalent stress value τ and the limit value $\bar{\tau}$, where flow occurs.

$$F: = \tau(\tau_{km}) - \bar{\tau} \, (e_{km}^{(p)}, \, w^{(p)}) \leq 0. \tag{8}$$

The flow rule of Prandtl-Reuss is defined with an unknown constant of proportionality $d\lambda$:

$$de_{km}^{(p)} = d\lambda \, \frac{\partial F}{\partial \tau_{km}}. \tag{9}$$

With existence of first derivatives from F follows:

$$\frac{\partial \tau}{\partial \tau_{km}} \, d\tau_{km} - \frac{\partial \bar{\tau}}{\partial e_{km}^{(p)}} \, de_{km}^{(p)} - \frac{\partial \bar{\tau}}{\partial W^{(p)}} \, dW^{(p)} = 0.$$ (10)

With

$$dW^{(p)} = \tau_{km} \, de_{km}^{(p)}$$ (11)

results:

$$\frac{\partial \tau}{\partial \tau_{km}} \, d\tau_{km} - [\frac{\partial \bar{\tau}}{\partial e_{km}^{(p)}} + \frac{\partial \bar{\tau}}{\partial W^{(p)}} \, \tau_{km}] \, de_{km}^{(p)} = 0.$$ (12)

For $d\lambda$ of Equation (9) it is obtained by insertion in Equation (12):

$$d\lambda = \frac{\partial \tau}{\partial \tau_{ij}} \, d\tau_{ij} \, / \, [\frac{\partial \bar{\tau}}{\partial e_{lp}^{(p)}} + \frac{\partial \bar{\tau}}{\partial W^{(p)}} \, \tau_{lp}] \, \frac{\partial \tau}{\partial \tau_{lp}} .$$ (13)

So, for the strain rates of Equation (9) it is obtained:

$$\dot{e}_{km}^{(p)} = \frac{\partial \tau}{\partial \tau_{km}} \, \frac{\partial \tau}{\partial \tau_{ij}} \, \dot{t}_{ij} \, / \, [\frac{\partial \bar{\tau}}{\partial e_{lp}^{(p)}} + \frac{\partial \bar{\tau}}{\partial W^{(p)}} \, \tau_{lp}] \, \frac{\partial \tau}{\partial \tau_{lp}}.$$ (14)

This means in short-hand notation with a plastic compliance tensor $\tilde{E}_{kmij}^{-1}$:

$$\dot{e}_{km}^{(p)} = \tilde{E}_{kmij}^{-1} \, \dot{t}_{ij}.$$ (15)

The inverse formulation of Equation (5) is inserted in Equation (15):

$$\dot{e}_{km}^{(p)} = \tilde{E}_{kmij}^{-1} \, E_{ijlp} \, \dot{e}_{lp}^{(e)}$$ (16)

and results with Equation (4) in:

$$\dot{e}_{km}^{(p)} = D_{kmlp} \, \dot{e}_{lp}.$$ (17)

The plastic strain rates are directly proportional to the total strain rates.

In addition to these kinematical formulations, a further static restriction has to be taken into account. Equilibrium in the absence of volume forces demands

$$\dot{t}_{km} = \dot{t}_{mk}$$ (18)

and

$$\dot{t}_{km,k} = 0.$$ (19)

BASIC EQUATIONS

The starting point is the Gaussian theorem:

$$\int_{\Gamma} G\,(x_m)\,n_m\,d\Gamma = \int_{\Omega} G_{,m}\,(x_m)\,d\Omega, \tag{20}$$

where $G(x_1)$ is a continuously differentiable tensor field and n_m is the exterior normal vector. With definition

$$G(x_m): = \dot{u}_k\,\dot{t}_{km} \tag{21}$$

it can be achieved:

$$\int_{\Gamma} \dot{u}_k\,\dot{t}_{km}\,n_m\,d\Gamma = \int_{\Omega} (\dot{u}_k\,\dot{t}_{km})_{,m}\,d\Omega \tag{22}$$

with

$$(\dot{u}_k\,\dot{t}_{km})_{,m} = \dot{u}_{k,m}\,\dot{t}_{km} + \dot{u}_k\,\dot{t}_{km,m} \tag{23}$$

and Equations (19), (18), and (4) follows:

$$\int_{\Gamma} \dot{u}_k\,\dot{t}_{km}\,n_m\,d\Gamma = \int_{\Omega} \dot{e}_{km}\,\dot{t}_{km}\,d\Omega, \tag{24}$$

which reads written with the traction field rate t_k:

$$\int_{\Gamma} \dot{u}_k\,t_k\,d\Gamma = \int_{\Omega} \dot{e}_{km}\,\dot{t}_{km}\,d\Omega. \tag{25}$$

From the physical point of view, this relation defines the principle of virtual work, demanding that the external virtual work ($\int_{\Gamma}$) must equal the internal virtual work ($\int_{\Omega}$). From the continuum, a subdomain Ω_n with the surface Γ_n will be cut out.

A stress field will be defined in rate form ($\dot{t}_{km}, t_k$) in Ω_n. With the material law of Equation (15) it follows the plastic strain rate $\dot{e}_{km}^{(p)}$, from which results the total strain rate $\dot{e}_{km}$ with the inverse of Equation (17). The solution of the partial differential equation system from Equation (3) gives the displacement rates.

A better way – and this is the only one practicable – is to formulate an integral equation. Starting with Betti's reciprocal work theorem, you get:

$$\int_{\Omega_n \backslash \Omega_n^*} \tau^*_{km}\,\dot{e}_{km}^{(e)}\,d\Omega_n - \int_{\Omega_n \backslash \Omega_n^*} e^*_{km}\,\dot{t}_{km}\,d\Omega_n = 0. \tag{26}$$

The functions (*) are developed from Kelvin's solution for a single force acting in an unbounded three-dimensional domain. A small spherical region Ω_n^* of radius ρ surrounding the load point, where Kelvin's solution diverges, is excluded from the volume integral.

In the further calculation, the Cauchy principal value is

used for the volume integral over Ω_n^*, demanding the limitation process $\rho \to 0$.

In the following equations, p,q are internal points, and P,Q are boundary points, the kernels T_{km}, V_{km}, and U_{km} originate from Kelvin's solution. For an internal point p of Ω^n with $\Omega_n^* \to 0$, the following equation is achieved (Swedlow[7], Mukherjee[8]):

$$\dot{u}_k(p) + \int\limits_{\Gamma_n} T_{km}(p,Q)\ \dot{u}_m(Q)\ d\Gamma_Q - \int\limits_{\Omega_n} V_{kmn}(p,q)\ \dot{e}_{mn}^{(p)}(q)\ d\Omega_q =$$

$$= \int\limits_{\Gamma_n} U_{km}(p,Q)\ \dot{t}_m(Q)\ d\Gamma_Q \qquad\qquad (27)$$

with $p \in \Omega_n$.

For a surface point P, it results in:

$$W_{km}(P)\ \dot{u}_m(P) + \int\limits_{\Gamma_n \backslash \{P\}} T_{km}(P,Q)\ \dot{u}_m(Q)\ d\Gamma_Q - \int\limits_\Omega V_{kmn}\ (P,q)\ \dot{e}_{mn}^{(p)}(q) \cdot$$

$$\cdot\ d\Omega_q = \int\limits_{\Gamma_n} U_{km}(P,Q)\ \dot{t}_m(Q)\ d\Gamma_Q \qquad\qquad (28)$$

with $P \in \Gamma_n$.

$W_{km}(P)$ is for a smooth boundary:

$$W_{km}: = \frac{1}{2}\ \delta_{km}. \qquad\qquad (29)$$

In a next step, the subdomain Ω_n will be discretized with finite elements Ω_e, see Figure 1.

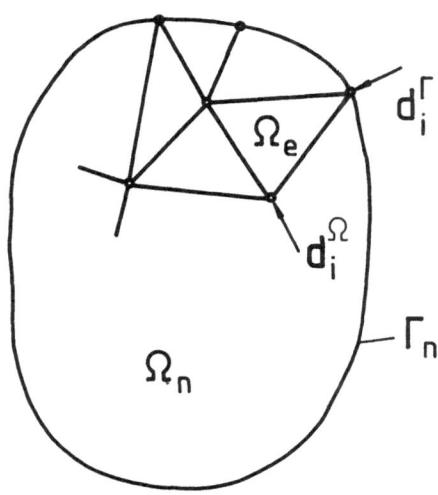

Figure 1.   Finite element discretization of Ω_n

So there are two sets of collocation points, defined by the nodes of the triangles in Figure 1. The first one is for the
boundary Γ_n, and the second one for the interior of Ω_n, i.e.
for $\Omega_n \backslash \Gamma_n$.

The traction rates of the rhs of Equations (27) and (28)
can be formulated by means of a form function R_{mk} and free parameters $\dot{f}_k$.

$$\dot{t}_m = R_{mk} \, \dot{f}, \quad \text{defined on } \Gamma_n \tag{30}$$

with the equilibrium constraints

$$\int_{\Gamma_n} \dot{t}_m \, d\Gamma_n = 0 \tag{31}$$

and

$$\int_{\Gamma_n} \varepsilon_{kij} \, r_i \, \dot{t}_j \, d\Gamma_n = 0. \tag{32}$$

The number of collocation points for Γ_n is m, and for $\Omega_n \backslash \Gamma_n$ it
is n. So, in $\mathbb{R}^3$, $\dot{f}_k$ of Equation (30), are 3m free parameters.

For the displacement field in Ω_e, the expression is defined:

$$\dot{u}_m = N_{mk} \, \dot{d}_k, \quad \text{defined in } \Omega_e, \tag{33}$$

where d_k, in $\mathbb{R}^3$, are $3(m+n)$ free parameters.

In addition, it is necessary to compute the plastic strain
rates. So, at first Equation (17) is transformed with the help
of Equation (3):

$$\dot{e}^{(p)}_{mn} = \frac{1}{2} \, (D_{mn l p} + D_{mnp l}) \, \dot{u}_{l,p} = \tilde{D}_{mn l p} \, \dot{u}_{l,p}. \tag{34}$$

Equation (33) results in:

$$\dot{u}_{l,p} = N_{lk,p} \, \dot{d}_k. \tag{35}$$

So, it is obtained for the plastic strain rates:

$$\dot{e}^{(p)}_{mn} = \tilde{D}_{mn l p} \, N_{lk,p} \, \dot{d}_k. \tag{36}$$

By use of Equations (30), (33), and (36), the integral equations (27) and (28) can be integrated. It results in a linear
equation system:

$$A_{mk} \, \dot{d}_k = B_{mn} \, \dot{f}_n. \tag{37}$$

From this, the displacement rates can be expressed as linear functions in the traction parameter rates:

$$\dot{d}_k = A_{k\ell}^{-1} B_{\ell n} \dot{f}_n = L_{kn} \dot{f}_n. \tag{38}$$

This dependence can also be shown to exist by theoretical consideration of the compliance relation Equation (15) with Equations (24), (25), and (30) on one hand, and the kinematical relation Equation (17) with Equations (3) and (33) on the other hand. For the displacement rates defined on Γ_n, it can be written with the help of Equation (33):

$$\dot{u}_m = N_{mk} L_{ki} \dot{f}_i = P_{mi} \dot{f}_i. \tag{39}$$

STIFFNESS RELATION

As a next step, the tangential stiffness matrix of **mixed type** is to be constructed. As shown in Figure 2, the subdomain Ω_n is connected to the classical finite elements Ω_f from the residual domain Ω_r with boundary Γ_r.

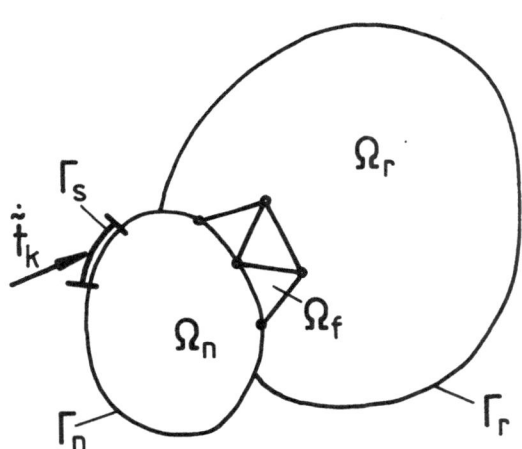

Figure 2.   Discretization of the whole structure

Therefore a displacement rate $\dot{\tilde{u}}_m$ is defined on Γ_n:

$$\dot{\tilde{u}}_m = M_{mk} \dot{r}_k. \tag{40}$$

On $\Gamma_r \cap \Gamma_n$ it means a discontinuity in the displacement rates from Ω_n and the adjacent finite elements Ω_f.

As known from nonconform methods, an integral weak soluti-

on is demanded for the compatibility. It can be constructed by the principle of virtual work, Equation (25):

$$\int_{\Gamma_n} \dot{t}_k \ddot{u}_k \, d\Gamma = \int_{\Omega_n} \dot{e}_{km} \dot{\tau}_{km} \, d\Omega \equiv \int_{\Gamma_n} \dot{t}_k \dot{u}_k \, d\Gamma, \tag{41}$$

from which follows:

$$\int_{\Gamma_n} \dot{t}_k (\dot{u}_k - \ddot{u}_k) \, d\Gamma = 0. \tag{42}$$

For a predetermined traction rate $\overset{*}{\dot{t}}_k$ defined on $\Gamma_s \subset \Gamma_n$ (see Figure 2), the equilibrium condition must be fulfilled in an in-
tegral weak formulation, too. With Equations (25) and (42) we
arrive at:

$$\int_{\Gamma_s} \overset{*}{\dot{t}}_k \ddot{u}_k \, d\Gamma = \int_{\Omega_n} \dot{e}_{km} \dot{\tau}_{km} \, d\Omega \equiv \int_{\Gamma_n} \dot{t}_k \ddot{u}_k \, d\Gamma, \tag{43}$$

and this results in:

$$\int_{\Gamma_s} \overset{*}{\dot{t}}_k \ddot{u}_k \, d\Gamma - \int_{\Gamma_n} \dot{t}_k \ddot{u}_k \, d\Gamma = 0. \tag{44}$$

The principle of virtual work demands no dependence between
statical values ($\dot{t}_k, \overset{*}{\dot{t}}_k$) and kinematical ones ($\dot{u}_k$, $\ddot{u}_k$) in Equations (42) and (44). As, for this elastoplastic behaviour, the
theory is described in an incremental form, it becomes a piecewise linear theory. So, for the equilibrium of the total structure with the adjacent finite elements, the principle of super-
position must be formulated.

By the insertion of the trial functions, defined in Equations (30), (39), and (40) into Equation (42), the matrix notation results in:

$$\overset{*}{\underset{\sim}{t}}_m \left(\underbrace{\int_{\Gamma_n} R^T_{mk} P_{kl} \, d\Gamma}_{H_{ml}} \dot{t}_l - \underbrace{\int_{\Gamma_n} R^T_{mn} M_{np} \, d\Gamma}_{S_{mp}} \dot{r}_p \right) = 0 \tag{45}$$

The parameters $\overset{*}{\underset{\sim}{t}}_m$ of $\overset{*}{\dot{t}}_k$ can be chosen independently ("$\sim$") of
$\dot{t}_l$, the parameters of $\ddot{u}_k$.
For det $H \neq 0$, and only this case leads to a solution, it follows:

$$\dot{t}_k = H^{-1}_{kl} S_{lm} \dot{r}_m. \tag{46}$$

Hence,

$$\left\langle \underbrace{\int_{\Gamma_s} \overset{*}{\dot{t}}_l M_{lk} \, d\Gamma}_{F_k} - \dot{f}_m \underbrace{\int_{\Gamma_n} R^T_{mn} M_{nk} \, d\Gamma}_{S_{mk}} \right\rangle \overset{*}{\underset{\sim}{r}}_k = 0 \tag{47}$$

leading to:

$$\dot{F}_m - S_{mk}\, \dot{f}_k = 0. \tag{48}$$

With Equation (46) follows:

$$\dot{F}_m = \underbrace{S_{mk}\, H_{kl}^{-1}\, S_{lp}}_{k_{mp}^t}\, \dot{r}_p. \tag{49}$$

As a result, a tangential stiffness matrix k_{mp}^t of mixed type is obtained.

CONCLUSION

Until today, for pure elasticity theory, a complete software system for two and three-dimensional notch and crack problems has been existing at the Institute of Solid Mechanics at Karlsruhe University. One example, treating a mixed notch-crack problem, shows the great advantage compared to the common algorithms for those problems.

Figure 3 shows the notch-crack problem, i.e. the crack is starting from the notch with a high stress field. Figure 4 shows the convergence curves compared with the common methods for computing the stress intensity factor for this problem. The convergence curves show, that the prescribed method in this paper is already working very effective for complicated problems in linear elasticity theory.

As, for the more complicated elastoplastic material behaviour, a piecewise linear theory is resulting, too, most part of the theoretical base has been at hand for existence and uniqueness of a solution. Therefore the theoretical and numerical base is given for software extension to problems in elastoplasticity.

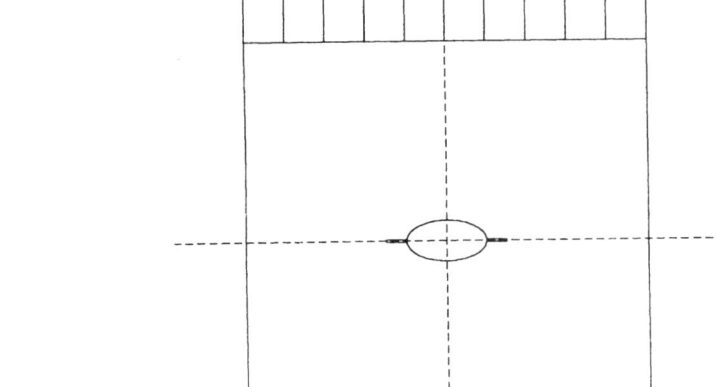

Figure 3.
Notch crack problem

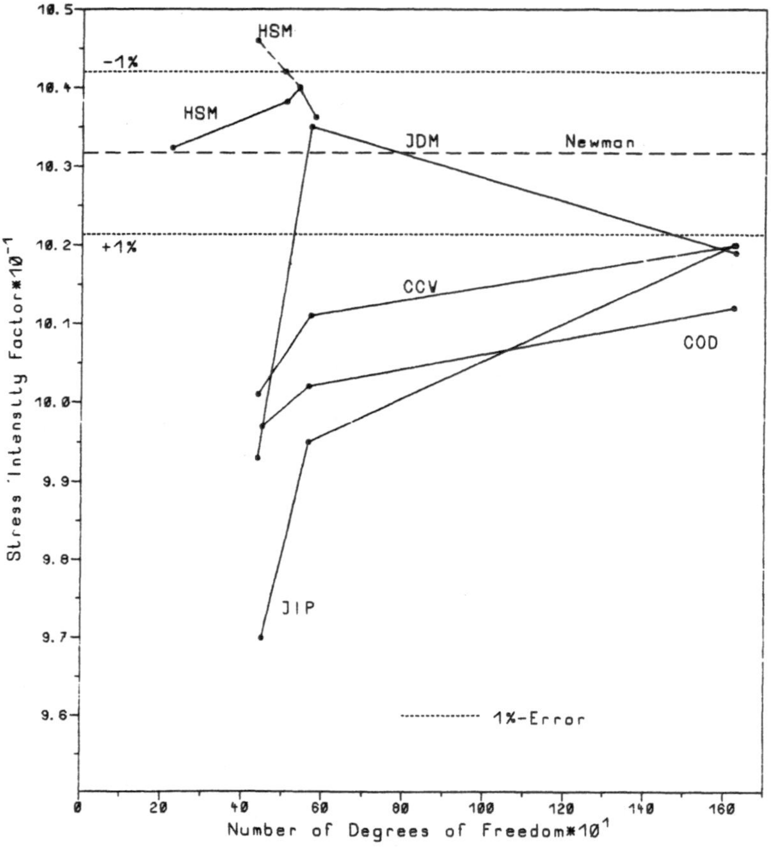

Figure 4.   Convergence curves for problem of Figure 3

JDM :   J-integral computation with classical displacement methods of FEM

CCW :   Crack-closure-work computation with classical displacement methods of FEM

JIP :   J-integral computation with isoparametric
elements

COD :   Crack-opening displacement concept

HSM :   Presented algorithm in this paper, called
'Hybrid Stress Method' with several meshes

REFERENCES

1. Brebbia C.A. (1980). The Boundary Element Method for Engineers,
   Pentech Press, London and Plymouth.

2. Brebbia C.A. (1981) Boundary Element Methods, (Ed. Brebbia
   C.A.), pp. 69-289, 397-622, Proceedings of the 3rd Int.
   Seminar, Irvine, California, 1981.

3. Zienkiewicz O.C. Kelly D.W. and Betess B. (1977), The
   Coupling of the Finite Element Method and Boundary Solution
   Procedures, Int. J. Num. Meth. Engng., Vol. 11, pp. 355-375.

4. Zienkiewicz O.C. Kelly D.W. and Betess B. (1979), Marriage
   à la mode. The best of both worlds (finite elements and
   boundary integrals). In: Energy Methods in Finite Element
   Analysis, (Ed. Glowinski R. Rodin E.Y. and Zienkiewicz
   O.C.), pp. 81-107, John Wiley & Sons, Chichester.

5. Schnack, E. (1985), Stress Analysis with a Combination of
   HSM and BEM. In: The Mathematics Of Finite Elements and
   Applications V, (Ed. Whiteman J.), pp. 273-282, Academic
   Press, London.

6. Wendland W.L. (1986) Asymptotic Error Estimates for the Combined Boundary and Finite Element Method, in Innovative Numerical Methods in Engineering (ed. Shaw R.P. Periaux J.
   Chaudouet A. Wu J. Marino C. and Brebbia C.A.), pp. 55-69,
   Proceedings of the 4th Int. Symp., Georgia Institute of
   Technology, Atlanta, Georgia, USA, 1986.

7. Swedlow J.L. and Cruse T.A. (1971), Formulation of Boundary Integral Equations for three-dimensional elasto-plastic
   Flow, Int. J. Solids Structures, Vol. 7, pp. 1673-1683.

8. Mukherjee S. (1982). Boundary Element Methods in Creep and
   Fracture, Applied Science Publishers. London and New York.

Special Shape Functions in Boundary Integral Method

J.B. Sellmeijer

Delft Soil Mechanics Laboratory, The Netherlands

1. INTRODUCTION

In hydraulic engineering the phenomenon of piping is known in the context of dikes, dams and weirs which are built on a sandy foundation. It generally occurs below a water retaining structure on top of granular material. The term piping refers to channels caused by seepage flow. It starts at the downstream side of the structure, where the flow lines through the aquifer converge, thus causing high seepage pressure. The erosion process develops backwards, forming irregular channels. This may result in loss of soil stability and ultimately in collapse of the structure.

Design rules to date are empirical and appropriate to quite different geographical locations. They are therefore often inadequate. Because of this piping is studied in depth with the aim to understand how it comes about. A two-dimensional model is set-up which includes erosion. This mathematical model has been worked out analytically. Design criteria are presented, based on the results. The minutiae will be published in a soil mechanics journal.

To describe the model use is made of a boundary integral method. Straightforward results, however, show unstable behaviour due to rather abrupt transitions on the boundaries. A remedy is found to smooth the calculation process. Special shape functions describing these transitions are the key to obtain reliable results. This paper deals with the derivation of these shape functions, explains their application in the computations and shows the results obtained.

2. CONCEPTUAL MODEL

The configuration studied is a water retaining structure built on granular material. The outflow area is of special interest, since this is where piping will be initiated. This region is schematically outlined in figure 1 . A graphical impression of the head is shown. The structure is simulated by line BCD ; line DAB represents a boundary with a free water table. The geometry far away is not important for the mechanism itself.

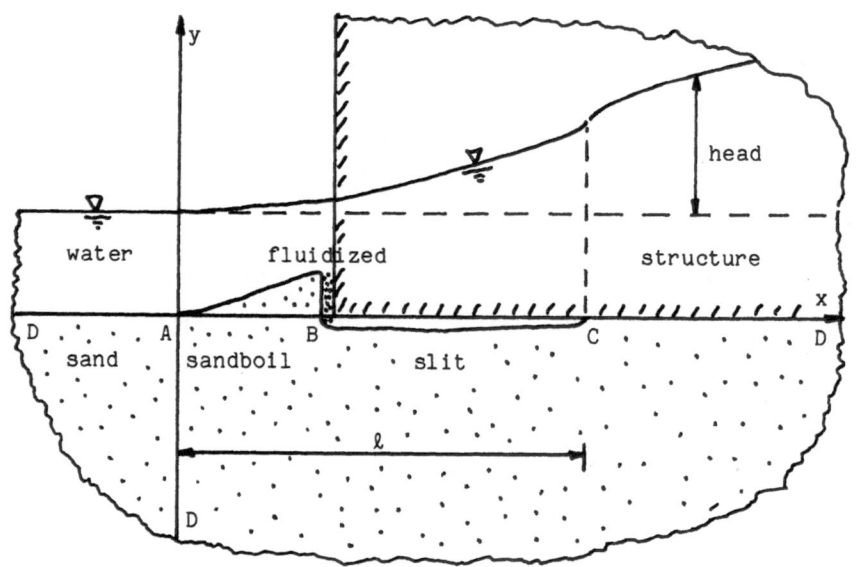

Fig. 1: geometry around the outflow region

Simple visual laboratory tests have shown that the geometry of
the outflow area changes for even small values of the hydraulic
head. Sandboils arise and slits appear. The sandboils consist
of material transferred from the slits. At the outflow point
the sand is fluidized, because water flows out of the slits at
relatively great speed.

Now that the geometry is understood a conceptual model for the
problem can be made. It involves two main aspects: groundwater
flow and limit equilibrium of sand particles. The limit
equilibrium is a special boundary condition for the flow along
the line ABC in figure 1 . The remaining boundary conditions
are: at DA a constant head and at CD no vertical discharge.

3. NUMERICAL ASPECTS OF THE SOLUTION

The conceptual model is worked out analytically resulting in a
set of integral equations and boundary conditions. The integral
equations contain an influence function which is yet to be
determined. In this paper the numerical aspects of the
calculation are dealt with. Unless precautions are taken the
computations show oscillations. To illustrate this it is not
necessary to show the complete solution merely that part
responsible for this unstable behaviour. Therefore this part
will be isolated and extensively investigated. In doing so the
following is obtained:

$$q(x) = \frac{1}{\pi} \oint_0^\ell p(r) \sqrt{\left(\frac{\ell-x}{\ell-r}\right)} \frac{dr}{r-x} \tag{1}$$

$$q + p \cot(\pi\theta) = 1 \tag{2}$$

Equation (1) results from the description of the groundwater
flow. It relates the vertical gradient q to the horizontal

one   p   along the line   ABC   in sandboil and slit. The erosion
length, being the total length of sandboil and slit, is denoted
by ℓ ;   the   horizontal coordinate by   x   and the integration
variable by   r . Since the pole   r = x   is involved integration
is   performed according to Cauchy's principle. It is noted that
the   integral   is   also valid for   x < 0   without having to use
Cauchy's principle.

The   second   equation   represents   the   condition of   limit
equilibrium   of   the sand grains. The angle Θ depends on soil
parameters which may range from   0   to $\frac{1}{2}$. In reality equation
(2)   is   more complicated, but the above form suffices to study
the   unstable behaviour. The unknown function   p   is determined
by the set of equations (1) and (2).

There   are   several   ways   of   evaluating   the   gradient   p ,
collocation   being   the   least complicated one. Straightforward
results, however, show unpleasant oscillations which are due to
the discontinuous transitions at the boundaries. The conditions
at   DA ,   ABC   and   CD   are   each of a different kind and a
mathematical effort is required in order to ensure a continuous
behaviour   of   the   head.   By   itself a method like collocation
cannot   cope with this problem, so a more powerful tool must be
used.

4. HYPERGEOMETRIC SHAPE FUNCTIONS

To   eliminate   the   oscillations use is made of knowledge about
the   true   behaviour   of   the gradient in the transition zones.
What   is known about this behaviour is the fact that there is a
linear   relation   between   p   and   q   according to equation (2)
and   it can be speculated that the gradient takes the form of a
power function. Since the character of   p(x)   is symmetrical
with the one of   q(ℓ-x) ,   only one transition zone, for example
the one around   x = 0 ,   needs to be studied. The true behaviour
of   p   will be supposed to be,

$$p_\varepsilon = \hat{p} \left(\frac{x}{\ell}\right)^\varepsilon = \hat{p} \; \chi^\varepsilon \tag{3}$$

The character $\hat{p}$ is a proportionality constant, also called weight of the contribution. The dimensionless variable χ is introduced for reasons of convenience. The corresponding behaviour of q follows by substitution of the assumed form of p into equation (1). Thus the behaviour of q then turns out to be a hypergeometric function. The theory of these functions, denoted by F , is adequately summarized in Abramowitch and Stegun (1968), chapter 15 . All formulas starting with 15. will refer to this book. According to formula 15.3.1 it may be written:

$$q_\varepsilon = -\frac{\hat{p}}{\pi} \frac{\sqrt{(1-\chi)}}{\chi} \frac{\Gamma(\tfrac{1}{2}) \; \Gamma(1+\varepsilon)}{\Gamma(\tfrac{3}{2}+\varepsilon)} \; F[1,1+\varepsilon;\tfrac{3}{2}+\varepsilon;1/\chi] - i \; \hat{p} \; \chi^\varepsilon \tag{4}$$

It must be noted that formula 15.3.1 contains a residue that has to be subtracted, since equation (1) is Cauchy integrated. The residue contains the imaginary unit $i = \sqrt{(-1)}$. As the present shape of the relation (4) is not very suitable for numerical interpretation, the complex term will initially be eliminated and the result rewritten. With the aid of formula 15.3.9 q may be written as follows:

$$q_\varepsilon = -\frac{\hat{p}}{\pi} \sqrt{(1-\chi)} \; \frac{\Gamma(-\tfrac{1}{2}) \; \Gamma(1+\varepsilon)}{\Gamma(\tfrac{1}{2}+\varepsilon)} \; F[1,\tfrac{1}{2}-\varepsilon;\tfrac{3}{2};1-\chi] \tag{5}$$

To arrive at equation (5) use was made of the fact that $F[\tfrac{1}{2}+\varepsilon,0;\tfrac{1}{2};1-\chi] = 1$ and that the value of the gamma function $\Gamma(\tfrac{1}{2})$ is equal to $\sqrt{(\pi)}$. The following steps are now taken:

- formula 15.3.6 is applied;
- the reflection formula for gamma functions is used:
  $\Gamma(\varepsilon) \; \Gamma(1-\varepsilon) = \pi \; \csc(\pi\varepsilon)$;
- use is made of the fact that according to formula 15.1.8 :
  $F[\tfrac{1}{2},1+\varepsilon;1+\varepsilon;\chi] = 1 \, / \, \sqrt{(1-\chi)}$.

It then follows that:

$$q_\varepsilon = \frac{\hat{p}}{\pi} \sqrt{(1-\chi)} \frac{\Gamma(\tfrac{1}{2})\ \Gamma(\varepsilon)}{\Gamma(\tfrac{1}{2}+\varepsilon)} F[1,\tfrac{1}{2}-\varepsilon;1-\varepsilon;\chi] - \hat{p}\ \chi^\varepsilon \cot(\pi\varepsilon) \qquad (6)$$

Finally the result may be polished by use of formula 15.3.3 ,

$$q_\varepsilon = \frac{\hat{p}}{\pi} \frac{\Gamma(\tfrac{1}{2})\ \Gamma(\varepsilon)}{\Gamma(\tfrac{1}{2}+\varepsilon)} F[-\varepsilon,\tfrac{1}{2};1-\varepsilon;\chi] - \hat{p}\ \chi^\varepsilon \cot(\pi\varepsilon) \qquad (7)$$

Indeed this is the hoped-for result. The behaviour of equation
(7) for small values of χ has the same character as equation
(3), provided that ε is small enough. As a matter of fact the
value of ε is determined, since the variation must satisfy
relation (2). This only can be achieved if $\cot(\pi\varepsilon) = \cot(\pi\Theta)$
or if $\varepsilon = \Theta$.

It turns out that the behaviour for small values of x can be
described very accurately by introducing a hypergeometric shape
function. Simply add equations (3) and (7) as part of the
solution. But   there   is a second transition zone around x = ℓ
where the obtained result may be applied too if χ is replaced
by $1-\chi$, Θ by $\tfrac{1}{2}-\Theta$ and $\hat{p}$ by a weight $\hat{q}$. In the real
piping problem the value of Θ around x = ℓ is different
from the one around x = 0 . But this is of no importance for
the stability of the calculation process.

5. APPLICATION OF SHAPE FUNCTIONS

The hypergeometric shape functions thus derived can be used
successfully in the calculation process. The procedure now
involves the use of these shape functions to build up the
horizontal and vertical gradient, as after all the system of
equations (1) and (2) is linear. In this simplified
representation of the piping problem it might be possible to
determine the hypergeometrical weights directly, but in the

true solution this is out of the question. The weights have to be calculated by a process of iteration.

Besides the hypergeometric functions a set of general shape functions is required in order to construct the solution. A straightforward way to obtain them is where discrete p_j values are introduced at a number of locations x_j. In between these values a linear behaviour may be assumed. And so the horizontal gradient may be defined,

$$p(\chi) = [p_j (x_{j+1}-\chi) + p_{j+1}(\chi-x_j)] / (x_{j+1}-x_j) \qquad (8)$$

This representation is valid between x_j and x_{j+1}. Where there is an N elements discretisation the index j ranges from 0 to N-1. Since the value of p is well defined by (8), the corresponding value of q can be determined from (1). If a linear grid is chosen, so that N $(x_{j+1}-x_j) = 1$, then the value of q can be found from,

$$q(\chi) = \sum_{j=1}^{N} \frac{2}{\pi}N \, p_j \, [f(x_{j+1}) - 2 \, f(x_j) + f(x_{j-1})] \qquad (9)$$

where,

$$f(\rho) = (\chi-\rho) \, arc^{tan}_{cot}h\{\sqrt{(\frac{1-\chi}{1-\rho})}\} + \sqrt{(1-\chi)} \, \sqrt{(1-\rho)}$$

In the auxiliary function f arctanh is implied if $\chi > \rho$; otherwise arccoth. Note that $p_0 = 0$; x_{N+1} is assumed 1.

Next a set of continuous shape functions is obtained. From the point of view of symmetry a conjugate set of equations can be defined, starting with $q(1-\chi)$ instead of $p(\chi)$. In this set the resulting $p(1-\chi)$ is defined by equation (9) and $q(1-\chi)$ itself by (8). The discrete values of q are indicated here by q_j, with $q_0 = 0$.

The solution of the problem is represented at this stage by 2N+2 different shape functions. They are:

-   linear discretisations of p with weight p_j ;
-   linear discretisations of q with weight q_j ;
-   hypergeometric behaviour around $\chi = 0$ with weight $\hat{p}$;
-   hypergeometric behaviour around $\chi = 1$ with weight $\hat{q}$;

The yet unknown weights of these shape functions can be determined, for instance by collocation.

6. RESULTS OF CALCULATION

It was noted that a computation of the set of equations (1) and (2) generates unpleasant oscillations, if no precautions are taken. It was suggested that these oscillations can be avoided by introducing the true behaviour of the solution into the transition zones. This was worked out and of course only from results of computation can it be shown that the method works. Therefore the formulas of the previous sections are gathered in a computer code. The program is as follows.

Two representations are composed, one consisting of discretisations p_j and the hypergeometric functions, the other of discretisations q_j and the same hypergeometric functions. Each consists therefore of N+2 shape functions. Collocation is applied for both representations in the discrete locations χ_j . This procedure yields two systems of N+1 equations to determine the 2N+2 unknown weights. Examples of the result of computation are shown in figures 2 and 3 . The discrete determined values are connected by straight lines.

What happens when the hypergeometric shape functions are not applied? This is of interest and so calculations have been made omitting these functions. In order to compare results the same value of θ as in the previous computation has been used as well as the same number of elements. In figure 4 and 5 the outcome is shown. As can be seen undesirable oscillations were generated. And it was observed that the iteration procedure took a long time and was troublesome.

It is demonstrated that hypergeometric shape functions are a powerful tool in the numerical procedure in the problem to hand. For a relatively small amount of 8 elements as shown in figure 2 a very accurate solution is obtained after only a few iterations. But 48 elements as shown in figure 5 still produce oscillations if no hypergeometric functions are applied.

It is interesting to note that the oscillations appear at the left side of the graphs where the abrupt behaviour is very severe due to the value of $\Theta = 0.1$. At the right side of the graphs, where the value of $\frac{1}{2}-\Theta$ equals 0.4 , oscillations are lacking, apparently since the abrupt behaviour there is more modest.

7. CONCLUSIONS

The result of a straightforward collocation procedure was compared with one using hypergeometric shape functions. The latter produced a smooth answer whereas the former suffered from counter-productive effects such as oscillations and was in addition much slower. A further advantage is the fact that far fewer shape functions are needed to describe the ultimate solution.

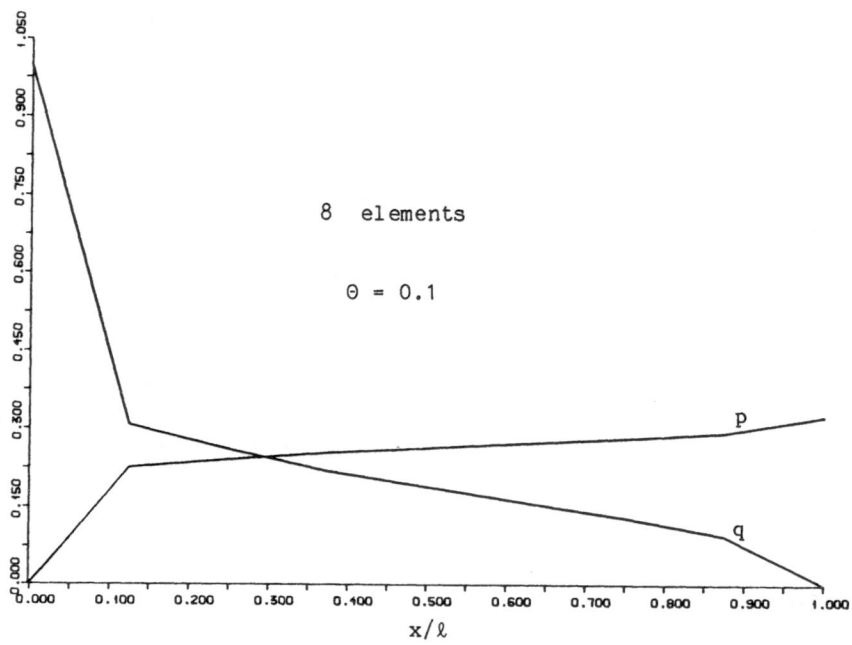

Fig 2: results of calculation applying hypergeometric functions

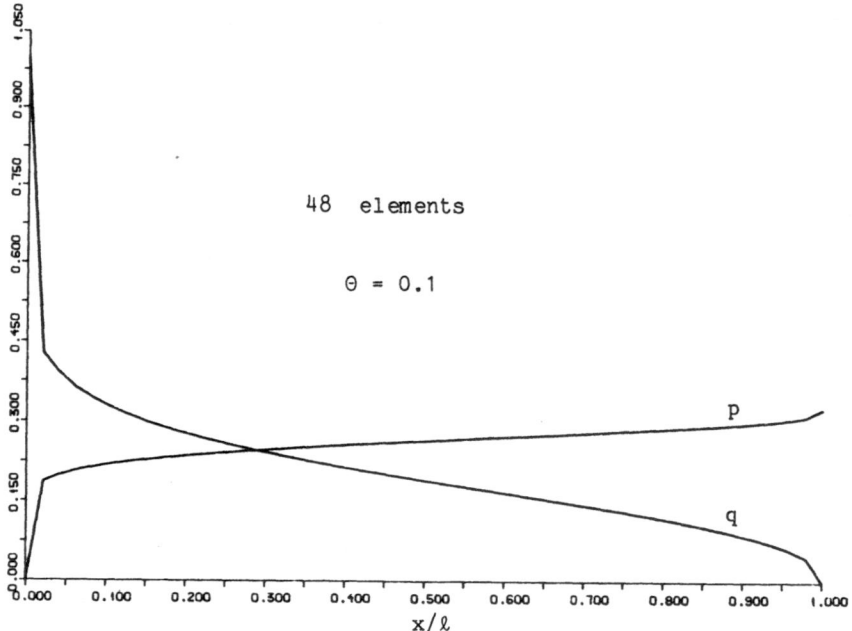

Fig 3: results of calculation applying hypergeometric functions

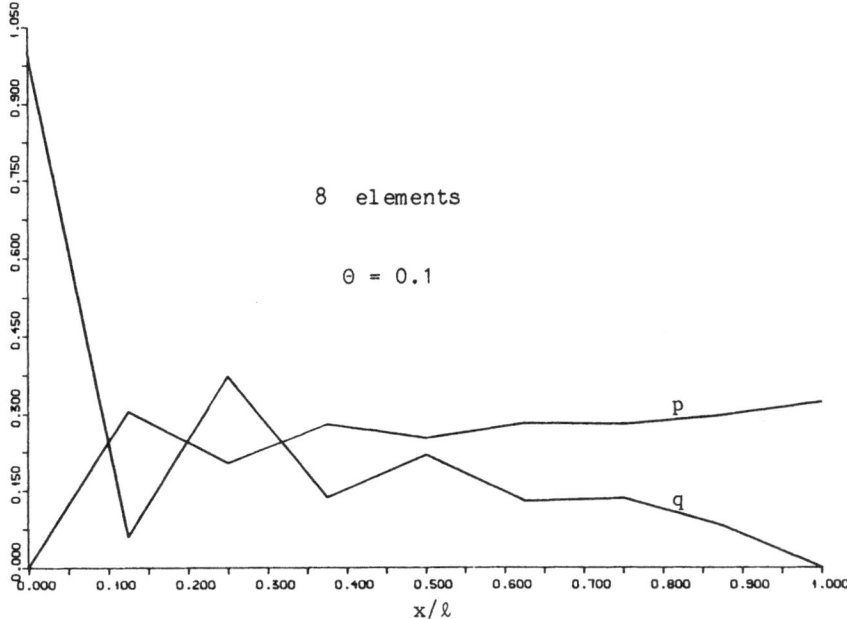

Fig 4: results of calculation without hypergeometric functions

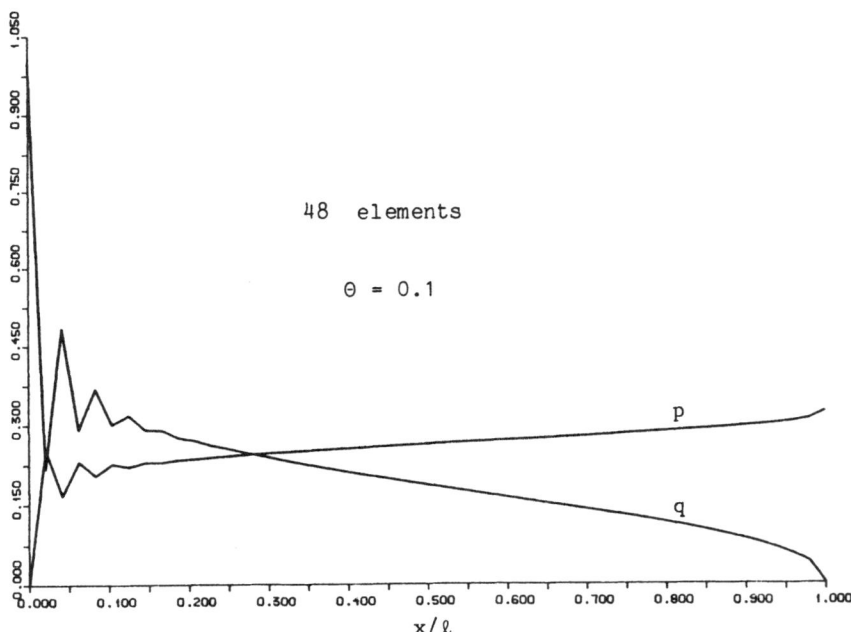

Fig 5: results of calculation without hypergeometric functions

LIST OF SYMBOLS

N   : number of elements

f   : auxiliary function

i   : imaginary unit $\sqrt{(-1)}$

ℓ [m] : erosion length

p [m] : horizontal gradient

p_j [m] : discrete value of horizontal gradient

$\hat{p}$ [m] : weight of hypergeometric function around $\chi = 0$

q [m] : vertical gradient

q_j [m] : discrete value of vertical gradient

$\hat{q}$ [m] : weight of hypergeometric function around $\chi = 1$

r [m] : integration variable

x [m] : horizontal coordinate

Θ : parameter depending on soil characteristics

ε : auxiliary parameter

χ : normalized horizontal coordinate

REFERENCE

(1968)   Abramowitz, M. and Stegun, I. A.
         Handbook of mathematical functions
         Dover publications, inc. , New York

A New Approach to Singular Kernel Integration for General Curved Elements

M. Koizumi, M. Utamura
Energy Research Laboratory, Hitachi Ltd., 1168 Moriyama-cho, Hitachi-shi, Ibaraki-ken 316, Japan

INTRODUCTION

Numerical accuracy in boundary element method depends on the errors appearing in the approximation of geometry, interpolation of functions and integration scheme except loss of digits inherent in computer hardware. Among them, the integration scheme takes an essential part and may significantly deteriorate numerical accuracy in ordinary boundary element formulation because integration in Cartesian coordinate includes singular kernel. It is known that direct application of numerical quadratures fails to predict the value of unknown function near the integration points. To overcome this difficulty, analytical integration method (Kuwabara and Takeda[1]), double exponential numerical integration method (Higashimachi et al.[2]) and Robmerg numerical integration method (Takahashi et al.[3]) have been attempted. However, in analytical method not only an element shape is limited to a piecewise plane element but also applicable integrand is limited to $1/r$ or $\log r$ type function. Thus it cannot be in general usage. On the other hand, double exponential integration and Romberg integration methods generally require much cpu time because many integration points are needed to acquire sufficient numerical accuracy. In recent years, a method has been practiced in two-dimensional problems, which divides one element into many sub-elements to each of which normal Gauss-Legendre integration scheme has been applied (Cox and Shugar[4]). However, this method cannot be extended to three-dimensional problems, because the number of sub-elements may increase which results in much CPU-time.

By the way, to calculate coefficient matricies in BEM, if the observation nodal point is located in an integrated element, a following method has been developed, that is, a numerical integration in the element is carried out in polar coordinate system in which coordinate origin is set to the nodal point (Rizzo and Shippy[5]; Utamura and Koizumi[6]). This method gives accurate numerical integration results because degree of singularity is reduced.

The purpose of this paper is to present a new integration method for general curved element having a singular kernel by the generalizing polar coordinate numerical integration method, in order to guarantee numerical accuracy in calculation of the potential and its gradient at any inner point near the boundary.

NUMERICAL INTEGRATION METHOD

The potential ϕ_i and its gradient $\partial \phi / \partial x_\alpha$ at the observation point i may be calculated by the following boundary integrals using potential ϕ and flux $\partial \phi / \partial n$ given on the boundary.

$$\phi_i = \int \frac{\partial \phi}{\partial n} \phi^* \, d\Gamma - \int \phi \frac{\partial \phi^*}{\partial n} \, d\Gamma \tag{1}$$

$$\frac{\partial \phi_i}{\partial x_\alpha} = \int \frac{\partial \phi}{\partial n} \frac{\partial \phi^*}{\partial x_\alpha} \, d\Gamma - \int \phi \frac{\partial^2 \phi^*}{\partial n \partial x_\alpha} \, d\Gamma \tag{2}$$

where ϕ^* is a weighting function.

In a conventional BEM, integration is approximated by summation of element-wise integration and in each element the integration has been done using Gauss-Legendre numerical formula. However, in the case that the observation point is located near the boundary, it is well known that the above numerical integration causes significant error because Eq.(1) and (2) include singular kernel. This difficulty may be overcome by a numerical integration method as follows.

First, as shown in Fig.1, a foot of the perpendicular (ξ_1, η_1) on the boundary which is nearest to the observation point was calculated by Eq.(3) as local coordinate (ξ, η) in an element.

$$(r - r_i, \frac{\partial r}{\partial \xi}) = 0$$
$$(r - r_i, \frac{\partial r}{\partial \eta}) = 0 \tag{3}$$

where r is a location vector on a boundary element and r_i is that of an observation point i. The above equation is solved by Newton-Raphson method.

Second, the boundary element in an isoparametric space was divided into four triangular regions and local Cartesian coordinate system was transformed to a polar coordinate system (ρ, θ) whose origin was denoted by (ξ_1, η_1). Numerical integration in each triangular region was carried out under this polar coordinate system as follows,

$$\int f(x,y,z)d\Gamma = \int f(\xi,\eta)Jd\xi d\eta = \int f(\rho,\theta)J\rho d\rho d\theta \qquad (4)$$

where J was Jacobian due to transformation coordinate system from global space (x,y,z) to the local isoparametric space (ξ,η).

Integration formuli and local coordinates of integration points in each region are displayed in table 1. In this table, ξ' and η' are integration point between from -1 to 1 and w_l and w_k are weight of integration point.

Combination of these formuli with Gauss-Legendre integration formula give finite integration results because the origin of the coordinate is not included in integration point.

CALCULATION RESULTS

Taking potential problem in a cubic as an example the present method was tested by varying observation point i as illustrated in Fig.2 and compared with conventional method.

Calculation conditions were as follows; potential ϕ =1 on one surface of the cubic ϕ =0 on opposite surface and flux $\partial\phi/\partial n$ =0 on other side surfaces. Observation point was set near the integration point given by the conventional method. Generally, large integration error appears in the case that observation point approaches the integration point. The number of integration points in each triangular region was selected 8x4 (8 for ρ and 4 for θ) for the present method and that in an element was 16x16 for the conventional method.

The calculation results of potential are shown in Fig.3. Solid line shows analytical solution. Integration error by conventional method begins to increase, when the distance between observation point and boundary surface decreases less than 0.1 times as much as element size. On the other hand, integration error by the present method increases, when the distance decreases less than 0.025 times as much as element size. Distinct difference between the present and the conventional is that upper bound of error is finite in the present method. Figure 4 displays the calculation result of potential gradient. In this case, singularity becomes stronger than that of the above case, because differential of weighting function has been included in the kernel. Due to this fact, calculation error becomes larger than that of above case. However, the error is 0.1-0.01 times less than that by conventional method, even if the number of integration point in an element by the present method is half of the conventional.

IMPROVEMENT OF INTEGRATION ERROR

Integration Error of Potential

As described in latter section, the integration error has remained by only transforming to the polar coordinate system.

First term of Eq.(1) has $1/r$ singularity at the highest and it will disappear by coordinate transformation. On the other hand the second term has $1/r^2$ singularity and this cannot disappear by coordinate transformation. This is the cause of numerical integration error.

When observation point i is located on the boundary, Eq.(1) becomes the folling equation because of including singularity in kernel.

$$c_i \phi_i = \int \frac{\partial \phi}{\partial n} \phi^* d\Gamma - \int \phi \frac{\partial \phi^*}{\partial n} d\Gamma \tag{5}$$

where c_i is constant and defined by following equation by using a solid angle ω viewing computational region from the observation point i.

$$c_i = \frac{\omega}{4\pi} \tag{6}$$

This c_i is also calculated by using Eq.(5). If potential is assumed to be uniform in analytical region, equation (5) is reduced as following boundary integration.

$$c_i = - \int \frac{\partial \phi^*}{\partial n} d\Gamma \tag{7}$$

In case that observation point is located in analytical region the solid angle becomes 4 , and c_i should be unity. However, if the integration of Eq.(7) is carried out numerically, c_i will not become unity, because numerical integration has error which come from singularity of kernel. Therefore, if the difference between exact c_i and numerically evaluated c_i can be known, the numerical error of the potential could be evaluated. Figure 5 depicts calculation results of c_i near the boundary. Exact c_i has discontinuity on the boundary, to the contrary, numerical result changes continuously and become equal to exact c_i (=0.5) on the boundary. Comparing Fig.3 with Fig.5, the behavior of numerical c_i is similar to that of numerical potential. The behavior of c_i shows that boundary region has a non-zero thickness numerically. Therefore, to calculate inner potential, it is not better to put c_i into unity in Eq.(5) but to transform Eq.(5) to the following equation.

$$\phi_i = \frac{1}{c_i} (\int \frac{\partial \phi}{\partial n} \phi^* d\Gamma - \int \phi \frac{\partial \phi^*}{\partial n} d\Gamma) \tag{8}$$

Numerical results of inner potential evaluated by Eq.(8) are shown in Fig.6. Numerical results agreed with exact solution much better and calculation error is 0.01 less than that of former results (shown in Fig.3).

Figure 7 shows comparison of relative errors calculated by the present method with those by the double exponential numerical integration method. In this figure the results of double exponential integration method have also been corrected by c_i. The present integration method gives convex curve of error distribution and maximum error becomes 1.5×10^{-3} in the case of 8x4x4 integration point. On the other hand, to obtain the same accuracy by double exponential numerical integration method the number of integration point needs 301x301. Relative CPU-time of both method are displayed in table 2. Hitac M-200H has been used for the calculation. As in table 2. CPU time by double exponetial integration method is 100 times as much as that by the present method.

Error of Potential Gradient

As mentioned in the foregoing section, large error of the potential gradient has remained near the boundary, unless local Cartesian coordinate system is transformed to polar coordinate system. This is because weighting function is differentiated and kernel has $1/r^3$ singularity. Therefore we have considered similar correction method as mentioned above to obtain accurate potential gradient.

As in Fig.5, numerical c_i changes continuously near the boundary. We have assumed that c_i is continuous and differentiatable in the region including boundary. Equation 5 is differentiated by x , the following equation has been obtained.

$$\frac{\partial \phi_i}{\partial x_\alpha} = \frac{1}{c_i} \left\{ \int \frac{\partial \phi}{\partial n} \frac{\partial \phi^*}{\partial x_\alpha} d\Gamma - \int \phi \frac{\partial^2 \phi^*}{\partial n \partial x_\alpha} d\Gamma - \frac{\partial c_i}{\partial x_\alpha} \phi_i \right\} \qquad (9)$$

where

$$\frac{\partial c_i}{\partial x_\alpha} = - \int \frac{\partial^2 \phi^*}{\partial n \partial x_\alpha} d\Gamma \qquad (10)$$

Analytically $\partial c_i/\partial x_\alpha$ is 0 and Eq.(9) is equal to Eq.(2). However, numerically, c_i could not be constant and $\partial c_i/\partial x_\alpha$ could not become 0. Accordingly potential gradient should be calculated by Eq.(9) numerically.

Calculation results by Eq.(9) is shown in Fig.8. Maximum error has been 25% near the boundary, comparing the results of Fig.4, numerical accuracy has been exceptionally improved. As the number of integration point increased, the region where numerical solution agrees with exact one spreads but maximum error does not decrease any further. In the case of 16x4x4 integration points, a distance between a point where maximum error has occurred and the boundary is 0.01 less than the boundary element size.   In the region where observation point is located far from the boundary, calculation error becomes less than 1.5%.   Therefore, Eq.(9) is useful in this region.

CONCLUSION

To improve numerical accuracy in inner potential and its gradient near the boundary, a new integration method has been developed which is featured by,
(1) The local Cartesian coordinate system in isoparametric boundary element is transformed to the polar coordinate system whose origin is the foot of the perpendicular of inner observation point and Gauss-Legendre formula is applied under this coordinate system.
(2) Coefficient c_i which appears in boundary integral equation is used to correct the error in numerical integration.
To study an availability of the present method, inner potential and its gradient of a cubic has been calculated under the following boundary condition i.e. potential ϕ =1 on a surface of the cubic ϕ =0 on opposite surface and flux $\partial\phi/\partial n$ =0 on other sufaces, following results have been obtained.
(1) Upper bound of error in the potential calculation has been less than 0.15% near the boundary where the error by conventional numerical integration goes infinite.
(2) Calculation error of the potential gradient has been less than 1.5% at the location where the distance from the surface is one -hundredth of the representation element size.
(3) CPU time by the present method is 0.01 less than that by double exponential numerical integration method   under the same numerical accuracy condition.

Table 1 Integration Formuli and Integration Points in Triangular Region

	Region 1	Region 2	Region 3	Region 4
Interval of Integration	$0 \leq \theta \leq \pi - (\theta_1 + \theta_2)$ $0 \leq \rho \leq \frac{1+\eta_l}{\sin(\theta+\theta_1)}$ $\theta_1 = \arctan(\frac{1+\eta_l}{1+\xi_l})$ $\theta_2 = \arctan(\frac{1+\eta_l}{1-\xi_l})$	$0 \leq \theta \leq \theta_2 + \theta_3$ $0 \leq \rho \leq \frac{1-\xi_l}{\cos(\theta_2 - \theta)}$ $\theta_3 = \arctan(\frac{1-\eta_l}{1-\xi_l})$	$0 \leq \theta \leq \pi - (\theta_4 + \theta_3)$ $0 \leq \rho \leq \frac{1-\eta_l}{\sin(\theta+\theta_3)}$ $\theta_4 = \arctan(\frac{1-\eta_l}{1+\xi_l})$	$0 \leq \theta \leq \theta_1 + \theta_4$ $0 \leq \rho \leq \frac{1+\xi_l}{\cos(\theta_4 - \theta)}$
Local Coordinate of Int. Point	$\theta_l = \frac{\pi - (\theta_1+\theta_2)}{2}(\eta'_l + 1)$ $\rho_k = \frac{1+\eta_l}{2\sin(\theta_l+\theta_1)}(\xi'_k+1)$ $\xi_{kl} = \xi_1 - \rho_k \cos(\theta_l+\theta_1)$ $\eta_{kl} = \eta_1 - \rho_k \sin(\theta_l+\theta_1)$	$\theta_l = \frac{(\theta_2+\theta_3)}{2}(\eta'_l + 1)$ $\rho_k = \frac{1-\xi_l}{2\cos(\theta_2-\theta_l)}(\xi'_k+1)$ $\xi_{kl} = \xi_1 + \rho_k \cos(\theta_l-\theta_2)$ $\eta_{kl} = \eta_1 + \rho_k \sin(\theta_l-\theta_2)$	$\theta_l = \frac{\pi - (\theta_4+\theta_3)}{2}(\eta_l + 1)$ $\rho_k = \frac{1-\eta_l}{2\sin(\theta_l+\theta_3)}(\xi'_k+1)$ $\xi_{kl} = \xi_1 + \rho_k \cos(\theta_3+\theta_l)$ $\eta_{kl} = \eta_1 + \rho_k \sin(\theta_3+\theta_l)$	$\theta_l = \frac{\theta_1+\theta_4}{2}(\eta_l + 1)$ $\rho_k = \frac{1+\xi_l}{2\cos(\theta_4-\theta)}(\xi'_k+1)$ $\xi_{kl} = \xi_1 - \rho_k \cos(\theta_l-\theta_4)$ $\eta_{kl} = \eta_1 - \rho_k \sin(\theta_l-\theta_4)$
Formula of Numerical Int.	$\int f(\xi,\eta) J\, d\xi\, d\eta$ $= \Sigma\Sigma f(\xi_{kl},\eta_{kl}) J \times$ $\left(\frac{1+\eta_l}{2\sin(\theta_1+\theta_l)}\right)^2 (\xi_k+1) \times$ $\frac{(\pi - (\theta_1+\theta_2)}{2} W_k W_l$	$\int f(\xi,\eta) J\, d\xi\, d\eta$ $= \Sigma\Sigma f(\xi_{kl},\eta_{kl}) J \times$ $\left(\frac{1-\xi_l}{2\cos(\theta_2-\theta_l)}\right)^2 (\xi_k+1) \times$ $\frac{(\theta_2 + \theta_3)}{2} W_k W_l$	$\int f(\xi,\eta) J\, d\xi\, d\eta$ $= \Sigma\Sigma f(\xi_{kl},\eta_{kl}) J \times$ $\left(\frac{1-\eta_l}{2\sin(\theta_l+\theta_3)}\right)^2 (\xi_k+1) \times$ $\left(\frac{\pi - (\theta_4+\theta_3)}{2}\right) W_k W_l$	$\int f(\xi,\eta) J\, d\xi\, d\eta$ $= \Sigma\Sigma f(\xi_{kl},\eta_{kl}) J \times$ $\left(\frac{1+\xi_l}{2\cos(\theta_4-\theta_l)}\right)^2 (\xi_k+1) \times$ $\frac{(\theta_1 + \theta_4)}{2} W_k W_l$

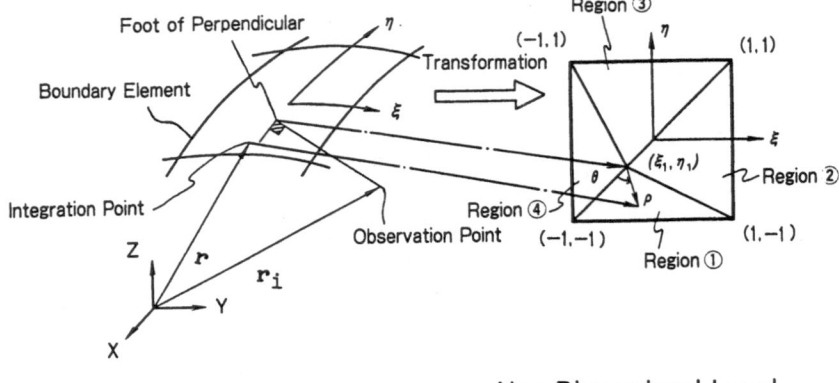

Fig.1 Concepts of Transformation of Local Cartesian Coordinate System to Polar Coordinate System in an Isoparamentric Element for Numerical Integration

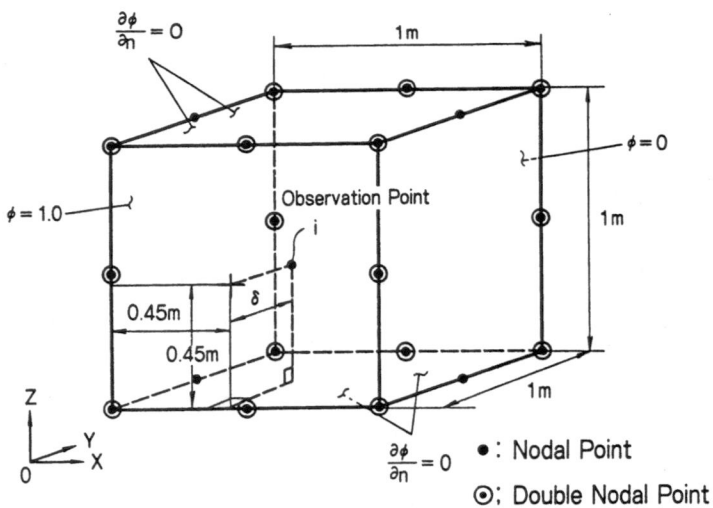

Fig.2 Calculation Model and Calculation Conditions

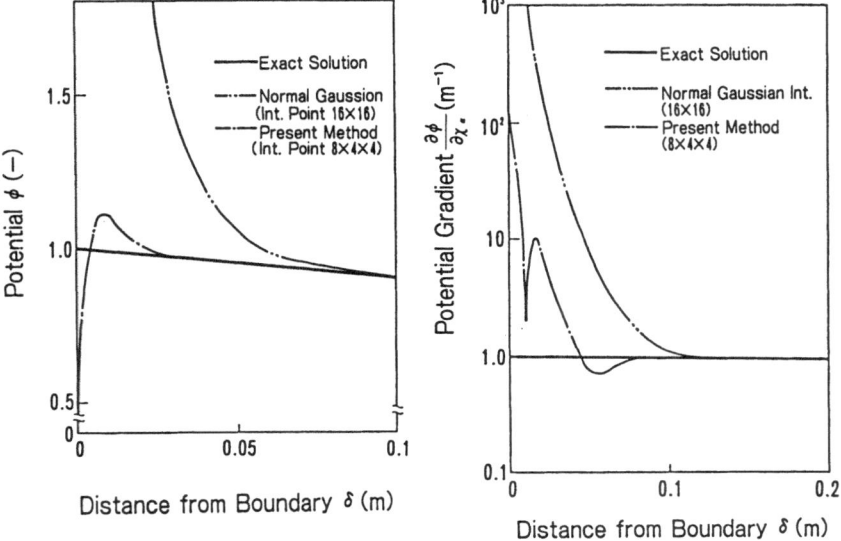

Fig.3 Change of Potential
near the Boundary

Fig.4 Change of Potential
Gradient near the Boundary

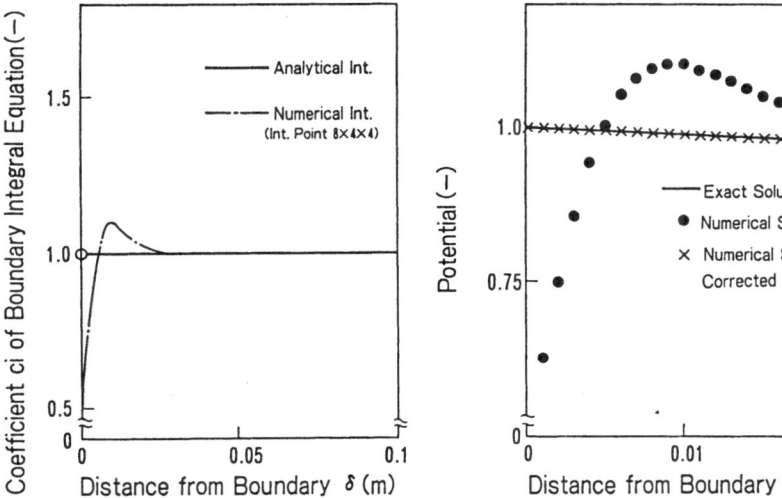

Fig.5 Change of Coefficient
c_i which appears in Boundary
Integral Equation near
the Boundary

Fig.6 Calculation Results of
Potentials Corrected by c_i

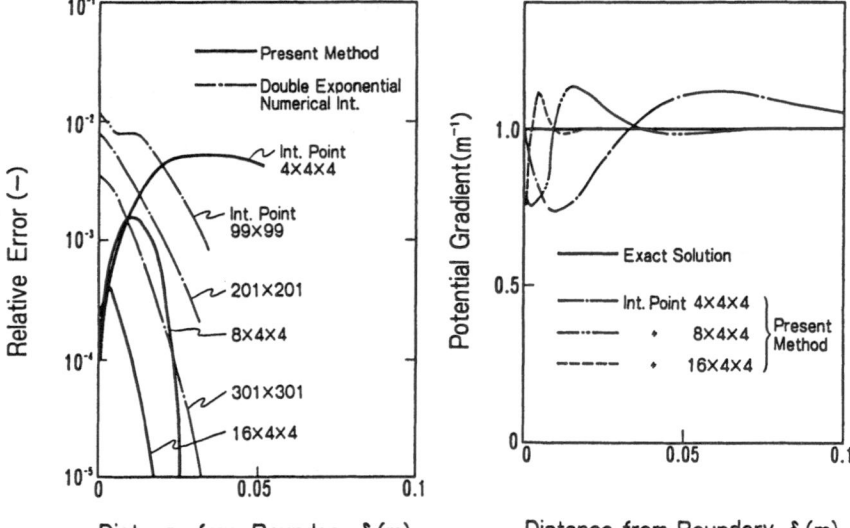

Fig.7 Comparison of Calculation
      Errors of the Present
      Numerical Integration Method
      with Double Exponential
      Numerical Integration Method

Fig.8 Calculation Results of
      Potential Gradient Corrected
      by c_i near the Boundary

Table 2 Comparison of Relative CPU-Time of the Present with
        Double Exponential Numerical Integration Method

Method	Number of Int. Point in an Element	Relative CPU Time
Present Method	4×4×4	0.56
	8×4×4	1.0
	16×4×4	2.5
Double Exponential Integration Method	99×99	13
	201×201	53
	301×301	117

REFERENCES
1. T. Kuwabara and T. Takeda (1985) Calculation of Potential and Potential Gradient for 3-Dimensional Field by B.E.M using Analytical Integration,pp 73 to 82 Proceeding of Conf. on Rotary Machines and Stational Apparatus, SA-85-47,JEC
2. T. Higashimachi et al.(1983) Interactive Structure Analysis System Using the Advanced Boundary Element Method, Boundary Elements Procceding of 5th Int. Conf. Springer-Verlag
3. K. Takahashi et al. (1985) On the Evaluation of Various Numerical Integrations in Boundary Element Method, PP 91 to 99 Procceding of Conf. on Rotary Machines and Stational Apparatus, SA-85-49,JEC
4. J.V. Cox and T.A. Shugar (1985) A Recursive Integration Technique for Boundary Element Method in Elastoatatics, Advanced Topics in Boundary Element Analysis AMD-Vol.72, The winter meeting of ASME
5. F.J. Rizzo and D.J. Shippy (1977) An Advanced Boundary Integration Method for Three-Dimensional Thermoelasticity, Int. J. Numer. Methods Eng., Vol.11 pp 1753 to 1768
6. M. Utamura and M. Koizumi (1985) Development of Computer Program for Analyzing Three-Dimensional Pressure Field in Pressure Supression System, J. Nucl. Sci. Technol.,Vol 22,No9, pp. 733 to 741

Application of the Cauchy Integral to the Displacement Discontinuity Method

J.P. Henry, A. Bouhadanne
Laboratoire de Mécanique de Lille – E.U.D.I.L. – Université des Sciences et Techniques de Lille, 59655 Villeneuve d'Ascq Cédex, France
E. Morel
Ecole des Mines de Douai, 241 rue Charles Bourseul, 59046 Douai, France

INTRODUCTION

In order to study isotropic elastic structures within cracks, the Displacement Discontinuity Method has been successfully developed and used for its simplicity and its consistency with the problem. (Crouch[1] ; Crawford and Curran[2,3] ; Wiles and Curran[4] ; Crouch and Starfield[5]). In this study, this method is considered in a more general formulation for which the fundamental solutions are determined in infinite plane. The approach is given by the complex potentials. The general solution of the Displacement Discontinuity Method for linear discretization is then given using Cauchy's integral formultation. This approach in the complex field also allows to develop solutions for circular elements which are more convergent. By using conformal mapping, we give the general solution for elliptical and corner elements. The method is then extended to the anisotropic behaviour.

THE DISPLACEMENT DISCONTINUITY METHOD IN COMPLEX FIELD - SOLUTION FOR LINEAR ELEMENTS.

In the complex field represented by the variables $z = x + iy$ and $\overline{z}$, the complex potentials $\phi(z)$ and $\psi(z)$ are connected to stresses and displacements applying the well known formulas of Kolosov-Muskhelishvili[6]

$$\sigma_{xx} + \sigma_{yy} = 2[\phi'(z) + \overline{\phi'(z)}]$$
$$\sigma_{yy} - \sigma_{xx} + 2i\sigma_{xy} = 2[\overline{z}\,\phi''(z) + \psi'(z)]$$
$$2\mu(u_1 + iu_2) = 2\mu D = k\,\phi(z) - z\,\overline{\phi'(z)} - \overline{\psi(z)}$$

$$(1)$$

with $k = 3 - 4\nu$ in plane strain and $k = (3 - \nu)/(1 + \nu)$ in plane stress. A new function $\Omega(z)$ is now introduced

$$\Omega(z) = -[z\,\phi'(z) + \psi(z)]$$

$$(2)$$

then the boundary conditions on each side of a crack lying on the x - axis ($x \in [-I, +I]$, fig. 1) are :

$$(\sigma_{22} - i\sigma_{22})^G(t) = \phi'^G(t) - \Omega'^D(t)$$
$$2\mu\, D^G(t) = k\,\phi^G(t) - \Omega^D(t) \tag{3}$$

in which $t \in [-I, +I] - G$ indicates the left side of the crack and D the right side with respect to the orientation of the x - axis. In this case we can impose a stress discontinuity $\hat{T}$

$$\hat{T} = T^G + T^D = (\sigma_{yy} + i\sigma_{xy})^G - (\sigma_{yy} + i\sigma_{xy})^D$$

and(or) a displacement discontinuity $\hat{D}$ ($D = D^G - D^D$) on the crack. The different possibilities for giving simple boundary conditions on the linear crack are :

1) $\quad 2\mu\,\hat{D}(x) = (k\,\phi - \Omega)^G(t) - (k\,\phi - \Omega)^D(t)$
   $\quad\quad O \quad\quad = (\phi' + \Omega')\,(t) - (\phi' + \Omega')^D(t)$ $\tag{4}$

2) $\quad \hat{T}(t) \quad = (\phi' + \Omega')^G(t) - (\phi' + \Omega')^D(t)$
   $\quad\quad O \quad\quad = (k\,\phi - \Omega)^G(t) - (k\,\phi - \Omega)^D(t)$

3) $\quad 2\mu\,\hat{D}(t) = (k\,\phi - \Omega)^G(t) - (k\,\phi - \Omega)^D(t)$
   $\quad\quad O \quad\quad = (k\,\phi + \Omega)\,(t) + (k\,\phi + \Omega)(t)$

4) $\quad \hat{T}(t) \quad = (\phi' + \Omega')^G(t) - (\phi' + \Omega')^D(t)$
   $\quad\quad O \quad\quad = (\phi' - \Omega')^G(t) + (\phi' - \Omega')^G(t)$

5) $\quad \hat{T}(t) \quad = (\phi' + \Omega')^G(t) - (\phi' + \Omega')^D(t)$
   $\quad\quad \hat{D}(t) \quad = (k\,\phi - \Omega)^G(t) - (k\,\phi - \Omega)^D(t)$

The solutions for these Plemelj equations are given by Bouhadanne[7]. Case 1-corresponds to the Displacement Discontinuity Method developed below. Case 2 - corresponds to the Fictitious stress Method (Crouch and Starfield[5])

Equation 4 is a Plemelj equation. The general solution is given by a Cauchy integral. Assuming stresses and displacements vanish at infinity, the fundamental solution is then

$$\phi(z) = -\Omega(z) = \frac{\mu}{i\pi(k+1)} \int_{-I}^{+I} \frac{\hat{D}(t)}{t-z}\,dt \tag{5}$$

This formulation for the fundamental solution has a great advantage to obtain simple analytical solutions when $\hat{D}(t)$ is assume to be a polynomial with complex coefficients. The complex coefficients are the linearization parameters of the method.

SOLUTION FOR A CIRCULAR ELEMENT.

In the polar system, stresses and displacements are as follows :

$$\sigma_{rr} + \sigma_{\theta\theta} = 2[\phi'(z) + \overline{\psi'(\bar{z})}]$$
$$\sigma_{\theta\theta} - \sigma_{rr} + 2i\,\sigma r_\theta = e^{2i\theta}[z\,\phi''(z) + \psi'(z)]$$
$$2\mu\,(u_r + iu_\theta) = 2\mu\,D_p = e^{-i\theta}[k\,\phi(z) - z\,\overline{\phi(\bar{z})} - \overline{\psi(\bar{z})}] \tag{6}$$

we now introduce the new function

$$\overline{\Omega}\left(\frac{a^2}{z}\right) = -\left[\frac{a^2}{z}\,\phi'(z) + \psi(z)\right] \tag{7}$$

where a is the circle radius. The boundary conditions are given on this circle along $A_1\,A_2$ (fig. 2). Then the Plemelj equations are :

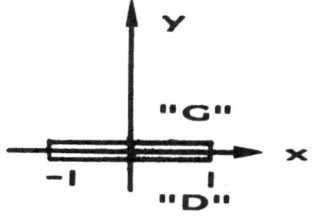

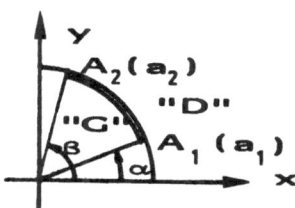

Figure 1   Figure 2

$$(\sigma_{rr} + i\sigma_{r\theta})^G(t) - (\sigma_{rr} + i\sigma_{r\theta})^D(t) = (\phi' + \Omega')^G(t) - (\phi' + \Omega')^D(t) = 0$$

$$2\mu\,(D_p^G - D_p^D)(t) = 2\mu\,\hat{D}_p(t) = \left[(k\,\phi - \Omega)^G(t) - (k\,\phi - \Omega)^D(t)\right]\,e^{-i\theta} \tag{8}$$

with $t = ae^{i\theta}$, $\theta\,\varepsilon[\alpha,\beta]$. To determine the solution of these equations, it is necessary to observe that the complex potentials $\phi'(z)$ and $\psi'(z)$ are holomorphic functions in the complete infinite plane, resulting in the fact that the function $\Omega'(z)$ has a second order pole at the origin. However, as stresses vanish at infinity and the resultant of the boundary stresses acting along both sides of the crack is equal to zero, the general solution is :

$$\phi(z) = \frac{\mu}{i\pi(k+1)}\int_{a_1}^{a_2}\frac{e^{i\theta}\,\hat{D}_p(t)\,dt}{t - z} \tag{9}$$

$$\Omega(z) = -\phi(z) + \phi'(0)$$

The integral (8) can be easily calculated if we chose $\hat{D}_p$:

$$\hat{D}_p = b_0 + \sum_{n=1}^{N} b_n\,\cos n\theta + \sum_{n=1}^{N} b_n'\,\sin n\theta$$

with $b_0,\,b_n,\,b_n'\,\varepsilon\,\mathbb{C}$

NUMERICAL EXAMPLE

The numerical example which follows is designed to show a better convergence of the circular element when the boundary itself is circular. Linear elements with one nodal point will be compared to circular element, again with onenodal point, which we just studied. In the comparison, the size of the influence matrix will be identical. We already notice that in the case of a hole in an infinite plane or in the case of a disc, both uniformally stressed, the previous solutions for circular elements

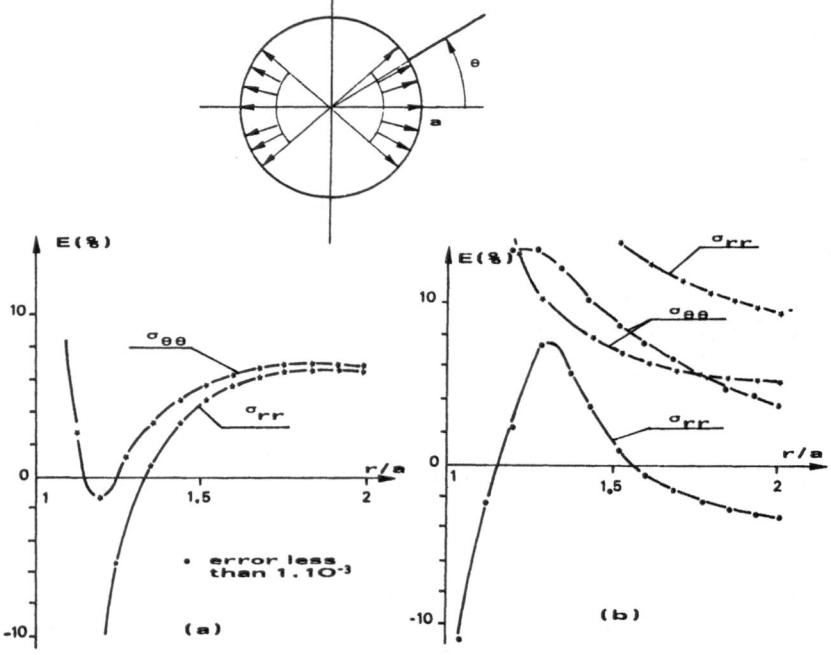

Figure 3
Error on stresses along $\theta = \pi/4$ and $\theta = 3\pi/16$ axes
✱linear element   • circular element

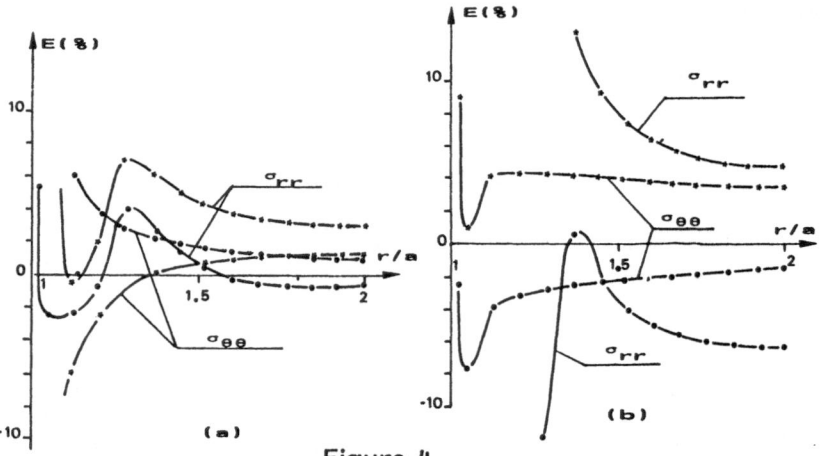

Figure 4
Error on stresses along $\theta = \dfrac{3\pi}{16}$ and $\theta = \dfrac{\pi}{2}$ axes.
✱linear element   • circular element

gives the right solution, whatever discretization is chosen. This proves the better adaptation of this method when used to solve circular boundary problems.

The example refers to an infinite plane with a hole. An uniform and perpendicular stress is applied along two opposite arcs whose opening equals $\pi/2$. The results for a discretization of 16 and 64 identical elements respectively are shown below (figure 3 and 4). It is noticed the relative error $E(\%)$ with respect to the analytical solution. Figure 3a gives the stress distribution $(\sigma_{rr}, \sigma_{\theta\theta})$ along the $\theta = \pi/4$ axis and Figure 3b along the $\theta = 3\pi/6$ axis. Figure 4a give the same stresses distribution along the $\theta = 3\pi/6$ axis and Figure 4b for the $\theta = \pi/2$ axis.

The programmating study was carried in complex field. The main difficulty was to solve the fact that the complex potentials have a multiform expression. It was chosen $\alpha = \pi/2 - \lambda$ $\beta = \pi/2 + \lambda$ in the fundamental solution (9) to simplify the problem.

This example shows a better convergence of the circular element.

CORNER ELEMENTS

Suppose the conformal mapping $z = \omega(\zeta)$ which links each point of the real plane "z" to each point of plane "ζ" (auxiliary plane for calculus) so that the corner boundary A_1 A_2 has an original defined by $|x| \le 1$ (fig.4). We note

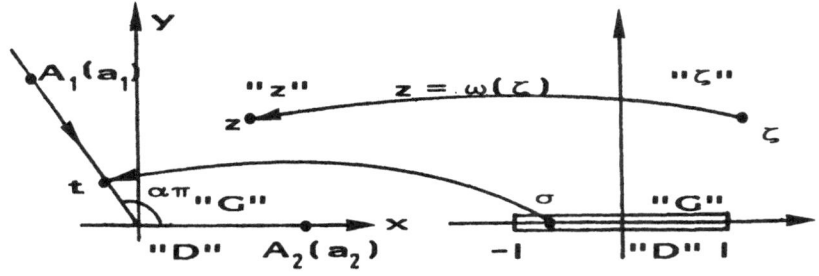

Figure 5

$\zeta = \rho e^{i\theta}, \ z = \omega(\zeta) = \zeta^{\alpha}$
$\psi(z) = \phi(\omega(\zeta)) = (\zeta)$
$\psi(z) = \psi(\omega(\ )) = \Psi(\zeta)$

we now change functions

$$\Omega(\overline{\zeta}) = - \left[\frac{\omega}{\overline{\omega}^\prime}\frac{(\overline{\zeta})}{(\zeta)}\overline{\phi}^\prime(\overline{\zeta}) + \overline{\Psi}(\overline{\zeta})\right]$$ (10)

The Kolosov Muskhelishvili relations now become

$$\sigma_{\eta\eta} - i\sigma_{\eta\xi} = \frac{\Phi'(\zeta)}{\omega'(\zeta)} - \frac{\zeta}{\zeta} \frac{1}{\overline{\omega'(\zeta)}} \Omega'(\zeta) + \frac{\overline{\Phi}'(\overline{\zeta})}{\overline{\omega'(\zeta)}} [1 - \frac{\zeta}{\zeta} \frac{\omega'(\overline{\zeta})}{\omega'(\zeta)}]$$

$$- \frac{\zeta}{\zeta} \frac{\overline{\omega}''(\overline{\zeta})}{\omega'(\zeta)} \frac{\overline{\Phi}'(\overline{\zeta})}{[\overline{\omega'(\zeta)}]^2} [\omega(\zeta) - \omega'(\zeta)] +$$

$$+ \frac{\zeta}{\zeta} \frac{\Phi''(\overline{\zeta})}{\omega'(\zeta) \overline{\omega'(\zeta)}} [\omega(\zeta) - \omega(\overline{\zeta})] \qquad (11)$$

$$2\mu(u_{\xi} + iu_{\eta}) = \frac{\zeta \overline{\omega'(\overline{\zeta})}}{|\zeta \omega'(\zeta)|} [k \phi(\zeta) - \frac{\overline{\Phi}(\overline{\zeta})}{\omega'(\zeta)} [\omega(\zeta) - \omega(\overline{\zeta})] + \Omega(\overline{\zeta})]$$

we can then obtain the Plemelj equations :

$$O = (\phi' + \Omega')^G(\sigma) - (\phi' + \Omega')^D(\sigma)$$

$$2\mu \frac{\sigma \omega'(\sigma)}{|\sigma \omega'(\sigma)|} \hat{D}_c = (k \phi - \Omega)^G(\sigma) - (k \phi - \Omega)^D(\sigma) \qquad (12)$$

in which $\hat{D}_c = (u_{\xi} + iu_{\eta})^G(t) - (u_{\xi} + iu_{\eta})^D(t)$. The solution for these equations (12) is :

$$\phi'(\zeta) = -\Omega'(\zeta)$$

$$k \phi(\zeta) - \Omega(\zeta) = \frac{\mu}{i\pi} \int_{I}^{O} \frac{e^{i\alpha\pi} \hat{D}_c(t)}{\sigma} \frac{1}{\zeta} d\sigma + \frac{\mu}{i \pi} \int_{O}^{I} \frac{\hat{D}_c(t)}{\sigma - \zeta} d\sigma \qquad (13)$$

cohere $t = \sigma^{\alpha}$. The above integral is easily calculated if $\hat{D}_c$ is a polynomial in ζ plane.

CURVILINEAR ELEMENTS - THE ELLIPTICAL ELEMENT

Suppose the conformal mapping $z = \omega(\zeta)$ which links each point of the real plane "z" to each point of plane ζ (auxiliary plane for calculus. So that the boundary ϕ has a circular original defined by $|\zeta| = a$ (fig. 6). We note

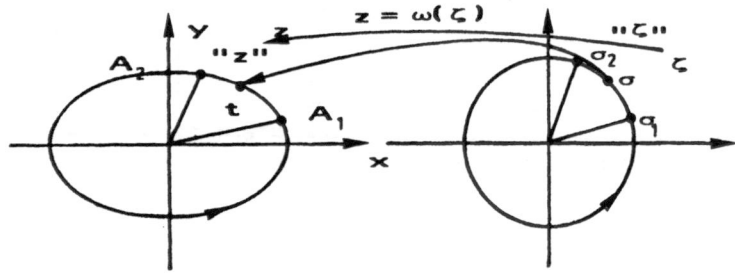

Figure 6

$$\zeta = e^{\xi + i\eta}$$
$$\phi(z) = \phi(\omega(\zeta)) = \phi(\zeta)$$
$$\psi(z) = \psi(\omega(\zeta)) = \Psi \zeta$$

We now change   functions

$$\bar{\Omega}\ (\frac{a^2}{\zeta}) = - [\frac{\bar{\omega}(\frac{a^2}{\zeta})}{\omega'(\zeta)}\ \Phi'(\zeta) + \Psi(\zeta)\ ] \tag{14}$$

The Kolosov-Muskhelishvili relations now become

$$\sigma_{\xi\xi} + i\ \sigma_{\eta\xi} = \frac{\Phi'(\zeta)}{\omega'(\zeta)z} - \frac{a^2}{\zeta\bar{\zeta}}\ \frac{1}{\omega'(\zeta)}\ \Omega'(\frac{a^2}{\zeta}) + \frac{\overline{\Phi'(\zeta)}}{\overline{\omega'(\zeta)}}\ [1 - \frac{a^2}{\zeta\bar{\zeta}}\frac{\omega'(a^2/\bar{\zeta})}{\omega'(\zeta)}]$$

$$+ \frac{\bar{\zeta}}{\zeta}\ \frac{\bar{\omega}''(\bar{\zeta})}{\omega'(\zeta)}\ \frac{\overline{\Phi'(\zeta)}}{[\overline{\omega'(\zeta)}]^z}\ [\omega(\zeta) - \omega(a^2/\bar{\zeta})]$$

$$- \frac{\bar{\zeta}}{\zeta}\ \frac{1}{\omega'(\zeta)}\ \frac{\Phi''(\zeta)}{\overline{\omega'(\zeta)}}\ [\omega(\zeta) - \omega(\frac{a^2}{\zeta})\ ] \tag{15}$$

$$2\mu(u_\xi + iu_\eta) = \frac{\bar{\zeta}\ \bar{\omega}'(\bar{\zeta})}{|\zeta\ \omega'(\zeta)|}\ [k\ \Phi(\zeta) - \frac{\overline{\Phi'(\zeta)}}{\omega'(\zeta)}(\omega(\zeta) - \omega(a^2/\bar{\zeta}))$$

$$- \Omega(a^2/\bar{\zeta})\ ]$$

and we can then obtain the Plemelj equations

$$O = (\phi' + \Omega')^G(\sigma) - (\phi' + \Omega')^D(\sigma) \tag{16}$$

$$2\mu\ \frac{\sigma\ \omega'(\sigma)}{|\sigma\ \omega'(\sigma)|}\ \hat{D}_e = (k\ \phi - \Omega)^G(\sigma) - (k\ \phi - \Omega)^D(\sigma)$$

where σ belongs to $A_1\ A_2$ and $\hat{D}_c = (u_\xi + iu_\eta)^G(t) - (u_\xi + iu_\eta)^D$
In the case of elliptical boundary :

$$z = \omega(\zeta) = h\zeta - d/\zeta$$

thas $\Phi(\zeta)$ and $\Psi(\zeta)$ have a second order pole at the origin.
The funtions $\Phi'(\zeta)$ and $\Psi'(\zeta)$ being holomorphic functions at
the infinity, function $\Omega'(\zeta)$ shows a second order pole at
the origin (equation 14). However, as stresses and displacements are supposed to vanish at infinity, this pole disappears and Ω' is equal to a constant at the origin. Moreover,
the resultant of the stresses acting on both sides of arc A_1
A_2 is equal to zero, this leads to the solution that follows :

$$\Phi'(\zeta) = A + \frac{B}{\zeta} + \frac{C}{\zeta^2} + I'(\zeta) \qquad A = \frac{\phi'(O)}{\omega'(O)} = -\bar{D}$$

$$\Omega'(\zeta) = D - I'(\zeta)$$

$$I(\zeta) = -\frac{\mu}{i\pi(k+1)}\int_{\sigma_1}^{\sigma_2}\frac{\sigma\ \omega'(\sigma)}{|\sigma\omega'(\sigma)|}\ \frac{\hat{D}_c}{\sigma - \zeta}\ d\sigma \qquad C = dA/n$$

$$[B = \lim_{\zeta \to O}\ [\ \frac{\phi'(\zeta)}{\omega'(\zeta)} - \frac{\phi'(o)}{\omega'(o)}\ ]\ \frac{1}{\zeta}$$

The above integral is easily calculated if $\hat{D}$ is equal to :

$$\hat{D}_p = \sqrt{\ C_\xi^2\ \cos\eta + C_1^2\ \sin\eta\ [\ b_o + \sum_{n=1}^{N}\ b_n\ \cos n\ \eta\ +}$$

$$\sum_{n=1}^{N}\ b'_n\ \sin n\ \eta] \tag{17}$$

with b_o, b_n, b_n' ε C . We are now in the process of progra
ming corner and elliptical elements. C_1 , C_2 principal axis of
the ellipse.

ANISOTROPIC MEDIUM

In anosotropic medium, the stresses are written in complex
field, introducing two potentials ϕ_1 and ϕ_2 (Lekhnitskii[8])

$$
\begin{aligned}
\sigma_{xx} &= 2Re \ [s_1^2 \ \Phi_1'(z_1) + s_2^2 \ \Phi_2'(z_2)] \\
\sigma_{yy} &= 2Re \ [\Phi_1' \ (z_1) + \Phi_2' \ (z_2) \ ] \\
\sigma_{xy} &= 2Re \ [s_1 \ \Phi_1' \ (z_1) + s_2 \ \Phi_2'(z_2) \ ]
\end{aligned}
\tag{1}
$$

In these relations we note $z_j = x + s_j \ y$, where $s_j (j = 1,2)$
are the complex roots of the characteristic equations calculated from the compatibility relation (Sih and Liebowitz[9]). The
displacements are :

$$
\begin{aligned}
u_x &= 2Re \ [\ p_1 \ \Phi_1(z_1) + p_2 \ \Phi_2(z_2)] \\
u_y &= 2Re \ [\ q_1 \ \Phi_1' z_1) + q_2 \ \Phi_2(z_2 \ ] \\
p_j &= b_{11} \ s_j^2 + b_{12} - b_{16} \ s_j \\
q_j &= \frac{1}{s_j} \ (b_{12} \ s_j^2 + b_{22} - b_{26} \ s_j)
\end{aligned}
\qquad j = 1,2
\tag{19}
$$

The b_{ij} terms are the coefficients of the elasticity matrix.
We cannot easily formulate the Displacement Discontinuity Method from the Kolosov-Muskhelishvili relations. It has been
found better to separate mode I and mode II and then super-
pose both solutions.

If we consider mode I, for example, and look for fundamental
solutions for a linear crack lying on the x-axis ($|x| \leq 1$ fig.1),
the boundary conditions can be written as follows :

$$
\begin{aligned}
u_y^G (x,o) - u_y^D(x,o) &= \hat{u}_y(x) \\
\sigma_{yy}^G(x,o) - \sigma_{yy}^D(x,o) &= 0 \qquad -1 < x < + 1 \\
\sigma_{xy} \ (x,o) &= 0 \ \forall \ x
\end{aligned}
\tag{20}
$$

Unknows $\hat{u}_y(x)$ are the linearization parameters of the system. The potentials will then verify the following system :

$$
\begin{aligned}
\rho_j \ \Phi_j \ (z_j) - \bar{\rho}_j \ \bar{\Phi}_j \ (z_j) &= \frac{1}{2i\pi} \ \int_{-1}^{+1} \frac{\hat{u}_y(x)}{x - z_j} \ dx \\
m_j \ \Phi \ (z_j) - \bar{m}_j \ \bar{\Phi}_j \ (z_j) &= 0 \qquad\qquad j = 1,2
\end{aligned}
$$

$$
m^z = \frac{s_1 - s_2}{s_1} \qquad \rho_1 = \frac{q_1 \ s_2 - q_2 \ s_1}{s_2} \qquad \rho_2 = \frac{q_2 \ s_1 - q_1 \ s_2}{s_1}
\tag{21}
$$

with $m_1 = (s_2 - s_1)/s_2$. As was the case with isotropic medium
the integral can be easily calculated, choosing $\hat{u}_y$ as a polynomial. When the structure really contains crack, these solutions lead to convergence defect at the crack tip, as was the

case for the previous solutions for isotropic media. We then decide on parabolical discontinuities, the summit of which coincides with the crack tip $(-2,0)$:

$$\hat{u}_y = \gamma \sqrt{1 + x} \qquad\qquad \gamma \ \varepsilon \ \mathbb{R}$$

The corresponding potentials ϕ_j are :

$$\phi_j (z_j) = \frac{\overline{m}_j}{\overline{p}_j\, \overline{m}_j - \overline{p}_j\, m_j} \quad \frac{\gamma}{2i\,\pi} \ [2\sqrt{2l} + \sqrt{z}_j + l \ \text{Log} \ \frac{\sqrt{z_j + l}\,.-\sqrt{2l}}{\sqrt{z_j + l} + \sqrt{2l}}]$$

$$(22)$$

In mode II, the boundary conditions are

$$u_x^G (x,o) - u_x^D (x,o) = \hat{u}_x (x)$$
$$\sigma_{xj}^G (x,o) - \sigma_{xj}^D (x,o) = 0 \qquad\qquad -1 < x < 1$$
$$\sigma_{yy} (x,o) = 0 \ \forall x$$

The solution is :

$$\phi_1 (z_1) = \frac{-1}{4\pi \ b_{11} \ (s_1 \cdot s_2) \ I_m \ (s_1 + s_2)} \int_{-1}^{+1} \frac{\hat{u}_x (t)}{t - z_1} \ dt$$

$$(23)$$

$$\phi_2 (z_2) = \frac{-1}{4\pi \ b_{11} \ (s_2 - s_1) \ I_m \ (s_1 + s_2)} \int_{-1}^{+1} \frac{\hat{u}_x (t)}{t - z_2} \ dt$$

By superposing (21) and (23), we can solve plane elastic anisotropic problems.

NUMERICAL EXAMPLE

Given a linear crack (length 2L) lying on the x-axis, symmetrically to O and stressed by and internal constant pressure. The crack is divided into 10 equal elements. Two degenerated parabolical discontinuities are located at each end of the crack. On the other 8 linear elements we chose second order discontinuities. The charactreristics used are $E_1 = 1,2 \ 10^5$ SI $E_2 = 0,6 \ 10^5$ SI, $\nu_{12} = 0,071$, $G_{12} = 0,07 \ 10^5$ SI. Figure 7. shows the evolution of the crack deformation when the orthotropic axes are rotated.

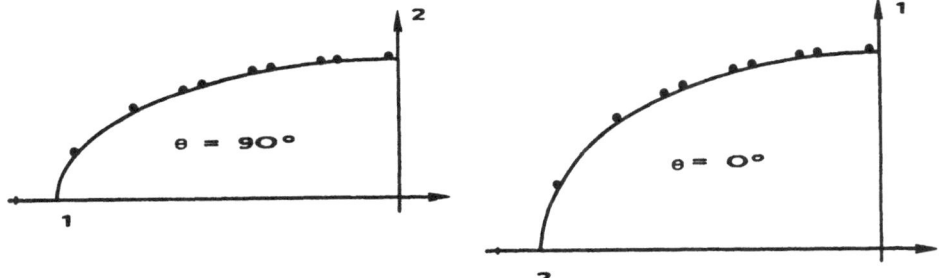

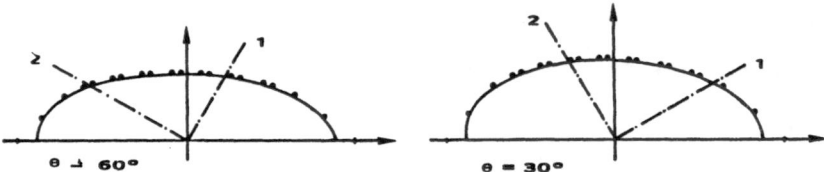

Figure 7
Deformation of a crack under pressure in orthotropic material
(axes 1 and 2)
• numerical solution ────── analytical solution

CONCLUSION

For solving plastic problems with cracks, the complex potentials give a new approach of the Displacement Discontinuity
Method. The Cauchy Integral allows to find fundamental solutions for curvilinear   elements (circular, corner, elliptical..)
The use of these solutions in the influence method give then
more convergent results.

BIBLIOGRAPHY

1. Crouch S.L. (1976), Solution of plane elasticity problems
   by the displacement discontinuity method, Int. J. Num.
   Meth. In Eng., Vol. 10, pp 301-343

2. Crawford A.M. and Curran J.H. (1982), Higher order functional variation displacement discontinuity elements, Int. J.
   Rock Mech. Min. Sci., Vol. 19, pp 143-148

3. Crawford A.M. and Curran J.H. (1983), A displacement
   discontinuity method approaching to modelling the creep
   behaviour of rocks and its discontinuties, Int. J. Num.
   Methods in Geomech., Vol. 7, pp.245-268

4. Wiles T.D. and Curran J.H. (1984), A general 3-D Displacement Discontinuity Method, Proceedings of the 4 th. Int.
   Conf. Num. Methods in Geomech, Edmonton, Canada, Vol.1,
   pp 103-111

5. Crouch S.L. and Starfield A.M. (1983), Boundary element
   methods in Geomechanics, George Allen and Unwin Ed.

6. Muskhelishvili N.I. (1963), Some basic problems of mathematical theory of elasticity, Noordhoff Ed., Groningen.

7. Bouhadanne A. (1986), Utilisation de l'intégrale de Cauchy
   dans les méthodes d'influence. Application à la méthode des
   discontinuités de déplacement, These de Doctorat (à paraî-
   tre).

8. Lekhnitskii S.G. (1963), Theory of an anisotropic elastic
   body, Holden Day Inc.

9. Sih G.C. and Liebowitz H.(1968), Rectilinearly Anisotropic
   bodies with cracks, in Fracture, Academic Press, Vol. 2,
   pp 108-123

A System of Boundary Integral Equations for the Problem of Acoustic Scattering by an Absorbing Wall

T. Ha Duong
Centre de Mathématiques Appliquées, Ecole Polytechnique, 91128 Palaiseau Cedex, France

1. INTRODUCTION.

We consider here a boundary integral equations (BIE) method for the exterior Helmholtz equation with the Robin boundary condition :

$$\zeta \frac{\partial u}{\partial n} + i\omega\, u = 0$$

where ζ is a complex function defined on the boundary, with a positive real part. This function describes the acoustic properties of the wall and is called the specific acoustic impedance at the frequency ω (we have taken the unities of length and time such that the wave velocity equals one). The case of a perfectly reflecting wall ($\zeta = 0$ or ∞) is not considered here.

This problem is well known, but according to ANGEL and KLEINMAN (1982) it was not solved by the BIE method. Thèse authors proposed a system of two BIE equivalent to the problem. Later, ANGEL and KRESS (1984) obtained a composite BIE based on the ansatz used by BURTON and MILLER (1971) for the Dirichlet and the Neumann problems.

Using an energy argument, we associate to this exterior problem an appropriate interior problem, and obtain another system of two BIE, equivalent to these problems. Compared with other BIE, our system presents some definite advantages with respect to numerical purposes : firstly, we can give an equivalent variational formulation of this system, with the usual properties of continuity and coercivity. Moreover, the sesqui-linear form in this variational problem can be written out with only weakly singular integrals. Finally we give a Galerkin method to approximate the solution of our BIE.

2. THE PROBLEM AND ITS ASSOCIATED BIE.

Let Ω^+ be an exterior domain in $\mathbb{R}^n (n = 2,3)$ with closed regular boundary $\Gamma = \partial\Omega$. We denote by Ω^- the complement of $\overline{\Omega}^+$ and orientate the unit normal vector ν on Γ from Ω^- to Ω^+. We consider the problem of calculating the (outgoing) scattered wave when Γ is an absorbing wall :

$$(P_+) \quad \begin{cases} - (\Delta + \omega^2)\, u = 0 & (1) \\[2mm] \lim_{|x|=r\to+\infty} r^{(n-1)/2} \left(\frac{\partial u}{\partial r} - i\omega\, u \right) = 0 & (2) \\[2mm] \zeta \frac{\partial u^+}{\partial\nu} + i\omega\, u^+ = g^+ \quad \text{on} \quad \Gamma & (3) \end{cases}$$

where u^+ is the trace of u on Γ, $\frac{\partial u^+}{\partial\nu}$ its normal trace and $g^+ = - (\zeta \frac{\partial u^I}{\partial\nu} + i\omega\, u^I)$ is given by the incident wave u^I. We denote by :

$$G(x,y) = \begin{cases} \frac{i}{4} H_0^1 (\omega\,|x-y|) & \text{if } n = 2 & (4) \\[2mm] \frac{e^{i\omega|x-y|}}{4\pi\,|x-y|} & n = 3 & (5) \end{cases}$$

the outgoing fundamental solution of the Helmholtz equation. We recall that any solution of (1) and (2) is representable as a sum of layer potentials on Γ :

$$\varepsilon u(x) = \int_\Gamma \{\frac{\partial G}{\partial\nu_y} (x,y)\, u_+(y) - G(x,y) \frac{\partial u_+}{\partial\nu}(y) \}d\sigma(y) \quad (6)$$

where $\varepsilon=1$ if $x \in \Omega_+$, 0 if $x \in \Omega_-$ and $\frac{1}{2}$ if $x \in \Gamma$.

In [1], where $\frac{i\omega}{\zeta} = \sigma$ and $g_+ = - (\frac{\partial u^I}{\partial\nu} + \sigma u^I)$, the authors simply replace in (6) the normal derivative $\frac{\partial u_+}{\partial\nu}$ by its value $g_+ - \sigma u_+$ from the boundary condition (3). They obtain the following BIE with the unknown function $\varphi = u_+$:

$$(- \frac{1}{2} I + S\sigma + K') \varphi = Sg_+ \quad (7)$$

$$(\frac{1}{2} \sigma + K\sigma + D) \varphi = (\frac{1}{2} I + K) g_+ \quad (8)$$

we have adopted in these equations the following notations, slightly different from those used in [1] : for $\underline{x \in \Gamma}$,

$$S\varphi(x) = \int_\Gamma G(x,y)\, \varphi(y)\, d\sigma_y \quad (9)$$

$$K\,\varphi(x) = \int_\Gamma \frac{\partial G}{\partial\nu_x} (x,y)\, \varphi(y)\, d\sigma_y \quad (10)$$

$$K'\varphi(x) = \int_\Gamma \frac{\partial G}{\partial\nu_y} (x,y)\, \varphi(y)\, d\sigma_y \quad (11)$$

$$D\varphi(x) = \int_\Gamma \frac{\partial^2 G}{\partial\nu_x\,\partial\nu_y}\,(x,y)\,\varphi(y)\,d\sigma_y \qquad (12)$$

In the last relation, the integral has a hypersingular kernal and should be understood in an appropriate distribution sense.

In [2], using the ansatz of Burton and Miller, the authors obtain the following BIE :

$$\frac{1}{2}\,(1 - i\eta\sigma)\varphi + (K + S\sigma + i\eta D + i\eta K'\sigma)\varphi = Sg_+ - i\eta(g - K'g) \qquad (13)$$

with $\eta \neq 0$, $\eta \mathrm{Re}\omega \geqslant 0$. While, in both cases, the authors can give an existence and uniqueness theorem for their BIE, we observe that these BIE are rather inadequate for an effective calculus of φ : the hypersingular integral operator D is difficult to be handled with in this form, and the presence of the operators K and D with the same variable φ prevents a variational treatment.

Next, we can note that the Robin boundary condition is <u>not</u> a totally reflecting boundary condition. One part of the energy passes   across the boundary and propagates into the interior. When ζ does not depend on ω, we can return to the wave propagation problem and obtain by Fourier transform, the condition $\zeta\frac{\partial u}{\partial n} + \frac{\partial u}{\partial t} = 0$ on Γ. From that condition, we can see that the variation of energy is the same for   the propagation of waves in Ω_- when it obeys to the boundary condition $\zeta\,\frac{\partial u-}{\partial n} - \frac{\partial u-}{\partial t} = 0$.

Thus, to deal with the exterior problem $(\dot{P}_+)$, it is natural to associate the following interior problem :

$$\begin{cases} -\,(\Delta + u^2)\,u = 0 \qquad \text{in } \Omega_- & (14) \\[2mm] \zeta\,\dfrac{\partial u-}{\partial n} - i\omega\,u_- = g_- \qquad \text{on } \Gamma & (15) \end{cases}$$

Now, a solution of (1), (2) and (14) is usually represented by layer potentials with the jumps $\varphi = u^- - u^+$ and $p = \dfrac{\partial u^-}{\partial\nu} - \dfrac{\partial u^+}{\partial\nu}$ of u and of its normal derivative as densities :

$$\begin{cases} u(x) = \int_\Gamma [\,G(x,y)\,p(y) - \dfrac{\partial G}{\partial\nu_y}\,(x,y)\,\varphi(y)\,]\,d\sigma_y \quad \forall x \in \Omega_-\cup\Omega_+ & (16) \\[2mm] =: Lp - M\varphi \end{cases}$$

Therefore, by using the well known formulas giving the traces of u as functions of φ and p, we simply deduce from (3) and (15) the necessary and sufficient relations to be satisfied by φ and p insuring that (16) yields the solution of (P_+) and (P_-) :

$$-\,i\omega\varphi + \zeta\,(\frac{\partial u^+}{\partial\nu} + \frac{\partial u^-}{\partial\nu}) = -\,i\omega\varphi + 2\zeta(Kp - D\varphi) = g_- + g_+ \qquad (17)$$

$$-\,i\omega(u^- + u^+) + \zeta p = -\,2i\omega(Sp - K'\varphi) + \zeta p = g_- - g_+ \qquad (18)$$

Or, with obvious notations, the final BIE equivalent to (P_+) and
(P_-) :

(Q)
$$\sigma\varphi - 2(Kp - D\varphi) = f_1 \tag{19}$$
$$2(Sp - K'\varphi) + \frac{1}{\sigma} p = f_2 \tag{20}$$

we proceed in the next section to the analysis of the BIE problem (Q).

3. THE VARIATIONAL FORMULATION OF (Q)

We denote by $\tilde{u} = (\varphi, p)$ the unknown of (Q), and by $\tilde{v} = (\psi, q)$ a
test-function. That is, $\tilde{v} = (v^- - v^+, \frac{\partial v^-}{\partial \nu} - \frac{\partial v^+}{\partial \nu})$ where $v = L\psi - Mq$
is a solution of (1), (2) and (14).
The variational formulation of (Q) in then, as usually :

(Q')
$$\text{To find } \tilde{u} \text{ such that}$$
$$b(\tilde{u}, \tilde{v}) = f(\tilde{v}) \text{ for all } \tilde{v} .$$

where
$$b(\tilde{u}, \tilde{v}) = \int_{\Gamma} [\, (\sigma\varphi - 2(Kp - D\varphi))\overline{\psi} + (2(Sp - K'\varphi) - \frac{1}{\sigma} p)\overline{q}\,] \tag{21}$$

$$f(v) = \int_{\Gamma} (f_1 \overline{\psi} + f_2 \overline{q}) \tag{22}$$

3.1.
We begin with the task of transforming the sesquilinear form b
such that all the involved integrals can be written with only
weakly singular kernels. The sole problem concerns the operator
D. But its treatment is now widely known (see Hamdi (1981)),
and we can write   in the 3-dimensional case :

$$\int_{\Gamma} (D\varphi)\overline{\psi}d\sigma = \int\int_{\Gamma\times\Gamma} G(x,y)\, \{\omega^2 \nu_x \cdot \nu_y \varphi(y)\overline{\psi}(x) - \overrightarrow{rot}_{\Gamma}\varphi(y) \cdot \overrightarrow{rot}_{\Gamma}\psi(x)\}\, d\sigma_y d\sigma_x$$

where $\overrightarrow{rot}_{\Gamma}\varphi = \nu \wedge \nabla\varphi$ is the tangential rotational of φ.

In the 2-d case, $\overrightarrow{rot}_{\Gamma}\varphi$ will be replaced by the tangential derivative of φ. Thus, we have in the 3-d case :

$$b(\tilde{u}, \tilde{v}) = \int_{\Gamma} (\sigma\varphi\overline{\psi} - \frac{1}{\sigma} p\overline{q})\, d\sigma_x + \int\int_{\Gamma\Gamma} \frac{e^{i\omega|x-y|}}{2\pi|x-y|} p(y)\overline{q}(x)\, d\sigma_y d\sigma_x$$

$$+ \int\int_{\Gamma\Gamma} \{\frac{\partial}{\partial\nu_x} (\frac{e^{i\omega|x-y|}}{2\pi|x-y|}) p(y)\overline{\psi}(x) + \frac{\partial}{\partial\nu_y} (\frac{e^{i\omega|x-y|}}{2\pi|x-y|})\varphi(y)\overline{q}(x)\}\, d\sigma_y d\sigma_x$$

$$+ \int\int_{\Gamma\Gamma} \frac{e^{i\omega|x-y|}}{2\pi|x-y|} \{\omega^2 \nu_x \cdot \nu_y \varphi(y)\overline{\psi}(x) - \overrightarrow{rot}_{\Gamma}\varphi(y) \cdot \overrightarrow{rot}_{\Gamma}\overline{\psi}(x)\}\, d\sigma_y d\sigma_x \tag{23}$$

The integrals are ordinary ones if $p, q \in L^2(\partial\Omega)$ and $\varphi, \psi \in H^1(\Gamma)$.

3.2. CONTINUITY AND COERCIVITY OF b.

First, we must precise the functional frame adapted to the problems
(Q) and (Q'). As the trace and normal derivative trace of locally
finite energy functions, φ and p can be taken in $H^{1/2}(\Gamma)$ and $H^{-1/2}(\Gamma)$
But the presence of the isolated term of p in equation (20) -and
of the integral $\int_\Gamma \frac{1}{\sigma} p\bar{q}$ in (21)- oblige us to restrict to the
square integrable p. So we pose the problems (Q) and (Q') for
$\tilde{u} \in V = H^{1/2}(\Gamma) \times L^2(\Gamma)$. The operators S, K, K are, as it is well
known, pseudo-differential operators of order -1, and the operator
D of order 1, so the functions g_+, g_- should be given such that
$f_1 \in H^{-1/2}(\Gamma)$ and $f_2 \in L^2(\Gamma)$. These conditions are clearly satisfied if g_+, $g_- \in L^2(\Gamma)$ and $\sigma \in L_\infty(\Gamma)$. In the other side, the
integral $\int_\Gamma (D\varphi)\bar{\psi}\, d\sigma_x$ should be taken in the sense of anti-duality
of $H^{1/2}(\Gamma)$ and $H^{-1/2}(\Gamma)$.
These preliminaries also make clear that the form b is continuous
on $V \times V$. And the main difficulty is to prove its coerciveness on
V. We do it by writing :

$$b(\tilde{u},\tilde{u}) = \int_\Gamma (\sigma\,|\varphi|^2 - \frac{1}{\sigma}\,|p|^2)\,d\sigma_x$$

$$- \int_\Gamma \{(\frac{\partial u^-}{\partial\nu} + \frac{\partial u^+}{\partial\nu})(u^- - u^+) - (u^- + u^+)(\frac{\partial u^-}{\partial\nu} - \frac{\partial u^+}{\partial\nu})\}\,d\sigma_x \qquad (24)$$

(cf (17) and (18)), then Green formula yields :

$$Jmb(\tilde{u},\tilde{u}) = Jm\int_\Gamma (\sigma\,|\varphi|^2 - \frac{1}{\sigma}\,|p|^2)\,d\sigma_x + 2\omega\,\|A\|^2 \qquad (25)$$

where A is the far fields pattern of u. From (25) the coercivity
of b on the subspace $L^2(\Gamma) \times L^2(\Gamma)$ of V is evident under the hypo-
thesis

$$Jm\ \sigma \geqslant \sigma_0 > 0$$

which is a consequence of the positivity of the real part of the impedance ζ.

In the other side, by a technique of MIKHLIN, we can prove that,
if $\varphi \in L^2(\Gamma)$, the double layer potential of density φ is locally
square integrable in $\Omega_- \cup \Omega_+$. This result permits us to obtain
the coercivity of b on V by a contradiction argument.

The well-posedness of problems (Q) and (Q') follows, and the solution of (P_+) and (P_-) is then calculated by (16).

3.3. That above results are exposed in a joint work of A.BAMBERGER
and the author to be published in the journal of Math.Meth. in Appl
Sci (1987). We also learned after having submitted our work that
HAMDI in his thesis (1982) had used the same BIE system as our's

to treat the case of an open surface (a crack) when the boundary conditions are $\zeta \frac{\partial u}{\partial n} \pm i\omega u = 0$ on the two sides of the surface. But he didn't give the mathematical results as we present here. In the case of a rectilinear crack (in R^2), we have also given, in the above cited work, a Fourier analysis of the problem, and pointed out the fact that the functional space should be $V = H_{00}^{1/2} (\Gamma) \times L^2(\Gamma)$. In this case, the problem is moreover de-coupled to two independant problems for φ and p respectively. One can also find in this work a proof of the uniqueness for problem (P_+), and a regularity theorem.

4. THE DISCRETIZED PROBLEM (Q_h).

The variational problem (Q') can be discretized as did Nedelec (1977) for the BIE of Laplace's equation :

a) Given a net of points in Γ, we can approximate the surface by a more simpler surface (Γ_h) made of triangles with these points as vertices.

If h, the maximal dimension of these triangles, is sufficiently small, a one to one projection exists between Γ and Γ_h, and it can be used to estimate the geometrical errors.

b) We construct then a finite dimensional space V_h on Γ_h, with P_1-functions for φ_h and P_0 functions for p_h. The functions of V_h are in bijection with those of a subspace of V. We consider the approximate problem in (V_h) :

(Q_h) $\begin{cases} \text{To find } \tilde{u}_h \in V_h \text{ such that :} \\ b_h(\tilde{u}_h, \tilde{v}_h) = f_h(\tilde{v}_h) \text{ for all } \tilde{v}_h \in V_h \end{cases}$

where b_h and f_h are obtained from b and f by replacing the integrals on Γ by integrals in Γ_h.

(Q_h) is then equivalent to a matrix problem $A_h \tilde{u}_h = f_h$, with a symmetric complex, **dissipative** matrix A_h (that is $Jm(A_h \tilde{u}_h, \tilde{u}_h) > 0$ for all $\tilde{u}_h \neq 0$). This last property allows us to solve the equation by the Cholesky decomposition.

c) Finally, as did Nedelec (1977), we obtain an O(h) error esti-mates for $\|\tilde{u} - P\tilde{u}_h\|_V$, where $P\tilde{u}_h$ is the inverse image of $\tilde{u}_h$ by the projection from Γ to Γ_h, when the regularity condition

$$\varphi \in H^{3/2}(\Gamma) \quad , \quad p \in H^1(\Gamma) \tag{25}$$

is satisfied.

5. CONCLUSION

We emphasize the fact that, by the energy argument, the problems (P_+) and (P_-) should be solved in the same time. And this association gives rise to a system of BIE with better properties than those of the BIE obtained when the problem (P_+) is solely considered.

BIBLIOGRAPHY

1. ANGEL T.S.§ KLEIMAN R.E. (1982). Math.Meth. in the Appl. Sci, vol 4, p 164-193.

2. ANGEL T.S.§ KRESS R. (1984) same journal, vol 6, p 345-352.

3. BAMBERGER A.§ HA DUONG T. (1987) same journal, vol 9.

4. HAMDI M.A. (1981) C.R.A.S. série II, vol 292, p.17-20.

5. HAMDI M.A. (1982) Thesis, University of Compiègne, FRANCE.

6. NEDELEC J.C. (1977) Cours EDF-CEA-INRIA, FRANCE.

On the Numerical Implementation of BEM for Axisymmetric Elasticity

M. Guiggiani, P. Casalini
Dipartimento di Costruzioni Meccaniche e Nucleari, Universita' di Pisa, via Diotisalvi 2, 56100 Pisa, Italy

INTRODUCTION

The extension of the Boundary Element Method (BEM) to axisymmetric elastic problems was first investigated by Mayr [1] and in particular by Cruse et al.[2], who extended the 'fictitious load' approach to the more general BEM approach by using the fundamental solution developed by Kermanidis [3]. A rather comprehensive description of the BEM for axisymmetric elastic problems has been recently presented in Brebbia, Telles and Wrobel [4] and in Telles [5].

However, on account of a rather exhaustive formulation of the theory, a few particulars are available in the literature concerning the numerical aspects involved in the axisymmetric version of BEM. This paper reports some of the particulars inherent in the axisymmetric formulation for both hollow and solid bodies of revolution. The important issue of the computation of the integrals involved is examined. In particular the authors suggest a method for the integration over elements on which the singular point is located. Finally two examples are provided to show the effectiveness of the procedure.

BOUNDARY INTEGRAL EQUATION

As described in Brebbia et al.[4], the boundary integral equation for axisymmetric problems, in (r,z) coordinates, is (see Fig. 1)

$$c_{ij}(P)u_j(P) + 2\pi \int_\Gamma T_{ij}(P,Q)u_j(Q)r(Q)d\Gamma(Q) =$$
$$= 2\pi \int_\Gamma U_{ij}(P,Q)t_j(Q)r(Q)d\Gamma(Q), \qquad i,j = r,z \tag{1}$$

where Γ is the one dimensional curve generator of the axisymmetric surface of the body. The kernel functions $U_{ij}(P,Q)$ and $T_{ij}(P,Q)$ are the displacements and the tractions in the direction j, at Q (field or observation point), due to unit ring load in the i direction, applied at P (source point or pole). By unit ring load (see Fig. 2) we intend a circular shape load applied on the circumference of radius $r(P)$ and with intensity F so that $2\pi Fr(P)=1$. Both P and Q are on Γ. Finally $u_j(Q)$ and $t_j(Q)$ are the displacement and traction components on the boundary of the body. The explicit expressions of the kernel functions U_{ij} are listed below:

$$U_{rr} = \frac{C_1}{\sqrt{Rr}} \left\{ \left[(3-4\nu)\frac{2-k^2}{k} + \frac{\bar{Z}^2}{2Rr}k \right] K(k) - \left[(3-4\nu)\frac{2}{k} + k\frac{\bar{Z}^2}{4Rr}\frac{2-k^2}{1-k^2} \right] E(k) \right\}$$

$$U_{rz} = \frac{C_1}{\sqrt{Rr}}\frac{\bar{Z}}{r}\frac{k}{2}\frac{r}{R} \left\{ -\frac{k}{2}K(k) + \left[\frac{k}{4}\frac{2-k^2}{1-k^2}\frac{2}{k^2}(\frac{r}{R}-1-\frac{2}{k^2}) - \frac{r}{R}\frac{1}{k} \right] E(k) \right\}$$

$$U_{zr} = -\frac{C_1}{\sqrt{Rr}}\frac{\bar{Z}}{r}\frac{k}{2} \left\{ K(k) - (\frac{2}{k^2}-1-\frac{r}{R}) \frac{k^2}{2(1-k^2)} E(k) \right\}$$

$$U_{zz} = \frac{C_1}{\sqrt{Rr}} k \left\{ K(k)(3-4\nu) + \frac{k^2}{1-k^2} E(k) \frac{\bar{Z}^2}{4Rr} \right\}$$

(2)

where $C_1 = 1/(16\pi^2 G(1-\nu))$, ν is the Poisson ratio, G is the shear modulus, and, with reference to Fig. 1,

$$R = r(P); \quad Z = z(P);$$

$$r = r(Q); \quad z = z(Q); \quad \bar{Z} = Z - z;$$

(3)

$$k = \sqrt{(4Rr/(\bar{Z}^2 + (R+r)^2))}, \text{ (note that } 0 \leq k < 1);$$

$K(k)$ and $E(k)$ are the complete elliptic integrals of the first and second kind, of modulus k.

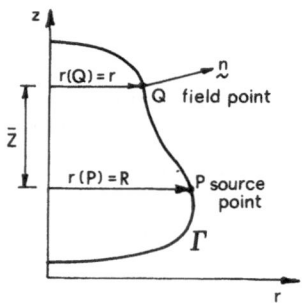

Figure 1. Geometry of the axisymmetric problem.

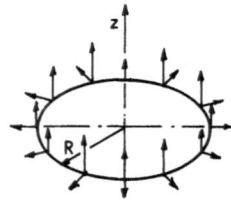

Figure 2.   Radial and axial
ring loads.

The expressions of the kernel functions T_{ij} can be achieved by deriving the U_{ij} with respect to r and z, considering the stress-displacement relations in cylindrical polar coordinates and then using the outward normal vector $\underset{\sim}{n} = (n_r, n_z)$ to Γ, in order to obtain the tractions on the boundary (Cruse et al.[2]). The explicit expressions of the derivatives of the U_{ij} are listed in the appendix. The coefficient $c_{ij}(P)$ of the so-called free term, when P is on Γ, depends on the local geometry of the boundary in P (Fig. 3)

$$\underset{\sim}{c} = \frac{1}{8\pi(1-\nu)} \left\{ \begin{array}{ll} 4(1-\nu)\,\bar{\theta} + & (\cos2\theta_2 - \cos2\theta_1) \\ + (\sin2\theta_1 - \sin2\theta_2) & \\ & \\ & 4(1-\nu)\,\bar{\theta} - \\ (\cos2\theta_2 - \cos2\theta_1) & - (\sin2\theta_1 - \sin2\theta_2) \end{array} \right\} \qquad (4)$$

where $\bar{\theta}$ is the absolute value of the internal angle ($\bar{\theta} < 2\pi$).

The above expressions hold in the case that point P does not lie on the z-axis. If R = 0 we must instead consider these limiting relations

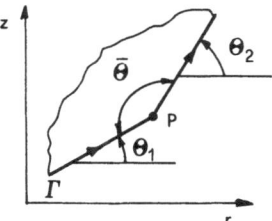

$U_{rr} = U_{rz} = 0$

$$U_{zr} = - \frac{C_2}{a} \frac{\bar{Z}r}{a^2}$$

$$U_{zz} = \frac{C_2}{a} [(3 - 4\nu) + \frac{\bar{Z}^2}{a^2}]$$

Figure 3. Definition of the local geometry

$T_{rr} = T_{rz} = 0$ $\qquad\qquad (5)$

$$T_{zr} = \frac{2C_2}{a^2} \frac{3\bar{Z}r}{a^2} [(n_r\frac{r}{a} - n_z\frac{\bar{Z}}{a}) - (1 - 2\nu)(n_r\frac{\bar{Z}}{a} + n_z\frac{r}{a})]$$

$$T_{zz} = - \frac{2C_2}{a^2} [(1 - 2\nu) + \frac{3\bar{Z}^2}{a^2}](n_r\frac{r}{a} - n_z\frac{\bar{Z}}{a})$$

in which $C_2 = 1/(16\pi G(1 - \nu))$ and $a = \sqrt{(r^2 + \bar{Z}^2)}$.

It is worth noting that if R = 0 (P on the z-axis), the radial components $u_r(P)$ and $t_r(P)$ are obviously zero and only one boundary integral equation (1) can be written, i.e. for i = z. Therefore, the only c_{ij} we need is (Telles[5])

$$c_{zz} = \{(1 - 2\nu)(1 - \cos\bar{\theta}) - \cos^3\bar{\theta} + 1\}/\{4(1 - \nu)\} \qquad (6)$$

where $\bar{\theta}$ is the internal angle between the z-axis and the tangent to the boundary at P. Note that if $\bar{\theta} = \pi/2$, $c_{zz} = 0.5$.

NUMERICAL IMPLEMENTATION

In the formulation here presented, the boundary Γ is represented
piecewise by quadratic shape functions $N^c(\xi)$ ($c = 1,2,3$) of the
intrinsic coordinate ξ, where $-1 \leq \xi \leq 1$ (see, e.g., Watson[6]),
and is divided in q elements Γ_b (each with three nodal points).
Note that the axis must not be discretised. The same quadratic
shape functions are also used to describe the displacement and
the traction in terms of nodal values. With these approximations, discretized forms of the equation (1) are obtained.
Furthermore, for a convenient representation of the boundary
values a normal-tangential-coordinate system, instead of a
global one, is used (Watson[6]; Mayr et al.[7]). The transformed
terms are marked with a prime.
By writing equation (1) for the node P and substituting the
parametric representation, we obtain

$$c'_{ij}(P)u'_j(P) + 2\pi \sum_b \sum_c u'^{b,c}_j \int_{\Gamma_b} T'_{ij}(P,Q)N^c(\xi)r(Q)J(\xi)d\xi =$$
$$= 2\pi \sum_b \sum_c t'^{b,c}_j \int_{\Gamma_b} U'_{ij}(P,Q)N^c(\xi)r(Q)J(\xi)d\xi \qquad (7)$$

where Γ_b is the b-th integration element of the meridian Γ, u'_j
and t'_j are the node values, $Q = Q(\xi)$ and $J(\xi)$ is the Jacobian
$d\Gamma/d\xi$. At this point the procedure is the same one of the
two-dimensional case, except for the computation of the
integrals along the actual boundary elements. The aim of this
paper is to exploit this important aspect.

Computation of Complete
Elliptic Integrals
From the expressions of the
kernel functions it clearly
arises that the first problem
to be solved is the
evaluation of E(k) and K(k)
at any value of the modulus k
between 0 and 1. As is well
known (Erdelyi[8]), both E(k)
and K(k) can be approximated
with a series of even powers
of the modulus k. This
procedure is computationally
useful if the value of k is
not very close to 1 because
the number of the series
terms for an assigned
precision rapidly increases.

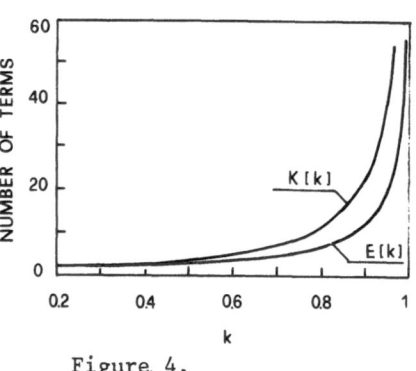

Figure 4.

We suggest the use of this technique in the range $0 < k < 0.99$
for E(k) and $0 < k \leq 0.96$ for K(k). The number of the series
terms to compute the integrals with an error lower than 0.1% is
plotted in Fig. 4. Because of the regular pattern of E(k), the
computation in the range $0.99 < k < 1$ can be easily accomplished

by using Gaussian quadrature formulas of the 5th order. The singular nature of K(k) as k approaches 1 requires some attention. Gaussian quadrature formulas can be successfully applied in the range $0.96 < k < 0.9969$, the order being: 5 if k < 0.97, 6 if $k < 0.987$, 7 if $k < 0.9945$ and 8 if $k < 0.9969$. For $k \geq 0.9969$, K(k) can be expanded as (Erdelyi[8])

$$K(k) = -(1/2k) \ln\{(1-k^2)/(16\ k^2)\}. \tag{8}$$

It is worth noting that to avoid loss of significant digits in the evaluation of $1-k^2$ when k is very close to 1 (and this occurs whenever Q approaches P), the computation should be performed in double precision. Following the previous advices the error is always lower than 0.1%.

Choice of the Order of Quadrature Formulas
The discretized version (7) of equation (1) allows the integrals to be approximated by the sum of the integrals over each element Γ_b. In non singular cases, i.e. when point P is not located on the integration domain, standard Gauss quadrature formulas can be used following the same scheme of 2D problems. Because of the kernel functions involved in axisymmetric problems, the only difference is in the evaluation of the proper order m of these formulas for rough uniform precision of integration without waste of cpu time. In 2D (and 3D) problems the integration order m is generally chosen (see Watson[6]), under certain simplifying assumptions, on the grounds of the 2m-th derivative of the integrands. The final simplified criterion suggests that the order should be chosen according to the product between the length L of the element Γ_b and the maximum value 1/d of a representation of the integrand in the same element (that is a measure of the rapidity of variation, if the integrand function is monotone).

For the axisymmetric formulation of BEM we suggest the application of a similar criterion. Consistently with (8), the function $1/\sqrt{(1-k^2)}$ is taken as a 'simulacrum' of the actual integrands, where k is the modulus of the complete elliptic integrals. It is important to note that

$$\frac{1}{\sqrt{(1 - k^2)}} = \frac{\sqrt{(\bar{Z}^2 + (R+r)^2)}}{d} \tag{9}$$

where $d = \sqrt{(\bar{Z}^2 + (R-r)^2)}$ is the distance between P and Q. Therefore our simulacrum has the same order of singularity of the kernels T_{ij}, but is much simpler. This function plays the same role of 1/d in the Watson criterion, and should be calculated in the node of the element under consideration that is nearest to P. In axisymmetric problems, a measure of the integration domain cannot be the simple length L. To take into account also the distance between the element and the z-axis, we should use, i.e., the ratio L/r_m, where r_m is the distance of the element middle node from the z-axis. Therefore the fundamental parameter for the choice of the integration order m could be the following

$$H = \frac{1}{\sqrt{(1 - k^2)}} \frac{L}{r_m} \qquad (10)$$

The parameter H is now to be related to the order m so as to provide the required criterion. By means of numerical experiments the authors established that a good, simple criterion for the choice of the proper order m of Gaussian formulas in axisymmetric elastic problems is

$$2^m \geq \frac{H}{C} > 2^{(m-1)} \qquad (11)$$

where C is a constant to be fixed in the light of experience. Fig.5 shows the effect of the application of (11), with C = 0.15, in two typical cases. If the ratio L/r_m does not exceed 1 and the adjacent elements are not too different in dimension, the order m is generally less than 5.

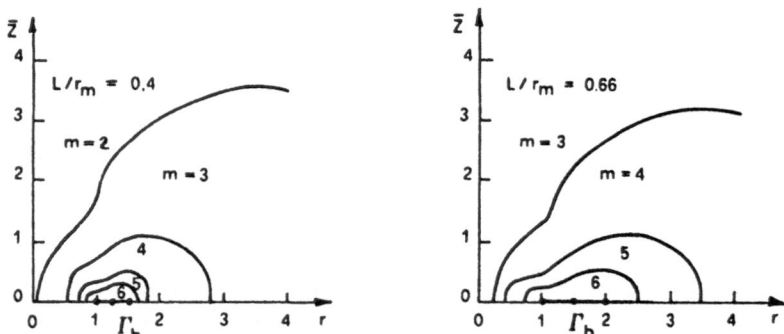

Figure 5. Order m chosen on the grounds of the position of source point P in two typical cases.

Computation of Singular Integrals

In the case $P \in \Gamma_b$ and $R \neq 0$ the kernels U_{ij} and T_{ij} contain singularities of the order $\ln(d)$ and $1/d$ respectively. However if P and the node whose relative values we are dealing with do not coincide, the integrand functions are regular because the shape function $N^c(\xi)$ tends to 0 as Q approaches P. On the contrary, the integrands are singular if P and the node under consideration are in the same position. Note that if $P \in \Gamma_b$ and R = 0, all integrand functions are either regular or need not be integrated because the corresponding nodal quantities are identically zero. For the computation of singular integrals containing U_{ij} one can use logarithmic Gaussian quadrature formulas.

This leaves us with the singular integrals containing $T_{ij}(P,Q)$. As pointed out by other authors, these integrals exist only in the sense of the Cauchy principal value. In 2D and 3D problems it is not necessary to evaluate the p.v. because of the application of the well-known rigid-body movement technique which provides directly the sum of the p.v. and the coefficient c_{ij} of the free term. In axisymmetric problems the rigid movement does not exist in radial direction. Rizzo and Shippy have recently demonstrated that p.v. integrals can be removed from equation (1) by applying kernels associated with ring sources of 'class 1' (namely bending problems, see Mayr et al.[7]). It is also possible to use rigid-body motion in the z direction ($u_r = 0$, $u_z = 1$) and inflating mode in the radial one ($u_r = r$, $u_z = 0$) (Sarihan and Mukherjee[10]).

However, according to our opinion, the application of finite-part integration is the most effective way of evaluation of principal value intergrals in the BEM for axisymmetric elastic problems. Considering the Cauchy principal value integral

$$I = \fint_a^b \frac{g(x)}{(x-s)}\,dx = \lim_{\varepsilon \to 0} \left\{ \int_a^{s-\varepsilon} \frac{g(x)}{(x-s)}\,dx + \int_{s+\varepsilon}^b \frac{g(x)}{(x-s)}\,dx \right\} \quad (12)$$

where $g(x)$ is required to be continuous in an interval containing (a,b) and to be of C^1 in a neighbourhood of s, $a<s<b$. It is known that a p.v. integral can be computed as a sum of two f.p. integrals (see, e.g., Kutt[11]; Brebbia et al.[4])

$$I' = \fint_a^s \frac{g(x)}{(x-s)}\,dx; \quad I'' = \fint_s^b \frac{g(x)}{(x-s)}\,dx; \quad I = I'+I'' \quad (13)$$

where

$$\fint_s^b \frac{g(x)}{(x-s)}\,dx = (\text{Def.}) = g(s)\,\ln|b-s| + \int_s^b \frac{dx}{(x-s)} \int_s^x g'(y)\,dy \quad (14)$$

In (14) reflection or translation of the interval of integration are permitted in the standard form, but scaling to a unit length requires the following expression

$$\fint_s^b \frac{g(x)}{(x-s)}\,dx = g(s)\,\ln|b-s| + \fint_0^1 \frac{g[(b-s)t+s]}{t}\,dt \quad (15)$$

It is important to note that the last term in (14) is a standard integral. Furthermore we found convenient to let

$$\int_s^b \frac{dx}{(x-s)} \int_s^x g'(y)\ dy = \int_s^b \frac{[g(x) - g(s)]}{(x-s)}\ dx \qquad (16)$$

Our aim is the application of (14), (15) and (16) in order to numerically compute the p.v. of singular integrals such as

$$2\pi \int_{\Gamma_b} T'_{ij}(P,Q)\ N^c(\xi)\ r(Q)\ d\Gamma \qquad (17)$$

where $N^c(\xi)$ equals unity at P. The first step to numerical integration is generally letting $Q = Q(\xi)$ and $d\Gamma = J(\xi)d\xi$, so that the integral on the curve Γ_b becomes an integral on (-1,1). In our case the new integrand function has a singularity of order $1/d$, $d = |P-Q|$, for $\xi = 1,0,-1$ depending on the position of P ($c = 1,2,3$ respectively). If the element is divided into two subelements (see Fig.6), only one or both subelements contain the singular point P, depending on whether P is located at one extremity or in the middle of the element. If the singularity does not occur at the first extremum of the subinterval, as in (14) and (15), only a reflection is required. The integration over non-containing P subelements is carried out by standard Gauss formulas.

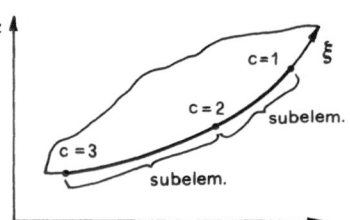

Figure 6. Subelements for integration.

For all three positions of P, the problem is now reduced to a single (c=1,3) or double (c=2) application of the last part of expression (15). By applying to it first the definition (14) and then (16), we obtain the important result that computation of this f.p. integral only requires the computation of a standard integral. To summarize the procedure, let us consider, for instance, the case $c = 1$ (i.e. $s = 1$)

$$\fint_0^1 T'_{ij}(P,Q(\xi))N^1(\xi)r(\xi)J(\xi)d\xi = \fint_0^1 f(\xi)\ d\xi = \fint_0^1 \frac{g(\xi)}{\xi - 1}\ d\xi =$$

$$= \int_0^1 \{f(\xi) - \frac{g(1)}{\xi - 1}\}\ d\xi \cong \frac{1}{2} \sum_{i=1}^m \{f(\frac{1+n_i}{2}) - \frac{g(1)}{(1+n_i)/2 - 1}\}\ w_i \qquad (18)$$

where n_i and w_i are the coordinates and weights of the <u>standard</u> Gaussian quadrature formula of order m (the sixth order is generally appropriate). Note that the last summation in (18) can even be avoided because the application of the analogous relation on the adjacent subelement provides a corresponding term of opposite value.

We shall now consider the logarithmic term in (15). Our aim is in fact the computation of a f.p. integral over a boundary element (see (17)), and this is not equivalent to the computation on the ξ interval (as in (18)). If the boundary element is a straight line of constant Jacobian $J(\xi)$, the coordinate transformation from Γ_b to ξ is a scaling. Thus the logarithmic term in (15) is equal to $g(s)\ln(J)$. For the (approximate) computation of $g(s)$ we suggest the calculation in double precision arithmetic of $f(\xi')(\xi'- s)$, with $|\xi'- s| = 0.00001$, so as to avoid the rather complicated and unnecessary task of isolating $g(\xi)$ from $f(\xi)$. However quadratic shape functions allow the use of curved and linear shifted-node elements of non-constant Jacobian. In this case the coordinate transformation is not a scaling and the application of (15) is not straightforward. Nevertheless we found that a formal equivalent expression can be used, but with the difference that the Jacobian must now be calculated in the singular abscissa s: the term $g(s)\ln(J(s))$ replaces the corresponding one of (15). Note that the computation of this term is unnecessary for c = 2. This method is based on the consideration that

$$\fint_s^b \frac{g(x)}{(x - s)}\, dx = \fint_s^{s+\varepsilon} + \int_{s+\varepsilon}^b \overset{\sim}{=} g(s)\ln|\varepsilon| + \int_{s+\varepsilon}^b \frac{g(x)}{(x - s)}\, dx \qquad (19)$$

for sufficiently small values of ε .

Therefore, the method we propose allows the direct computation of p.v. integrals, even in distorted quadratic boundary elements, by means of standard Gaussian formulas plus an additional logarithmic term. To our best knowledge, this procedure appears to be the most effective and efficient for axisymmetric elastic problems. Furthermore it could also be applied in 2D cases. A check with a similar procedure employing Kutt's formulas (Brebbia el al.[4]) showed no improvement in result accuracy.

EXAMPLES

In this section we present two examples, a hollow and a solid body, in order to verify the described integration schemes. Namely, we consider a tube under internal pressure and the classical circumferentially grooved bar (see also Shippy and Rizzo[12]). Isoparametric quadratic elements were used for both problems. The material properties were: Poisson's ratio $\nu = 0.3$ and Young's modulus $E = 210,000$ N/mm^2.

Tube under internal pressure
Fig. 8 shows the model employed for the stress analysis by the BEM of a tube of infinite length. The specified boundary conditions were $t_r =$ const. on the inner surface, $u_z = 0$ on the upper and lower side and zero traction otherwise. All calculated tractions agree with the exact value to within 1.7%, and all displacements show an error always lower than 0.4%. The cpu time was 4.5 s.

Grooved bar

Fig. 7 shows the BEM model used for the stress concentration
analysis of a grooved bar. Owing to symmetry, it was necessary
to discretise only half a generator of the bar. The mesh
employed 14 elements (the z-axis need not be discretised). The
specified boundary conditions (pure tension) were t_z = const. on
$z = 60$, $u_z = 0$ on $z = 0$, and zero traction otherwise. The
calculated stress concentration factor was equal to 2.16 that
differs from the exact value (2.18) of 0.9%. The cpu time was
6 s.

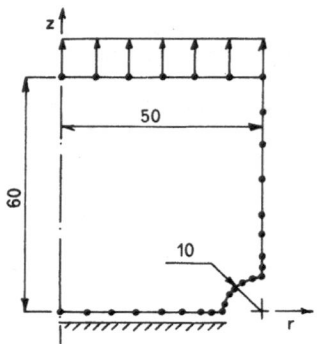

Figure 7. Grooved bar   Figure 8. Tube under
                                     internal pressure

Acknowledgements   The support by M.P.I. is gratefully
acknowledged.

REFERENCES

1.   Mayr M. (1976), On the Numerical Solution of Axisymmetric
     Elasticity Problems Using an Integral Equation Approach,
     Mech. Res. Comm., Vol.3, pp.393-398.

2.   Cruse T.A., Snow D.W. and Wilson R.B. (1977), Numerical
     Solutions in Axisymmetric Elasticity, Computers and
     Structures, Vol.7, pp.445-451.

3.   Kermanidis T. (1975), A Numerical Solution for Axially
     Symmetrical Elasticity Problems, Int. J. Solids Structures,
     Vol.11, pp.493-500.

4.   Brebbia C.A., Telles J.C.F. and Wrobel L.C. (1984), Boundary
     Elements Techniques, Springer-Verlag, Berlin and New York.

5.   Telles J.C.F. (1983), A Boundary Element Formulation for
     Axisymmetric Plasticity, in Boundary Elements (Ed. Brebbia
     C.A., Futagami T. and Tanaka M.), pp.577-588, Proceedings of
     the 5th Int. Conf., Hiroshima, Japan, 1983, Springer-Verlag,
     Berlin and New York.

6. Watson J.O. (1979), Advanced Implementation of the Boundary Element Method for Two- and Three-dimensional Elastostatic. Chapter 3, Developments in Boundary Element Method-1, (Eds. Banerjee P.K. and Butterfield P.), pp.31-63, Applied Science Publishers, London.

7. Mayr M., Drexler W. and Kuhn G. (1980), A Semianalytical Boundary Integral Approach for Axisymmetric Elastic Bodies with Arbitrary Boundary Conditions, Int. J. Solids Structures, Vol.16, pp.863-871.

8. Erdelyi A. et al. (1953), Higher Trascendental Functions, Vol.1, Bateman Manuscript Project, MacGraw-Hill, New York.

9. Rizzo F.J. and Shippy D.J. (1986), A Boundary Element Method for Axisymmetric Elastic Bodies. Chapter 3, Developments in Boundary Element Method-4, (Eds. Banerjee P.K. and Watson J.O.), pp.67-90, Elsevier Applied Science Publishers, Barking.

10. Sarihan V. and Mukherjee S. (1982), Axisymmetric Viscoplastic Deformation by the Boundary Element Method, Int. J. Solids Structures, Vol.18, pp.1113-1128.

11. Kutt H.R. (1975), The Numerical Evaluation of Principal Value Integrals by Finite-Part Integration, Numer. Math., Vol.24, pp.205-210.

12. Shippy D.J. and Rizzo F.J. (1982), On the Effectiveness of Three Boundary Integral Equation Formulations for Certain Axisymmetric Elastostatic Problems, Res Mechanica, Vol.4, pp.43-56.

APPENDIX

Derivatives of displacement kernels:

$$\frac{\partial U_{rr}}{\partial r} = -\frac{1}{2r} U_{rr} - \frac{C_1}{\sqrt{Rr}} \frac{\overline{Z}^2}{2Rr^2} k \left[K(k) - \frac{1}{2} \frac{2-k^2}{1-k^2} E(k)\right] + \frac{\partial U_{rr}}{\partial k} \frac{k}{2r} (1 - \frac{R+r}{2r} k^2);$$

$$\frac{\partial U_{rr}}{\partial z} = -\frac{C_1}{\sqrt{Rr}} \frac{\overline{Z}}{Rr} k \left[K(k) - \frac{1}{2} \frac{2-k^2}{1-k^2} E(k)\right] + \frac{\partial U_{rr}}{\partial k} \frac{\overline{Z}}{4Rr} k^3;$$

where:

$$\frac{\partial U_{rr}}{\partial k} = \frac{C_1}{\sqrt{Rr}} \{ \left[- (3-4\nu) \frac{k^2+2}{k^2} + \frac{\overline{Z}^2}{2Rr}\right] K(k) + \left[(3-4\nu) \frac{2-k^2}{k} + \frac{\overline{Z}^2}{2Rr} k\right] \frac{dK(k)}{dk} +$$

$$+ \left[(3-4\nu) \frac{2}{k^2} - \frac{\overline{Z}^2}{4Rr} \frac{2-k^2+k^4}{(1-k^2)^2}\right] E(k) - \left[(3-4\nu) \frac{2}{k} + k \frac{\overline{Z}^2}{4Rr} \frac{2-k^2}{1-k^2}\right] \frac{d E(k)}{dk} \}.$$

$$\frac{\partial U_{rz}}{\partial r} = -\frac{3}{2} \frac{1}{r} U_{rz} + \frac{C_1}{\sqrt{Rr}} \frac{\overline{Z}}{2Rr} k \left[K(k) - \frac{1}{2} \frac{2-k^2}{1-k^2} E(k)\right] + \frac{\partial U_{rz}}{\partial k} \frac{k}{2r} (1 - \frac{R+r}{2R} k^2);$$

$$\frac{\partial U_{rz}}{\partial z} = -\frac{C_1}{\sqrt{Rr}} \frac{1}{r} \{\frac{r}{R} \frac{k}{2} K(k) + \left[\frac{k}{4} \frac{2-k^2}{1-k^2} (\frac{2}{k^2} - 1 - \frac{r}{R}) - \frac{1}{k}\right] E(k)\} + \frac{\partial U_{rz}}{\partial k} \frac{\overline{Z}}{4Rr} k^3;$$

where:

$$\frac{\partial U_{rz}}{\partial k} = \frac{C_1}{\sqrt{Rr}} \frac{\overline{Z}}{r} \{\frac{1}{2} \frac{r}{R} \left[K(k) + k \frac{dK(k)}{dk}\right] + \left[\frac{k}{4} \frac{2-k^2}{1-k^2} (\frac{2}{k^2} - 1 - \frac{r}{R}) - \frac{1}{k}\right] \frac{dE(k)}{dk} +$$

$$+ \frac{1}{1-k^2} \left[\frac{1}{4} \frac{2-k^2+k^4}{1-k^2} (\frac{2}{k^2} - 1 - \frac{r}{R}) - \frac{1}{k^2}\right] E(k)\}.$$

$$\frac{\partial U_{zr}}{\partial r} = -\frac{3}{2} \frac{1}{r} U_{zr} - \frac{C_1}{\sqrt{Rr}} \frac{\overline{Z}}{4Rr} \frac{k^3}{1-k^2} E(k) + \frac{\partial U_{zr}}{\partial k} \frac{k}{2r} (1 - \frac{R+r}{2R} k^2) ;$$

$$\frac{\partial U_{zr}}{\partial z} = \frac{C_1}{\sqrt{Rr}} \frac{k}{2r} \{ K(k) - (\frac{2}{k^2} - 1 - \frac{r}{R}) \frac{1}{2} \frac{k^2}{1-k^2} E(k)\} + \frac{\partial U_{zr}}{\partial k} \frac{\overline{Z}}{4Rr} k^3;$$

where:

$$\frac{\partial U_{zr}}{\partial k} = \frac{-C_1}{\sqrt{Rr}} \frac{\overline{Z}}{r} \{\frac{1}{2} \left[K(k) + k \frac{dK(k)}{dk}\right] - (\frac{2}{k^2} - 1 - \frac{r}{R}) \frac{1}{4} \frac{k^3}{1-k^2} \frac{dE(k)}{dk} +$$

$$+ \frac{1}{1-k^2} \left[1 - \frac{k^2}{4} \frac{3-k^2}{2-k^2} (\frac{2}{k^2} - 1 - \frac{r}{R})\right] E(k)\} .$$

$$\frac{\partial U_{zz}}{\partial r} = -\frac{1}{2r} U_{zz} - \frac{C_1}{\sqrt{Rr}} \frac{\overline{Z}^2}{4Rr^2} \frac{k^3}{1-k^2} E(k) + \frac{\partial U_{zz}}{\partial k} \frac{k}{2r} (1 - \frac{R+r}{2r} k^2) ;$$

$$\frac{\partial U_{zz}}{\partial z} = -\frac{C_1}{\sqrt{Rr}} \frac{\overline{Z}^2}{2Rr} \frac{k^3}{1-k^2} E(k) + \frac{\partial U_{zz}}{\partial z} \frac{\overline{Z}}{4Rr} k^3;$$

where:

$$\frac{\partial U_{zz}}{\partial k} = \frac{C_1}{\sqrt{Rr}} \{(3-4\nu) \left[K(k) + k \frac{dK(k)}{dk}\right] + \frac{\overline{Z}^2}{4Rr} \frac{k^2}{1-k^2} \left[\frac{3-k^2}{1-k^2} E(k) + k \frac{dE(k)}{dk}\right]\} .$$

$$C_1 = \frac{1}{16\pi^2 G(1 - \nu)}$$

An Indirect Boundary Element Method in the Solution of the
Diffusion Equation
J. Zhu
Chongqing Institute of Architecture and Engineering, Chongqing, Sichuan, China

INTRODUCTION

The indirect boundary element method based on potential
theory constructs the solution of the problem in terms
of some potential functions. Although the representations of the solution are different in form from those
obtained by direct method, it's easy to bring out the
connections between them. For some problems, the indirect method has certain advantages. This method, which
has been widely used for numerical solution to elliptic partial differential equations, can be extended to
solve the linear parabolic differential equations.
In this paper, a potential approach is proposed for
the diffusion equation. The potentials are deduced by
considering both interior and its complementary exterior problems via Green's formula. For the problem
with Dirichlet boundary condition the solution of both
interior and exterior problems may be represented by a
simple layer potential. Since this kind of potential
is continuous, the equation for the unknown which is
the jump of flux through the boundary is an integral
equation of the first kind. It can be transformed into
an equivalent variational problem. Then, a finite
element discretization scheme for the spatial domain
and time region is applied to this variational formulation in which the unknowns are on the boundary only.
The problem with Neumann boundary condition can be
solved in terms of double layer potential in a similar
manner.
The variational approach of treating the integral
equations is rather involved since the coefficients
are computed through a time consuming double integrations, nevertheless, it offers a way of dealing with
problems with non-integrable singularities, its main

advantage is that the error analysis can be developed along the lines of finite element method for variational problems in Sobolev spaces.

BIE FORMULATION

Let Ω be an open bounded domain in R^2 with sufficient smooth boundary Γ. the complement of $\bar{\Omega}=\Gamma+\Omega$ is denoted by Ω'. Let t varies on a finite or infinite interval I: $0 < t < T \leqslant +\infty$. We consider the initial-boundary value problem for the diffusion equation with the Dirichlet boundary condition over Ω and Ω' simultaneously

$$(D) \begin{cases} \dfrac{\partial u(x,t)}{\partial t} = \Delta u(x,t) & (x,t)\epsilon(\Omega\cup\Omega')\times I \qquad (1) \\[2mm] u(x,t) = g(x,t) & (x,t)\epsilon\,\Gamma\times I \qquad (2) \\[2mm] u(x,0) = u_o(x) & x \epsilon \Omega \qquad (3) \\[2mm] u(x,0) = 0 & x \epsilon \Omega' \qquad (4) \end{cases}$$

The fundamental solution of the parabolic operator $\partial/\partial t - \Delta$ in $R^2\times I$ is of the form

$$u*(x,t;\zeta,\tau) = \frac{H(t-\tau)}{4\pi(t-\tau)}\exp(-\frac{|x-\zeta|^2}{4(t-\tau)}) \qquad (5)$$

where the Heaviside function $H(t-\tau)$ is included to emphasize the fact that u* is identically zero for $t<\tau$, u* satisfies

$$\frac{\partial u*}{\partial t} = \Delta u* \qquad \frac{\partial u*}{\partial \tau} = -\Delta u* \qquad \lim_{\tau\to t} u* = \delta(x-\zeta) \qquad (6)$$

therefore we have

$$\frac{\partial}{\partial \tau}(u\,u*) = u\,\frac{\partial u*}{\partial \tau} + u*\frac{\partial u}{\partial \tau} = u*\,\Delta u - u\,\Delta u*$$

Integrating this formula over the domain Ω and on the time region I, applying Green's theorem to the right-hand side, we obtain

$$\int_I\int_\Omega\frac{\partial}{\partial \tau}(u\,u*)d\zeta\,d\tau = \int_I\int_\Gamma(u*\,\frac{\partial u^-}{\partial n} - u^-\frac{\partial u*}{\partial n})d\zeta\,d\tau \qquad (7)$$

where n denotes the unitary exterior normal to Γ and the minus sign signifies that $x\to\zeta$ from Ω whilst the plus sign will signify that $x\to\zeta$ from Ω' . Similarly,

$$\int_I \int_\Omega \frac{\partial}{\partial \tau}(u\ u^*)d\zeta\, d\tau = -\int_I \int_\Gamma (u^* \frac{\partial u^+}{\partial n} - u^+ \frac{\partial u^*}{\partial n})ds_\zeta d\tau \qquad (8)$$

Adding each side of the two formulas above and interchanging the order of integrations on the left-hand side, we obtain from (2), (3), (4), (6) the integral representation of the solution of both interior and exterior problems

$$u(x,t)=\int_\Omega u^*(x,t;\zeta,0)u_o(\zeta)d\zeta + \int_I \int_\Gamma u^*(x,t;\zeta,\tau)q(\zeta,\tau)ds_\zeta\, d\tau$$

$$(x,t)\epsilon(\Omega\cup\dot\Omega)\times I \qquad (9)$$

where $u^- - u^+ = 0$, $q = \frac{\partial u^-}{\partial n} - \frac{\partial u^+}{\partial n}$ is the intermediate unknowns we are going to find.

Obviously this is an indirect formulation, It involves just one boundary integral in contrast to two for the direct formulation in the case of Dirichlet boundary conditions. Since u is continuouse through Γ, the following integral equation is obtained

$$g(x,t)=\int_\Omega u^*(x,t;\zeta,0)u_o(\zeta)d\zeta + \int_I \int_\Gamma u^*(x,t;\zeta,\tau)q(\zeta,\tau)ds_\zeta\, d\tau$$

$$(x,t)\epsilon\Gamma\times I \qquad (10)$$

which defines a continuous mapping $g\to q$ from $L^2(0,T; H^{\frac{1}{2}}(\Gamma))$ into $L^2(0,T; H^{-\frac{1}{2}}(\Gamma))$, $H^{\frac{1}{2}}(\Gamma)$ is the trace space on the boundary of the Sobolev space $H^1(\Omega)$ or $H^1(\Omega')$, its dual is $H^{-\frac{1}{2}}(\Gamma)$. The notation $L^2(0,T; H)$ denotes the functions $u(\cdot,t)$ taking values in Hilbert space H are square integrable for time t.

The discretization of Equation(10) can be done by the collocation method, but we present here a variational method.

VARIATIONAL FORMULATION AND DISCRETIZATION

Consider Equation(10) in the distributional sense, it has the following equivalent variational formulation :

For $u\epsilon L^2(\Omega)$, $g\epsilon L^2(0,T;H^{\frac{1}{2}}(\Gamma))$, find $q\epsilon L^2(0,T;H^{-\frac{1}{2}}(\Gamma))$ such that $\forall\ p\epsilon L^2(0,T;H^{-\frac{1}{2}}(\Gamma))$

$$a(q,p) = \int_0^T\int_\Gamma g(x,t)p(x,t)ds_x\, dt$$

$$-\int_0^T\int_\Gamma\int_\Omega u_o(\zeta)u^*(x,t;\zeta,0)p(x,t)d\zeta\, ds_x\, dt \qquad (11)$$

We denote by $a(q,p)$ the bilinear form

$$a(q,p) = \int_0^T \int_0^t \int_\Gamma \int_\Gamma q(\zeta,\tau)p(x,t)u^*(x,t;\zeta,\tau)ds_\zeta \, ds_x \, d\tau dt \quad (12)$$

The unique solvability of the variational problem(11) can be proved by the generalized Lax-Milgram theorem. We now present the numerical discretization scheme. The boundary Γ is discretized into N boundary elements Γ_i (i=1,$\cdots$,N), the domain Ω is divided into M cells e_m (m=1,$\cdots$,M) for the purpose of performing the domain integration. The time region(0,T) is divided into K time steps $\Delta t = T/K$ with $t_k = k\Delta t$ (k=1,$\cdots$,K).

Since q,p belong to $L^2(0,T;H^{-\frac{1}{2}}(\Gamma))$, we can assume that they are piecewise constant over each boundary element and on each time step. Then $L^2(0,T;H^{-\frac{1}{2}}(\Gamma))$ is approximated by a finite dimensional subspace $V_h = S_h \times T_h$

$$\begin{aligned} S_h &= \text{span}\{\,\Phi_i(x),\ i=1,\cdots,N\} \subset H^{-\frac{1}{2}}(\Gamma) \\ T_h &= \text{span}\,\{\,\theta_k(t),\ k=1,\cdots,K\} \subset L^2(0,T) \end{aligned} \quad (13)$$

with basis functions

$$\Phi_i(x) = \begin{cases} 1 & x \in \text{boundary element } \Gamma_i \\ 0 & \text{otherwise} \end{cases}$$

$$\theta_k(t) = \begin{cases} 1 & t \in (t_{l-1},\ t_l] \\ 0 & \text{otherwise} \end{cases} \quad (14)$$

The approximate solution $q(x,t) \in V_h$ of the variational problem(11) can be represented as the linear combination

$$q(x,t) = \sum_{j=1}^N \sum_{l=1}^K q_{jl}\Phi_j(x)\,\theta_l(t) \quad (15)$$

We have the discretized form of the variational problem

$$\sum_{j=1}^N \sum_{l=1}^K G_{ijkl}\,q_{jl} = F_{ik} \qquad \begin{matrix} i=1, & ,N \\ k=1, & ,K \end{matrix} \quad (16)$$

where

$$G_{ijkk} = \int_{t_{k-1}}^{t_k} \int_{t_{k-1}}^t \int_{\Gamma_i} \int_{\Gamma_j} u^*(x,t;\zeta,\tau)ds_\zeta \, ds_x \, d\tau dt \quad (17)$$

$$G_{ijkl} = \int_{t_{k-1}}^{t_k} \int_{t_{l-1}}^{t_l} \int_{\Gamma_i} \int_{\Gamma_j} u^*(x,t;\mathfrak{z},\tau)\,ds_{\mathfrak{z}}\,ds_x\,d\tau dt \qquad (18)$$

$$l = 1, \cdots, k-1$$

$$F_{ik} = \int_{t_{k-1}}^{t_k} \int_{\Gamma_i} g(x,t)\,ds_x\,dt$$

$$+ \int_{t_{k-1}}^{t_k} \int_{\Gamma_i} \int_{\Omega} u_o(\mathfrak{z})u^*(x,t;\mathfrak{z},0)\,d\mathfrak{z}\,ds_x\,dt \qquad (19)$$

We use a time marching process. The value q_{jk} at time t_k which is a vector with N components ($j=1,\cdots,N$) is obtained step by step by use of the already evaluated quantities q_{jl} ($l=1,\cdots,k-1$).

$$\sum_{j=1}^{N} G_{ijkk}\,q_{jk} = -\sum_{j=1}^{N} \sum_{l=1}^{k-1} G_{ijkl}\,q_{jl} + F_{ik} \qquad (20)$$

$$i = 1, \cdots, N \qquad k = 1, \cdots, K$$

The time integrals in (17), (18) can be carried out analytically

$$G_{ijkk} = -\frac{1}{4\pi} \int_{\Gamma_i} \int_{\Gamma_j} \left\{ (\Delta t + \frac{r^2}{4})\mathrm{Ei}(-\frac{r^2}{4\Delta t}) + \Delta t\,\exp(-\frac{r^2}{4\Delta t}) \right\} ds_{\mathfrak{z}}\,ds_x$$

$$(21)$$

$$G_{ijkl} = \frac{1}{4\pi} \int_{\Gamma_i} \int_{\Gamma_j} \left\{ (2(k-l)\Delta t + \frac{r^2}{2})\mathrm{Ei}(-\frac{r^2}{4(k-l)\Delta t}) \right.$$

$$- ((k-l-1)\Delta t + \frac{r^2}{4})\mathrm{Ei}(-\frac{r^2}{4(k-l-1)\Delta t})$$

$$- ((k-l+1)\Delta t + \frac{r^2}{4})\mathrm{Ei}(-\frac{r^2}{4(k-l+1)\Delta t})$$

$$+ 2(k-l)\Delta t\,\exp(-\frac{r^2}{4(k-l)\Delta t}) \qquad (22)$$

$$- (k-l-1)\Delta t\,\exp(-\frac{r^2}{4(k-l-1)\Delta t})$$

$$\left. - (k-l+1)\Delta t\,\exp(-\frac{r^2}{4(k-l+1)\Delta t}) \right\} ds_{\mathfrak{z}}\,ds_x$$

in which $r = |x-\mathfrak{z}|$ and $\mathrm{Ei}(\cdot)$ is the exponential integral function. When $j=i$, $i-1$, $i+1$, the diagonal coefficients G_{ijkk} contain integrals with a logarithmic singularity, They are integrable, we integrate them analytically. The rest of the coefficients are evaluated by using a standard Gaussian quadrature. It's worth noting, the expression(21) of G_{ijkk} is independent of k. Once G_{ij11} is calculated for the first time step, it can be kept in storage for the use of the following time steps. The matrix G_{ijkl} depends on the quantities $(k-l)\Delta t$. Only one new matrix G_{ijk1} need to be evaluated and to be kept in storage at each time step t_k for the other matrices

$$G_{ijkl} = G_{ij(k-l+1)1} \qquad l=1,\cdots,k-1$$

have been already obtained from previous time step, which will multiply the calculated values of q_{jl} to form the new right-hand side term of the linear system(20). Finally, the approximate solution $u(x,t)$ of the problem(D) at time t_k will be evaluated by the following expression

$$u(x,t_k) = \sum_{m=1}^{M} \int_{e_m} u^*(x,t_k;\zeta,0)u_o(\zeta)d\zeta$$

$$+ \frac{1}{4\pi}\sum_{j=1}^{N}\sum_{l=1}^{K} q_{jl} \cdot \int_{\Gamma_j}\left\{ \text{Ei}(-\frac{r^2}{4(k-l)\Delta t}) - \text{Ei}(-\frac{r^2}{4(k-l+1)\Delta t})\right\} ds_\zeta$$

$$(23)$$

FORMULATION FOR NEUMANN PROBLEM

Consider the problem with Neumann boundary condition

$$(N)\quad\begin{cases} \dfrac{\partial u(x,t)}{\partial t} = \Delta u(x,t) & (x,t)\in(\Omega\cup\Omega')\times I \\[2mm] \dfrac{\partial u(x,t)}{\partial n} = \overline{q}(x,t) & (x,t)\in\Gamma\times I \\[2mm] u(x,0) = u_o(x) & x\in\Omega \\[2mm] u(x,0) = 0 & x\in\Omega' \end{cases}$$

The solution of this problem admits a double layer representation of density $\mu = u^- - u^+$

$$u(x,t)=\int_{\Omega} u^*(x,t;\zeta,0)u_o(\zeta)d\zeta - \int_{\Gamma}\mu(\zeta,\tau)\frac{\partial u^*}{\partial n} ds_\zeta d\tau \quad (24)$$

in which

$$\frac{\partial u^*(x,t;\mathfrak{z},\tau)}{\partial n_\mathfrak{z}} = \frac{(r \cdot n_\mathfrak{z})}{8\pi(t-\tau)^2} \exp\left(-\frac{r^2}{4(t-\tau)}\right) \qquad (25)$$

and $\quad (r \cdot n_\mathfrak{z}) = (x_1-\mathfrak{z}_1)n_1(\mathfrak{z}) + (x_2-\mathfrak{z}_2)n_2(\mathfrak{z}) \qquad (26)$

The integral equation liking μ and $\bar{q}$ is then

$$\bar{q}(x,t) = \int_\Omega \frac{\partial u^*(x,t;\mathfrak{z},0)}{\partial n_x} u_0(\mathfrak{z})d\mathfrak{z}$$

$$- \int_0^T \int_\Gamma \mu(\mathfrak{z},\tau) \frac{\partial^2 u^*}{\partial n_x \partial n_\mathfrak{z}}(x,t;\mathfrak{z},\tau)ds_\mathfrak{z}\,d\tau \qquad (27)$$

The normal derivative above is not authentic integral operator but a pseudodifferential operator. Equation (27) is a finite part expression of divergent integral in the sense of the theory of distributions. We have for the distribution kernel

$$\frac{\partial^2 u^*(x,t;\mathfrak{z},\tau)}{\partial n_x \partial n_\mathfrak{z}} = \frac{(n_x \cdot n_\mathfrak{z})}{8\pi(t-\tau)^2} \exp\left(-\frac{r^2}{4(t-\tau)}\right)$$
$$- \frac{(r \cdot n_\mathfrak{z})(r \cdot n_x)}{16\pi(t-\tau)^3} \exp\left(-\frac{r^2}{4(t-\tau)}\right) \qquad (28)$$

Equation(27) has the following variational expression

Find $\mu \in L^2(0,T; H^{\frac{1}{2}}(\Gamma))$, such that $\forall \nu \in L^2(0,T; H^{\frac{1}{2}}(\Gamma))$

$$b(\mu,\nu) = \int_0^T \int_\Gamma \bar{q}(x,t)\,\nu(x,t)dxdt$$
$$- \int_0^T \int_\Gamma \int_\Omega u_0(x)\frac{\partial u^*(x,t;\mathfrak{z},0)}{\partial n_x}\,\nu(x,t)d\mathfrak{z}\,ds_x\,dt \qquad (29)$$

The approximation of the variational problem(29) using finite element techniques is similar to (11). Although the integrals of the diagonal coefficients

$$\int_{t_{k-1}}^{t_k} \int_{t_{k-1}}^t \int_\Gamma \int_\Omega \frac{\partial^2 u^*}{\partial n_x \partial n_\mathfrak{z}}(x,t;\mathfrak{z},\tau)ds_\mathfrak{z}\,ds_x\,d\tau dt \qquad (30)$$

Contain strong singularities, in the case of constant (or linear) elements, the expression(28) can be simplified due to the orthogonality between r and n, so that

$$(n_x \cdot n_\mathfrak{z}) = 1 \qquad (r \cdot n_x) = (r \cdot n_\mathfrak{z}) = 0 \qquad (31)$$

Then the evalution of integral(30) can be carried out in closed form with respect to time first, and to space secondly.

REFERENCES

1. Brebbia, C.A., Telles, J. and Wrobel, L.(1984). Boundary Element Techniques, Springer-Verlag.

2. Friedman, A. (1964). Partial Differential Equations of Parabolic Type. Prentice-Hall, Inc.

3. Lions, J.E., Magenes, E. (1972). Non-Homogeneous Boundary Value Problems and Applications II. Springer-Verlage.

4. Nedelec, J.C.(1982). Integral Equations with Non Integrable Kernels, Integral Equations and Operator Theory, Vol.5, pp.562-572.

5. Sgallari, F. (1985). A Weak Formulation of boundary integral equations for time dependent parabolic problems, Appl. Math. Modelling, Vol.9 pp. 295-301.

6. Tanaka, M., Tanaka, K.,(1980). Transient Heat Conduction Problems in Inhomogeneous Media Discretized by Means of Boundary-volume Element, Nuclear Engineering and Design 60, pp.381-387.

The Usefulness and the Limit of Direct Regular Method in Boundary Element Elastostatic Analysis

R. Yuuki, T. Matsumoto
Institute of Industrial Science, University of Tokyo, 22-1, Roppongi 7 Chome, Minato-ku, Tokyo, Japan
H. Kisu
Faculty of Engineering, Nagasaki University, Bunkyo-cho, Nagasaki, Japan

INTRODUCTION

A Boundary Element Method(BEM) has been developed as a new efficient numerical analysis method, and is applied widely to various fields of engineering problems[1,2] these days. The mainstream of them is the Direct (Singular) Method, in which a boundary integral equation is discretized directly. Very accurate results have come to be obtained by introducing the sophisticated discretization techniques of FEM[3]. In the Direct Method, a special care must be taken to carry out the singular integral, since both the load point and the object point of the fundamental solution are located on the boundary. Therefore much computation time is necessary for the numerical integration.

On the other hand, a new method is proposed recently by Patterson et al.[4,5], in which the singular integral can be avoided by locating the load point outside the domain. It is called a Direct Regular Method in contrast with the usual Direct (Singular) Method.

Some characteristics of this method are as follows;

(1) Singular integral can be avoided by locating the load point of the fundamental solution outside the domain and therefore the computing time of numerical integration can be reduced.

(2) No unknown freedoms are generated on the load point outside the boundary, so that both the number and the arrangement of the load points can be taken arbitrarily in theory.

(3) From the above point of view, we can obtain the simultaneous equations as many as necessary in no connection with the

number of nodes on the boundary.

(4) Besides putting the load point outside the domain, this method is the same as the usual Direct Method and therefore we can use the Direct Method programme without any change.

Applying the above merits, it may be expected to overcome some difficulties on numerical analysis and to make BEM more efficient. Patterson et al. applied this method to fluid flow problems[4] and elastostatic problems[5] and Honma et al. did to convected diffusion problems[6] and plasma problems[7]. But they confirmed the effectiveness of the method only in the case of very simple boundary value problems and not clarified whether it is also applicable effectively to the problem under complicated boundary conditions which is important in the practical case. It is not also clarified how the load points should be located, relevant to the above (2). Furthermore, no study can be found, noticing the merits (3), (4) mentioned above.

In the present paper, the Direct Regular Method based on the Somigliana internal reciprocal formula is applied to two-dimensional elastostatic problems. We investigate systematically both the accuracy and the convergence of solutions relating to the location of the load points and examine whether this method is applicable effectively to the intensive stress concentration problems. Furthermore, we propose a new method using the Direct Regular Method combined with the usual Direct Method, considering the merits of above-mentioned (3) and (4).

BASIC FORMULATIONS

We consider a boundary value problem as shown in Fig.1. Here Ω is the elastic region to be solved, Ω^* is the region outside Ω and Γ is the boundary of Ω. Locating the point p in the region Ω, following Somigliana's identity[8] is obtained from Betti's reciprocal theory.

$$u_i(p) + \int_\Gamma T_{ij}(p,Q)u_j(Q)d\Gamma = \int_\Gamma U_{ij}(p,Q)t_j(Q)d\Gamma$$

$$+ \int_\Omega U_{ij}(p,q)b_j(q)d\Omega$$

$(p,q \in \Omega, Q \in \Gamma)$ (1)
where $U_{ij}(p,Q)$, $T_{ij}(p,Q)$ are Kelvin's fundamental solutions and $b_j(q)$ is a body force vector given. After the displacement $u_j(Q)$ and the traction $t_j(Q)$ of all points on the boundary have been obtained, the displacement of the arbitrary point p in Ω is calculated by eq.(1).

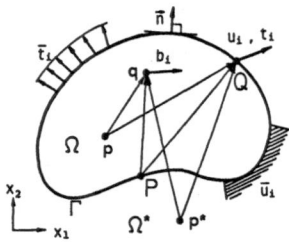

Fig.1 Boundary value problem for a elastic domain.

Taking the appropriate limit of $p \in \Omega$ to the boundary point $P \in \Gamma$, eq. (1) can be deduced to Somigliana's boundary identity as follows,

$$C_{ij} u_j(P) + \int_\Gamma T_{ij}(P,Q) u_j(Q) d\Gamma = \int_\Gamma U_{ij}(P,Q) t_j(Q) d\Gamma$$

$$+ \int_\Omega U_{ij}(P,q) b_j(q) d\Omega \qquad (P,Q \in \Gamma, \ q \in \Omega) \qquad (2)$$

Eq. (2) is the basis of the Direct Method. Because of the singularity of the kernel, the integral of eq. (2) becomes a singular integral.

Furthermore, moving P to the point $p^* \in \Omega^*$ out of Γ, the following internal reciprocal formula is obtained because C_{ij} in eq. (2) comes to be zero.

$$\int_\Gamma T_{ij}(p^*,Q) u_j(Q) d\Gamma = \int_\Gamma U_{ij}(p^*,Q) t_j(Q) d\Gamma + \int_\Omega U_{ij}(p^*,q) b_j(q) d\Omega$$

$$(p^* \in \Omega^*, Q \in \Gamma, q \in \Omega) \qquad (3)$$

The Direct Regular Method dicretizes eq. (3) directly instead of eq. (2). The integral of eq. (3) is no more a singular integral equation because the load point p^* of the fundamental solution is not on Γ. Besides the discussion whether the integral equation (3) is solvable or not, it has an obvious physical meaning. That is to say, the eq. (3) is no other than Betti's reciprocal theory consists of two kinds of stress field; one is that obtained by the fundamental solution without body force and the other is that of the solution of the boundary value problem with body force to be solved.

A corner point problem with two unknown tractions as shown in Fig.2 cannot be solved by the Direct Method based on eq. (2) with use of high order elements, because of the insufficiency of equations. That is, one can obtain only a equation for the double nodes as shown in Fig.2 when the load point is on the corner point. The problem can be solved by using non-conforming elements[9] there, but in this case, another problem comes out that the compatibility of the displacement is not assured in the similar manner to the case of the constant element discretization. But employing the eq. (3), it comes to be possible to obtain equations as many as necessary, since the number of load points is not necessarily subject to that of the nodes on the boundary. Therefore, the corner point problem mentioned above can be solved easily and one can get a compatible solution there.

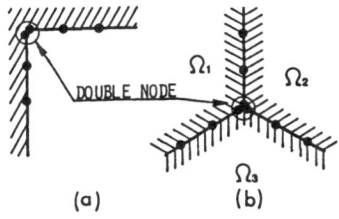

(a)   (b)

Fig.2 A Corner point where the tractions on both edges are unknown.

The Direct Regular Method is also a kind of Direct Methods, so that one can combine the former with the latter by putting some of the load points outside the domain and the other on the boundary.

DISCRETIZATION ANALYSIS

The Direct Regular Method has a great merit that one can use the usual Direct Method programme without any change, because it is just the same as the usual method discretizing analysis besides putting load points outside the domain. In the present study, we use the Direct Method programme reported previously[10,11] without any change, in which the quadratic isoparametric elements were used and the efficient numerical integration scheme was adopted based on the error estimation method by Lachat et al.[3]

No singular integral is required in the analysis of the Direct Regular Method because the load point is located outside the boundary. But if it is close to the boundary, the accuracy of the numerical integration is expected to be lowered by the behavior of the kernel.

To avoid the above difficulty, we have employed the following integration scheme to improve the accuracy of numerical integration. Namely, consider when the load point p^* is located outside the boundary as shown in Fig.3 (a). Let $R = min(R_1, R_2, R_3)$ be the minimum distance between p^* and each node of the element and L be the half length of the element. If $R \leq L$, the numerical integration is performed by dividing the element into a few sub-elements as shown in Fig.3. For example, the element is sub-divided like (c) in the case of $R = R_2$ and (d) in the case of $R = R_3$. This scheme is similar to that of the singular integration of the Direct Method.

After the whole boundary values of both displacement and traction are obtained by the discretization analysis, displacement and stress components at the internal point are calculated by eq.(3) in the same way as the Direct Method. However in this calculation, there also arises the same difficulty as stated previously if the internal point is close to the boundary. The sub-division scheme of the element stated above is also applicable to such a case.

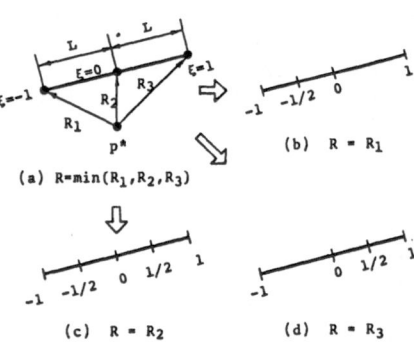

(a) $R=min(R_1,R_2,R_3)$

(b) $R = R_1$

(c) $R = R_2$

(d) $R = R_3$

Fig.3 Sub-element division scheme.

NUMERICAL RESULTS BY THE DIRECT REGULAR METHOD

In applying the Direct Regular Method, it is an important problem where the load point should be located. It is theoretically possible to locate it at an arbitrary point, but numerically, the farther it is located from the boundary, the less each of the simultaneous equations become distinguishable; therefore, the determinant of the coefficient matrix of the system of equations tends to become singular. Moreover, the close location of the load point to the boundary causes the low accuracy of the integration. Accordingly, we performed the Direct Regular BEM elastostatic analyses on the problems of a uniform stress field and also the problems of a hole and a crack with a intensive stress concentration, with changing widely the distance between the load point and the boundary. And we investigated numerically where to locate the load points for the accurate solutions.

Analyses of a Square Plate under Uniform Tension
First, we examined the relation between the accuracy and the distance of the load points from the boundary with regard to a square plate under uniform tensile stress σ as shown in Fig.4.

Fig.5 shows the results of the analyses of a shadowed quarter region of the plate in Fig.4 for the symmetry. Each edges are divided into two elements and the load point is located at a distance of d in the normal direction of the boundary. Absolute values of the relative errors of the stress obtained at the node Ⓐ are plotted against d/L: the ratio of d to the half length L of the element. Solid symbols in the figure denote the results obtained without using the sub-element integration scheme even if the load points are located close to the boundary. Very accurate results are obtained in the wide range of $d/L=0.5\sim10.0$, better than that of the usual Direct Method shown by the broken line. But in the range of $d/L\geqq10.0$, the appearance of the errors of

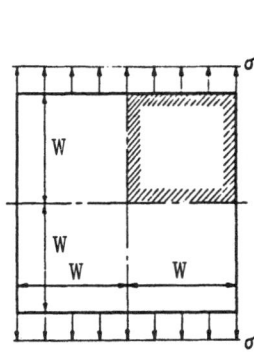

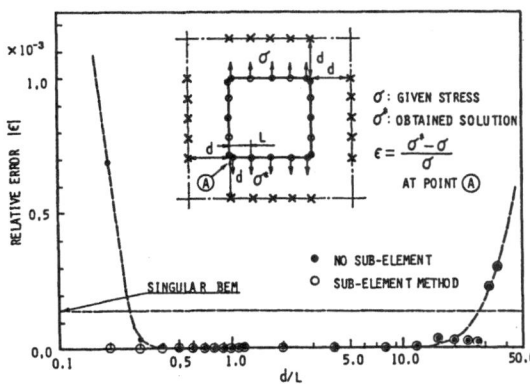

Fig.4 Square plate under uniform tension.

Fig.5 Relative error of obtained traction σ^* at A against d/L (8 boundary elements).

the solutions looks unstable and also in $d/L < 0.5$, erroneous
results are obtained. However, if the accuracy of the integration
is improved by employing the sub-element integration scheme, high
accurate results are obtained to $d/L \to 0$, as shown in Fig.6 by the
open symbols in Fig.6.

From the above results, it is found that in such a problem,
this method is more effective the usual Direct Method if the
location of load points is appropriately chosen. It is also found
that even when the load point is located very close to the
boundary, this method with sub-element integration scheme gives
very accurate results.

Analyses of a Circular Hole

In order to examine whether the Direct Regular Method is also
effective to the problems with complicated stress field, we
analyzed a quarter region of a square plate with a circular hole
as shown in Fig.6. Stress concentration factors were analyzed in
two cases of load point locations. First, all the load points
were located at a distance of constant d/L from the
boundary(Fig.7 (a)). Next, some of the load points were located
on the boundary of the circular hole where there was a intensive
stress concentration and the others were outside the
boundary(Fig.7 (b)). Fig.8 shows the results in the case where
all the load points were located outside the boundary, in
comparison with the reliable result obtained by the usual Direct
Method shown by the broken line. When the sub-element scheme is
not used even if the load point is close to the boundary, the
correct results can be obtained only in the very small range of
the relative distance, that is, about $d/L \fallingdotseq 0.25$. In the range of
$d/L \leqq 0.01$, all the results were smaller by 7% than the reliable
result, and in the range of $d/L = 0.75 \sim 2.5$, the obtained tractions
were oscillating, which means that the coefficient matrix of the
system of equations were in bad condition in that range; in fact,
it became singular in the range of $d/L > 2.5$. However, if we
employ the sub-element scheme when the load point is close to the
boundary, the correct results are obtained in the whole range of
$d/L < 1.0$ as plotted with the solid symbols in Fig.8.

Thus it is found that the Direct Regular Method can solve
such the stress concentration problem as a hole and, however, the
load point must be located close to the boundary. Even in that
case, accurate results are obtained by employing the sub-element
integration scheme, but the merit of avoiding the singular
integral is lost at this stage.

We have thus proposed a new method combining the Direct
Regular Method with the usual Direct Method. In the method, load
points are located on the boundary around the hole where an
intensive stress distribution is expected and the rest are
located outside the boundary as shown in Fig.7 (b). Fig.9 shows
the results of it. Compared with the case when all the load
points are located outside, rather accurate results(less than

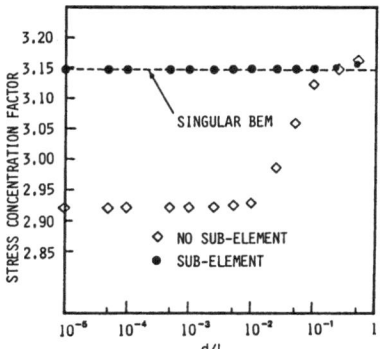

Fig.6 A quarter region of a square plate with a circular hole.

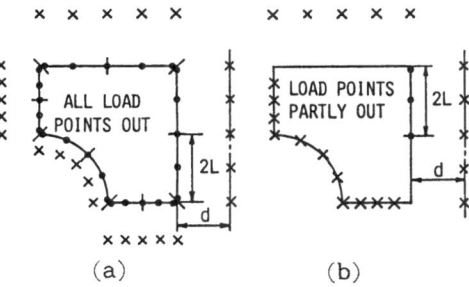

(a)   (b)

Fig.7 Arrangement of load points for Fig.6.

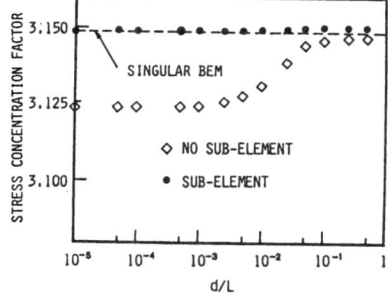

Fig.8 Stress concentration factor obtained when all load points located outside (19 boundary elements).

Fig.9 Stress concentration factor obtained when load points located partly on the boundary (19 boundary elements).

1%) are obtained without employing the sub-element scheme. It is found that this method can make good use of the merits of the Direct Regular Method and can give the stable solutions in accuracy.

Analyses of the Crack Problems

As an example of the more complicated stress field problems, we analyzed a center crack in a square plate under uniform tension as shown in Fig.10 and calculated the stress intensity factors K_I by the extrapolation method using the obtained displacements around the crack tip. For the symmetry of the problem, the shadowed quarter region are divided into 26 elements. The singular elements are placed at the crack tip.

Fig.11 shows the normalized stress intensity factor F_I obtained. In crack analyses, one has no choice but to locate the load points very near the boundary and therefore, the sub-element integration scheme must be employed in every case. When every load point are located outside the boundary, the range where correct results can be obtained is limited to $d/L < 10^{-3}$ and F_I becomes greater by degrees for $d/L > 10^{-3}$. The result for $d/L \fallingdotseq 1.0$

seems to be accurate but the obtained tractions on the ligament of the crack are oscillating, which means that the coefficient matrix of the system of equations is nearly singular. However, if we employ the combined method by locating the load points on the boundary of a cracked edge and the other load points outside, we can obtain the results as accurate as those obtained by the usual Direct Method in the whole range of d/L as shown in Fig.11.

Next, the above combined method is applied to an analysis of a mixed mode crack problem as shown in Fig.12. The similar results as that of the mode I crack are obtained as shown in Fig.13. The coefficient matrix became singular for $d/L > 0.05$. Therefore, it is found that the load points should be located more close to the boundary than in the case of the mode I crack. The combined method also gives the accurate results in the range $d/L \leq 0.25$.

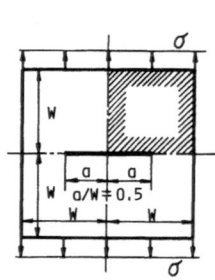

Fig.10 Center crack in a square plate under uniform tension.

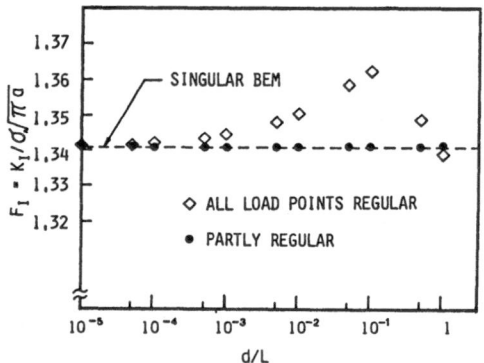

Fig.11 Stress intensity factor of the center crack.

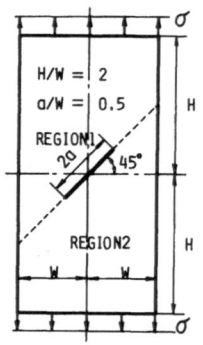

Fig.12 Slant crack in a rectangular plate under uniform tesion.

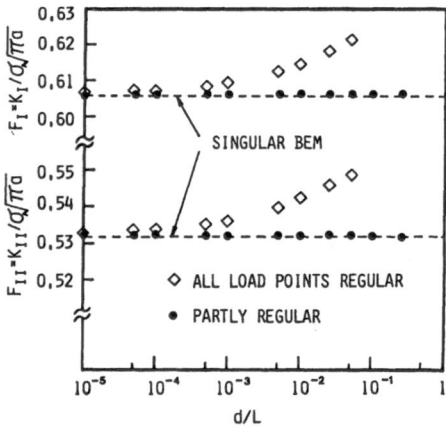

Fig.13 Stress intensity factors of the slant crack.

From these results, it becomes clear that the Direct Regular Method cannot give the correct result unless the load points should be located close to the boundary according to the intensity of the stress field of the problem. Therefore, the Direct Regular Method seems to be not necessarily effective from a practical point of view, since it needs a special numerical integration scheme like the sub-element integration scheme of the Direct Method when the load point is located near the boundary. However, the combined Direct Regular Method with the usual Direct Method, which is proposed in this paper, is expected to be a new efficient BEM method, because it makes a good use of both methods.

It should also be noted that this combined method is very easily applied to the problem with a corner point where the tractions on both edges are unknown, since we can easily make the additional equation by locating a load point just outside the corner[1,2].

CONCLUSIONS

The Direct Regular Method based on the Somigliana internal reciprocal formula is applied to two-dimensional elastostatic analysis. The usefulness and the limit of this method is discussed through various numerical examples, and the following conclusions are obtained.

(1) For the uniform stress field problem, the Direct Regular Method is useful and gives quite accurate results, even if the load points are located relatively far from the boundary in the normal directions.

(2) For the problem with intensive stress field like that of a hole or a crack, the load points must be located very close to the boundary, so that it is necessary to employ some special numerical integration scheme like the sub-element scheme in order to improve the accuracy of integration. Therefore the merit of the Direct Regular Method is lost in such problems.

(3) Converged accurate solutions can be obtained even in the problem, if the usual method is combined with the Direct Regular Method by locating some of the load points on the boundary where the stress changes intensively and the other load points outside.

(4) This combined method proposed in this study seems to be very useful and efficient. It can deal easily with the problem of a corner point where the traction on both edges are unknown.

REFERENCES

1. Brrebia. C.A.,Telles J.C.F. and Wrobel L.C. (1984), Boundary
   Element Techniques, Springer-Verlag. Berlin and New York.

2. Washizu. K., Tanaka. M. and Tanaka. K. (1982), Boundary
   Element Method -Fundamentals and Applications-, MARUZEN,
   Tokyo (in Japanese).

3. Lachat. J.C. and Watson. J.O. (1976), Effective Numerical
   Treatment of Boundary Integral Equations, Int. J. Num. Mech.
   Eng., 10, pp.991-1005.

4. Patterson. C. and Sheikh. M.A. (1981), Regular Boundary
   Integral Equations for Fluid Flow (Ed. Taylor et al.),
   Numerical Methods in Laminar and Turbulent Flow, Pineridge
   Press.

5. Patterson. C. and Sheikh. M.A. (1984), Applications of The
   Direct Regular Methods to Linear Elastic Fracture Mechanics
   (Ed. Brrebia. C.A.), Boundary Element, Springer Verlag.

6. Honma. T., Tanaka. Y. and Kaji. I. (1985), Three-Dimensional
   Analysis of Convective Diffusion Equation using Regular
   Boundary Element Method, Proc. 2nd Symp. on BEM's, JASCOM,
   pp.25-30, (in Japanese).

7. Honma. T., Tsuchimoto. M. and Kaji. I. (1985), An Analysis of
   MHD Equilibria of Cylindrical Plasmas using Boundary Element
   Methods, Proc. 2nd Symp. on BEM's, JASCOM, pp.275-280, (in
   Japanese).

8. Jaswon. M.A. and Symm. G.T. (1977), Integral Equation Methods
   in Potential Theory and Elastostatics, Springer Verlag.

9. Patterson. C. and Sheikh. M.A. (1981), Nonconforming Boundary
   Elements for Stress Analysis (Ed. Brrebia. C.A.), Recent
   Advances in Boundary Element Methods, Springer Verlag.

10. Kisu. H., Yuuki. R. and Kitagawa. H. (1985), The Analysis of
    Stress Intensity Factor for Surface Crack by Boundary Element
    Method, Trans. JSME, Vol.51, pp.660.

11. Yuuki. R., Kisu. H. and Matsumoto. T. (1985), On the Method to
    Determine the Stress Intensity Factors in Boundary Element
    Method, Proc. 1st Symp. on BEM's, JASCOM, pp.85-90, (in
    Japanese).

12. Matsumoto. T., Yuuki. R. and Kisu. H. (1985), Application of
    Direct Regular Method to Two-dimensional Elastostatic
    Analysis, Proc. 2nd Symp. on BEM's, JASCOM, pp.215-220, (in
    Japanese).

Stability Experiment of FDM, FEM and BEM in Convective Diffusion

M. Kanoh
Civil Engineering Department, Kyushu Sangyo University, Fukuoka 813, Japan
G. Aramaki
Civil Engineering Department, Saga University, Saga 840, Japan
T. Kuroki
Civil Engineering Department, Fukuoka University, Fukuoka 814-01, Japan

INTRODUCTION

It is important to know if the model has stability in numerical calculation.
Usually diffusion number, Courant number and cell Peclet number are used as stability criteria. For one-dimensional problems, stability has been studied theoretically in various schemes of finite difference method[1].
However theoretical approach is not easy in two-dimensional problems.
Numerical experiment on stability criteria may be another approach to study the appropriate mesh size and time increment under given physical constants.
The purpose of this paper is to study numerical stability criteria on two-dimensional transient convective analysis by weighted finite difference method, finite element method and boundary element method.
The first method is presented by the first author.
One of the advantages of this method is that it shows higher accuracy than simple finite difference and finite element method do.
As for boundary element method, we use the time dependent fundamental solution for thermal conducting problem. And constant element is used to discretise integral equations. Domain integration is evaluated by using eight noded cells.
Numerical experiments are carried out for two examples by varying diffusivity, time increment and mesh size.
The results showed that the stability criteria derived for one-dimensional problems are also useful in two-dimensional finite element and boundary element analysis. On the other hand weighted finite difference method was stable in a wide

range of cell Peclet number.

GOVERNING EQUATION

The governing equation is described below.

$$\dot{\phi} + u_i \, \phi_{,i} - k \, \phi_{,ii} = 0 \tag{1}$$

Here we consider the following boundary condition.

$$\phi = \bar{\phi} \quad \text{on } \Gamma_1 \quad (2) \quad ; \quad q = \frac{\partial \phi}{\partial n} = \bar{q} \quad \text{on } \Gamma_2 \tag{3}$$

BOUNDARY INTEGRAL EQUATION

By the application of the Green's theorem and time-integration, we obtain the equation with the weighting function ϕ^*

$$\gamma\phi(x_p, t_2) + k \int \phi_{,n} \cdot \left(-\frac{\partial r/\partial n}{2k\pi r} \right) \exp(-a) \, d\Gamma$$

$$= k \int \phi_{,n} \cdot \frac{1}{4\pi k} E_i(a) \, d\Gamma + \int \phi(x, t_1) \, \phi^*(r, t_1) \, d\Omega$$

$$- \int u_i \phi_{,i} \cdot \frac{1}{4\pi k} E_i(a) \, d\Omega \tag{4}$$

Here we consider weighting function ϕ^* as below

$$\phi^* = \frac{1}{4k\pi(t_2 - t)} \exp\left(-\frac{r^2}{4k(t_2 - t)} \right) \tag{5}$$

Eq. 4 can be discretized to Eq. 6 by using boundary constant element and internal eight noded cell for Gaussian integration.

$$\gamma\phi(x_p, t_2) + \sum \phi \sum_{k=1}^{M} \left(-\frac{\partial r/\partial n}{2\pi r} \right) \exp(-a)\Delta x \, W_k = \sum \phi_{,n} \sum_{k=1}^{M} \left(\frac{E_i(a)}{4\pi} \right)$$

$$\cdot \Delta x \, W_k + \sum_{k=1}^{M} \sum_{L=1}^{M} \phi(x, t_1) \frac{\exp(-a)}{4\pi D \, \Delta x^2} |J| W_k W_L - \sum_{k=1}^{M} \sum_{L=1}^{M} \frac{P_e}{4\pi \Delta x} \frac{\partial \phi/\partial x_1}{}$$

$$\cdot E_i(a) |J| W_k W_L \tag{6}$$

where $\quad E_i(a) = -\ln(a) - E_c - \sum \frac{(-a)^n}{n \cdot n!} \quad ; \quad E_c = 0.5772157$

$$a = r^2/(4D \, \Delta x^2) \, , \quad Pe = u_1 \Delta x_1/D \, , \quad u_2 = 0 \, , \quad D = d\Delta t/\Delta x^2$$

From Eq. 6 the stability and accuracy of boundary element method analysis depend on diffusion number(D), cell Peclet number (P_e) and size of element(Δx).

FEM SOLUTION BY IMPLICIT SCHEME

The one-dimensional domain is divided into linear elements. Then the application of the Galerkin method to one-dimensional convective diffusion equation and the replacement of the time differencial term with backward finite difference result in

$$[B_1] [\phi^t] = [B_2] [\phi^{t-\Delta t}] + \frac{\Delta t}{\Delta x} [P] \tag{7}$$

where $[\phi^t]$ is the vector of ϕ^t(ϕ of time t), then $[B_1]$ and $[B_2]$ are the coefficient matrices as described below.

$[B_1] =$

$$\begin{pmatrix} \frac{1}{3}+D & \frac{1}{6}+\frac{C}{2}-D & 0 & 0 & . & . & 0 \\ \frac{1}{6}-\frac{C}{2}-D & \frac{2}{3}+2D & \frac{1}{6}+\frac{C}{2}-D & 0 & . & . & . \\ 0 & \frac{1}{6}-\frac{C}{2}-D & \frac{2}{3}+2D & \frac{1}{6}+\frac{C}{2}-D & 0 & . & . \\ . & . & . & . & . & . & . \\ 0 & . & . & 0 & \frac{1}{6}-\frac{C}{2}-D & \frac{2}{3}+2D & \frac{1}{6}+\frac{C}{2}-D \\ 0 & . & . & . & 0 & \frac{1}{6}-\frac{C}{2}-D & \frac{1}{3}+D \end{pmatrix}$$

$$[B_2] = \begin{pmatrix} \frac{1}{3} & \frac{1}{6} & 0 & 0 & . & . & 0 \\ \frac{1}{6} & \frac{2}{3} & \frac{1}{6} & 0 & . & . & . \\ 0 & \frac{1}{6} & \frac{2}{3} & \frac{1}{6} & 0 & . & . \\ . & . & . & . & . & . & . \\ . & . & . & . & . & . & . \\ 0 & . & . & 0 & \frac{1}{6} & \frac{2}{3} & \frac{1}{6} \\ 0 & . & . & . & 0 & \frac{1}{6} & \frac{1}{3} \end{pmatrix} \tag{8}$$

Here $C = u\Delta t/\Delta x$ is Courant number.
From Eqs. 7 and 8 the stability and accuracy of finite element method analysis depend on Courant number and diffusion number.

WEIGHTED FINITE DIFFERENCE METHOD(WFDM)

One-dimensional WFDM

The procedure to decided the one-dimensional WFDM was given by Kanoh, Aramaki and Kuroki[2], then the one-dimensional weighted finite difference equation was shown as Eq. 9.

$$\phi(i,j) = Q_{-1}^0 \cdot \phi(i-1,j) + Q_1^0 \cdot \phi(i+1,j) + Q_{-1}^{-1} \cdot \phi(i-1,j-1)$$

$$+ Q_0^{-1} \cdot \phi(i,j-1) + Q_{-1}^{-2} \cdot \phi(i-1,j-2) \tag{9}$$

Determination of two-dimensional WFDM

We formulate the two-dimensional WFDM similar to the one-dimensional WFDM. The polynomials composed of x, y, t satisfying two-dimensional convective diffusion equation are given as

$$\phi^{(r)}(x,y,t) = \sum_{i=0}^{r/2} \left\{ \frac{(x-u_1 t + y - u_2 t)^{r-2i}}{(r-2i)!} \cdot \frac{(k_1 t + k_2 t)^i}{i!} \right\} \tag{10}$$

Here discretisation is introduced, namely when $\Delta x = G_1 h$, $\Delta y = G_2 h$ and $\Delta t = k$ are increments of x, y, t, we have

$$x = p_1 G_1 h, \quad y = p_2 G_2 h, \quad t = qRh^2 \tag{11}$$
$$(p_1, p_2, q: 0, \pm 1, \pm 2 \ldots; G_1, G_2 : \text{positive constant})$$

Subsequently if we set

$$C_* = u_1 \Delta t/\Delta x + u_2 \Delta t/\Delta y, \quad D_* = k_1 \Delta t/\Delta x^2 + k_2 \Delta t/\Delta y^2 \tag{12}$$

The two-dimensional infinite polynomial progression composed of superimposed polynomials in Eq. 10 is described below.

$$\phi^{(r)}(p_1 \Delta x, p_2 \Delta y, q \Delta t) = (\Delta x)^r \sum_{i=0}^{r/2} \left[\frac{(G_1 p_1 + G_2 p_2 - qC_*)^{r-2i}}{(r-2i)!} \cdot \frac{qD_*^i}{i!} \right] \tag{13}$$

In case the vicinity points in the model(referred herein as angle upwind-scheme model) shown in Fig. 1 are adopted, the FDM is

$$\phi(i,g,j)=S^0_{-1,0}\cdot\phi(i-1,g,j)+S^0_{-1,-1}\cdot\phi(i-1,g-1,j)+S^{-1}_{-1,0}\cdot\phi(i-1,g,j-1)$$

$$+S^{-1}_{0,0}\cdot\phi(i,g,j-1)+S^{-1}_{1,0}\cdot\phi(i+1,g,j-1) \tag{14}$$

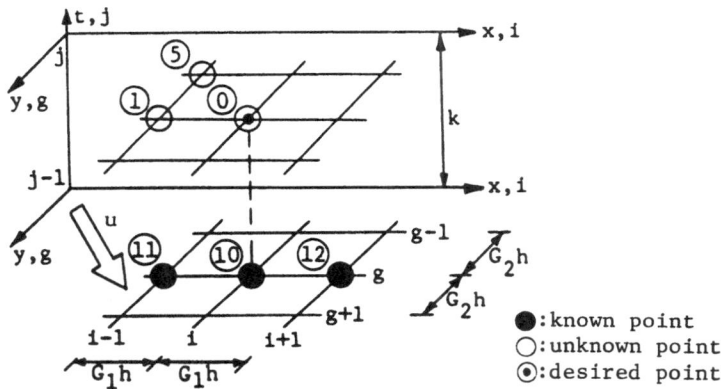

Figure 1 Angle upwind-scheme model

Five weights $S^0_{-1,0},\ldots,S^{-1}_{1,0}$ are decided from the following linear equations.

$$
\begin{bmatrix}
1 & 1 & 1 & 1 & 1 \\
-G_1 & -G_1-G_2 & C_*-G_1 & C_* & C_*+G_1 \\
G_1^2 & (G_1+G_2)^2 & (C_*-G_1)^2-2D_* & C_*^2-2D_* & (C_*+G_1)^2-2D_* \\
-G_1^3 & (-G_1-G_2)^3 & \begin{array}{c}(C_*-G_1)^3\\-6(C_*-G_1)D_*\end{array} & \begin{array}{c}C_*^3\\-6C_*D_*\end{array} & \begin{array}{c}(C_*+G_1)^3\\-6(C_*+G_1)D_*\end{array} \\
G_1^4 & (G_1+G_2)^4 & \begin{array}{c}(C_*-G_1)^4\\-12(C_*-G_1)^2D_*\\+12D_*^2\end{array} & \begin{array}{c}C_*^4\\-12C_*^2D_*\\+12D_*^2\end{array} & \begin{array}{c}(C_*+G_1)^4\\-12(C_*+G_1)^2D_*\\+12D_*^2\end{array}
\end{bmatrix}
\cdot
\begin{bmatrix}
S^0_{-1,0} \\
S^0_{-1,-1} \\
S^{-1}_{-1,0} \\
S^{-1}_{0,0} \\
S^{-1}_{1,0}
\end{bmatrix}
=
\begin{bmatrix}
1 \\
0 \\
0 \\
0 \\
0
\end{bmatrix}
\tag{15}
$$

NUMERICAL EXAMPLES

Two cases of transient convective diffusion problems shown in Fig. 2 are considered. In case one, the concentration along the left hand side boundary is fixed to unity under zero initial concentration. On the other hand, the concentration along the right hand side boundary is given by Eq. 17 in case two. For those problems, analytical solutions are given as follows;

case 1

$$\phi(x,t)=2k \sum_{n=1}^{\infty} [\frac{n\pi}{\lambda_n} \{1-\exp(-\lambda_n \cdot t)\}\exp(\frac{ux}{2k})\sin(n\pi x)] \qquad (16)$$

case 2

$$\phi(x,t)= \frac{1}{2} \{erfc(\frac{x-ut}{2\sqrt{kt}}) + \exp(\frac{ux}{k}) \ erfc(\frac{x-ut}{2\sqrt{kt}})\} \qquad (17)$$

where $\qquad \lambda_n = k \{(u/2k)^2 + (n\pi)^2\}$

$$erfc(x)=\frac{2}{\sqrt{\pi}} \int_{x}^{\infty} \exp(-T^2) \ dT$$

Numerical calculations are carried out as two-dimensional problems while fixing velocity to unity and varying steady diffusivity field.

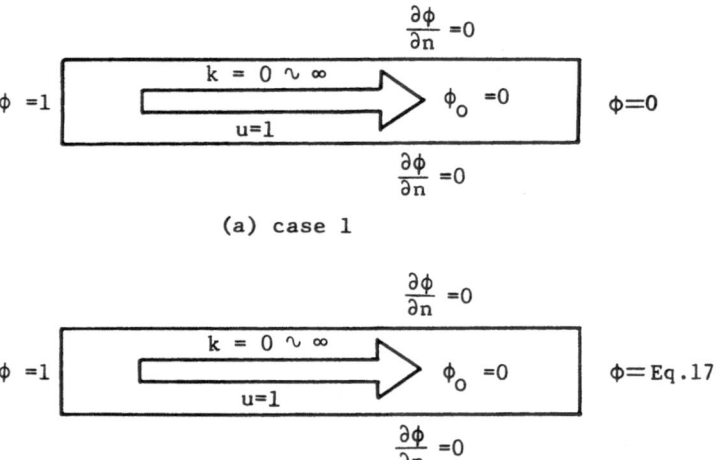

(a) case 1

(b) case 2

Figure 2 Convective diffusion geometry

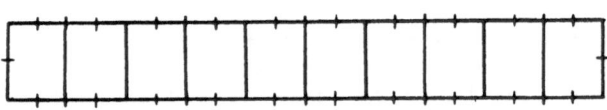

Figure 3 B E M mesh example ($\Delta x=1/20$)

RESULTS and DISCUSSION

To discuss the stability of the above discretised equations, we set following loose guide line for numerical experiments. The solutions may be stable, even fluctuate in the early stage of numerical calculation, if width of fluctuation has tendency to decrease as time passes.
If maximum deviation of numerical solution occurs less than five percent error, then we may say the solution is stable. For example Fig. 4 shows unstable and Fig. 5, Fig. 6 and Fig.7 show stable solutions.
In boundary element method, due to the characteristics of the fundamental solution, we have to take diffusion number in the range of one half ~ one. For large or small diffusion number, speed of numerical phenomena in computers deviates from the exact speed.
If we fix Courant number to one and diffusion number to one half, then we have two as cell Peclet number.
For weighted finite difference method and finite element method, there is more flexibility than boundary element method in choosing diffusion number.

For our examples, from Table 1, weighted finite difference method is stable as long as following conditions are satisfied;
    Courant number $\le$ 1 , cell Peclet number $\le$ 40

Finite element method is stable as long as following conditions are satisfied;
    Courant number $\le$ 1 , cell Peclet number $\le$ 2 ....in case1

    Courant number $\le$ 1 , cell Peclet number $\le$ 5 ....in case2

In boundary element method, it is necessary to decide time interval from the following relation due to the characteristics of the fundamental solution;
    diffusion number = $1/2 \sim 1$
If Courant number is less or equal to one using this time interval, stable solution can be obtained. In this case cell Peclet number is assured to be less or equal to two.

Table 1 Stability criteria of WFDM, FEM and BEM

D	WFDM case1		WFDM case2		FEM case1		FEM case2		BEM case1		BEM case2	
$\frac{1}{4}$	Pe≤40		Pe≤40			C≤1.0			Pe≤2.8			
$\frac{1}{10}$	Pe≤40	C≤1	Pe≤40	C≤1	Pe≤2	C≤1.5	Pe≤5	C≤1	Pe≤1.7	D=$\frac{1}{2}$	Pe≤2	$\frac{1}{2}$≤D≤1 and C≤1
$\frac{1}{20}$	Pe≤50		Pe≤45			C≤1.5			Pe≤2.0			

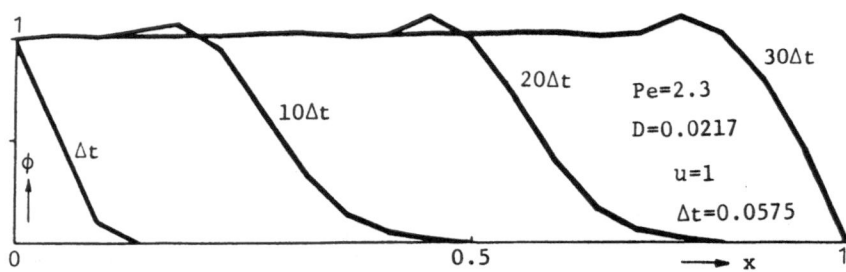

Figure 4 BEM solution

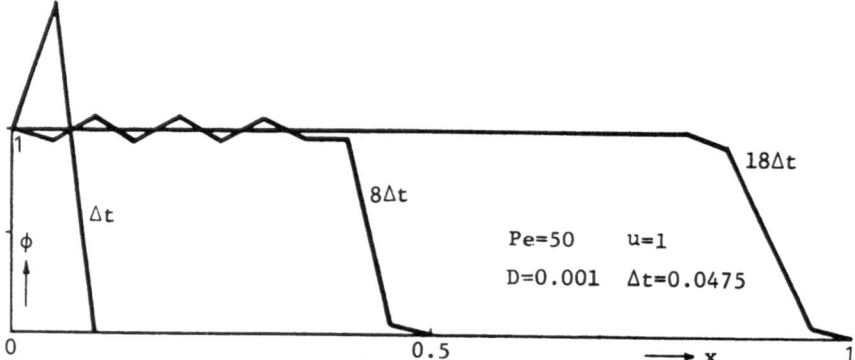

Figure 5 WFDM solution

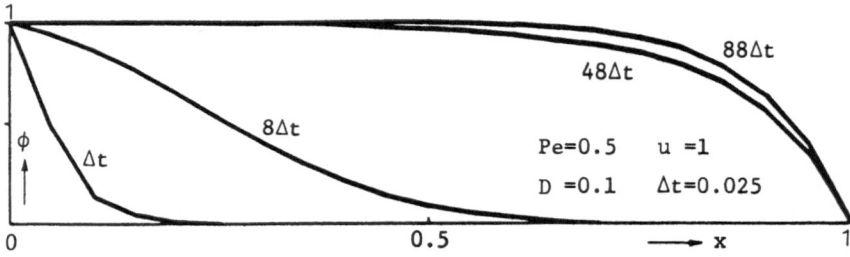

Figure 6 WFDM solution (case 1)

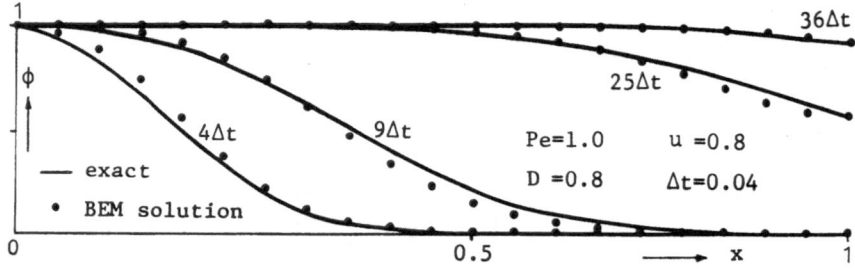

Figure 7 BEM solution (case 2)

CONCLUSIONS

Numerical stability tests on transient convective diffusion in two dimension were carried out by using weighted finite difference method, simple finite element method and boundary element method.

Stability was maintained as long as the following conditions were satisfied;
    diffusion number, Courant number ≤ 1
        cell Peclet number ≤ 2
It is recommendable to set diffusion number in the range of one half $\sim$ one in boundary element analysis. Weighted finite difference method was applicable to the model with cell Peclet number ≤ 40.

REFERENCES

1.   Roache P.J.(1976). Computational Fluid Dynamics, Hermosa Publisher.
2.   Kanoh M. Aramaki G. and Kuroki T.(1985), Boundary Element Solutions of Convective Diffusion compared with Finite Element and Weighted finite difference Method, in Boundary Elements VII(Ed. Brebia C.A.,Maier G.),pp.9-13 to 9-22, Proceedings of the 7th Int. Conf. Villa Olmo, Lake Como ,Italy, September 1985.

Boundary Element Modelling of Dilatant Interfaces

A.P.S. Selvadurai, M.C. Au

Department of Civil Engineering, Carleton University, Ottawa, Ontario, Canada

ABSTRACT

The present paper focusses on the boundary element modelling of material interfaces which exhibit dilatancy phenomena. Dilatancy phenomena at materials interfaces can occur as a result of movement at a fracture plane with asperities or by virtue of a dilatant filler material such as a granular material which can be present at a joint region. The boundary element modelling is used to examine the influence of such dilatant phenomena on the global irreversible responses that develop in two layer systems subjected to a flexural deformations.

INTRODUCTION

The group of problems which examine the mechanical behaviour of contact regions is an important branch of applied mechanics which has extensive engineering applications in areas such as composite materials, tribology, soil structure interaction, mechanical response of fractured and damaged interfaces etc. (see e.g., Gudehus[1]; Desai and Christian[2]; Johnson[3]; Selvadurai and Voyiadjis[4]; Selvadurai and Au[5]). For these studies the local behaviour at the interface exerts a dominant influence on the global response of the surfaces or bodies in contact. The mathematical modelling of these interfaces can be approached via diverse analytical and numerical schemes. The analytical studies lend themselves to only very simple interface constitutive laws, contact conditions and interface geometries. Therefore, numerical methods of stress analysis offer considerable prospects for the study of interface phenomena. In particular, the boundary element methods can be used quite successfully to examine the response of interfaces possessing frictional, plastic, and unilateral contact conditions (see e.g., Andersson and Allan Persson[6]; Paris and Garrido[7] and Selvadurai and Au[5]). In this paper we examine the category of interfaces in which dilatant phenomena exist by virtue of an interface with

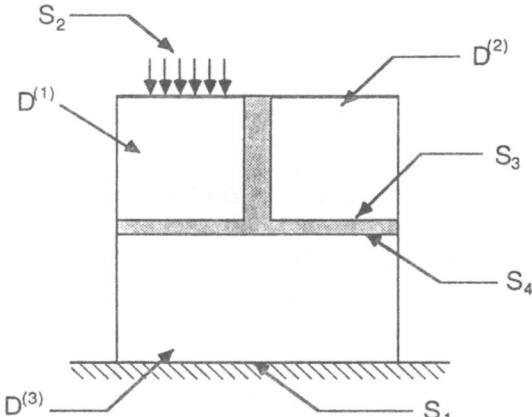

Fig (1)   A Multi-region representation

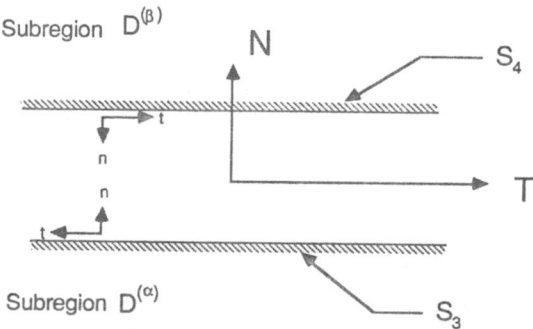

Fig (2)   Interface Description ($\alpha < \beta$)

asperities or by the presence of a dilatant filler material at the interface. Dilatant phenomena are of considerable interest to the modelling of rock interfaces which have saw-tooth or irregular structure or to   the modelling of rock interfaces which contain densely packed granular debris. The boundary element scheme developed in the paper is used to examine the bending behaviour of a two layer system with a dilatant interface and subjected to a self stressing compression. Numerical results presented in the paper illustrate the manner in which the flexural response of the two layer system is influenced by the degree of self stressing and the dilatancy characteristics at the interface.

BOUNDARY ELEMENT METHOD

This section presents a brief account of the two-dimensional boundary element formulation applicable to the multi-region problem shown in Figure (1). The boundary integral equation governing an individual region $D^{(\alpha)}$ given by Brebbia[8] can be written as

$$C_{ik} U_k^{(\alpha)} + \int_{\Gamma^{(\alpha)}} P_{ik}^{(\alpha)} u_k^{(\alpha)} d\Gamma = \int_{\Gamma^{(\alpha)}} u_{ik}^{(\alpha)} P_k^{(\alpha)} d\Gamma \quad \text{on } D^{(\alpha)} \quad (1)$$

where i,k = 1,2; $u_k^{(\alpha)}$ and $P_k^{(\alpha)}$ are the boundary displacements and tractions respectively; $U_{ik}^{(\alpha)}$ and $P_{ik}^{(\alpha)}$ are the corresponding fundamental solutions. The superscript (α) were $\alpha=1,2,3$ etc., indicates the subregion under consideration. The boundary conditions for the domain can be classified as follows:

(a)   Displacement prescribed boundary (S_1) where

$$u_i = \bar{u}_i \qquad \text{on } S_1 \qquad (2)$$

(b)   Traction prescribed boundary (S_2) where

$$P_i = \bar{P}_i \qquad \text{on } S_2 \qquad (3)$$

and finally

(c)   The interface conditions between two subregions which are represented by (S_3) and (S_4);

these will be defined in a subsequent section. After the usual discretization procedure, the boundary element matrix equation from (1) can be written as

$$[\underset{\sim}{H}^{(\alpha)}] \{\underset{\sim}{U}^{(\alpha)}\} = [\underset{\sim}{G}^{(\alpha)}] \{\underset{\sim}{P}^{(\alpha)}\} \qquad (4)$$

An equilibrium condition is also imposed to each subregion so that

$$\int_{\Gamma^{(\alpha)}} P_i^{(\alpha)} \, d\Gamma = 0 \qquad (i=1,2) \tag{5}$$

With the formulation given by Mustoe[9], (4) can be modified as

$$\begin{bmatrix} \underset{\sim}{H}^{(\alpha)} \\ \underset{\sim}{0} \end{bmatrix} \{ U^{(\alpha)} \} = \begin{bmatrix} \underset{\sim}{G}^{(\alpha)} & \underset{\sim}{Q}^{(\alpha)T} \\ \underset{\sim}{Q}^{(\alpha)} & \underset{\sim}{0} \end{bmatrix} \begin{bmatrix} \underset{\sim}{P}^{(\alpha)} \\ \underset{\sim}{\lambda}^{(\alpha)} \end{bmatrix} \tag{6}$$

where $[Q^{(\alpha)}]$ is the submatrix corresponding the discretization of equation (5) and $\{\lambda^{(\alpha)}\}$ is a 2x1 vector composed of the Lagrange multipliers for the imposed condition (5). The matrix equation (6) will be transformed further to reflect the local co-ordinate (n,t) of an element (normal and tangential direction), in order to study the problem at an interface.

ELASTIC INTERFACE PROBLEMS

In order to consider the local elastic behaviour at the interface between two subregions $D^{(\alpha)}$ and $D^{(\beta)}$ ($\alpha<\beta$), a common local system for the subregions is defined in the Figure (2), which indicates the tangential (T) and the normal (N) direction. (Note that (n,t) is different from (T,N)). In the present discussion, àn elastic interface is capable of experiencing a jump or discontinuity in displacements from one side of the interface to the other. This can be expressed in the form

$$\Delta_T = u_t^{(\alpha)} + u_t^{(\beta)}$$

and

$$-\Delta_N = u_n^{(\alpha)} + u_n^{(\beta)} \tag{7}$$

where Δ_T and Δ_N are the relative tangential and normal displacements respectively. The well-known interface-element approach in the finite element procedure (Goodman and St. John[10]; Heuze and Barbour[11]) requires a constitutive relation of the form

$$\begin{bmatrix} \sigma_T \\ \sigma_N \end{bmatrix} = \begin{bmatrix} K_{TT} & K_{TN} \\ K_{NT} & K_{NN} \end{bmatrix} \begin{bmatrix} \Delta_T \\ \Delta_N \end{bmatrix} \tag{8}$$

where K_{ij} (i,j = T,N) are the interface stiffnesses and σ_i are the relevant stress components. Considering the equilibrium at the interface, the stresses given in (8) can be related to the interface traction given by

$$-\sigma_T = P_t^{(\alpha)} = P_t^{(\beta)}$$

and

$$\sigma_N = P_n^{(\alpha)} = P_n^{(\beta)} \qquad (9)$$

Substituting (7) and (9) into (8) and rearranging the terms, the constitutive relation at the interface expressed in terms of the boundary values can be written as

$$\begin{bmatrix} P_n^{(\alpha)} \\ P_t^{(\alpha)} \end{bmatrix} = \begin{bmatrix} -K_{NN} & K_{NT} & -K_{NT} & K_{NT} \\ K_{TN} & -K_{TT} & K_{TN} & -K_{TT} \end{bmatrix} \begin{bmatrix} u_n^{(\alpha)} \\ u_t^{(\alpha)} \\ u_n^{(\beta)} \\ u_t^{(\beta)} \end{bmatrix}$$

$$= \begin{bmatrix} P_n^{(\beta)} \\ P_t^{(\beta)} \end{bmatrix} \qquad (10)$$

which will be applied as a coupling boundary condition given by S_3 on $D^{(\alpha)}$ and S_4 on $D^{(\beta)}$ defined in the previous section. The global system matrix resulting from all subregions will be assembled by the procedure given in the subsequent section.

There are some limiting cases for the values of K_{ij} so that one can consider the following conditions

(a) For full compatibility and equilibrium conditions, at an interface (bonded contact)

$$K_{TN} = K_{NT} = 0 \; ; \; K_{TT} \to \infty \text{ and } K_{NN} \to \infty \qquad (11)$$

(b) For a smooth interface contact condition (slipping-bilateral contact)

$$K_{TN} = K_{NT} = K_{TT} = 0 \; ; \; K_{NN} \to \infty \qquad (12)$$

and

(c) For uncoupled the interface conditions (separation) (slipping-unilateral contact)

$$K_{TN} = K_{NT} = K_{TT} = K_{NN} = 0 \qquad (13)$$

ELASTO-PLASTIC INTERFACE PROBLEMS

In order to consider the elasto-plastic behaviour at the interface, the solution will be studied incrementally. Hence any
quantity will be expressed as a sum of increments; say, $\dot{u}$ is an
incremental form of u. As the material within the subregion
remains elastic, the boundary influence coefficient matrices
will be constant for any step of increments. Therefore, the
governing equation for a subregion can be written as

$$
\begin{bmatrix} H^{(\alpha)} \\ 0 \end{bmatrix} \{\dot{u}^{(\alpha)}\} = \begin{bmatrix} G^{(\alpha)} & Q^{(\alpha)^T} \\ Q^{(\alpha)} & 0 \end{bmatrix} \begin{bmatrix} \dot{P}^{(\alpha)} \\ \dot{\lambda}^{(\alpha)} \end{bmatrix} \tag{14}
$$

which is an incremental form of (6). The elastic interface
constitutive relationship can be expressed in the form

$$
\dot{\sigma}_i = K_{ij}^{(e)} \dot{\Delta}_j^{(e)} \qquad (i,j = T,N) \tag{15}
$$

where the superscript (e) indicates the stress field which is
still in the elastic range. Such an elastic-stress field
satisfies the constraint

$$
F = F (\sigma_i, W) < 0 \tag{16}
$$

where σ_i are the interface stresses and W is the interface

plastic work. When F=0, the interface behaviour will be
expressed by an elasto-plastic relationship. The relative
displacements at the interface will be composed of a recoverable
elastic part and a non-recoverable plastic part. That is

$$
\dot{\Delta}_i = \dot{\Delta}_i^{(e)} + \dot{\Delta}_i^{(p)} \tag{17}
$$

In the theory of plasticity, the plastic deformation rate can
be given by a flow rule which is defined by

$$
\dot{\Delta}_i^{(p)} = \begin{bmatrix} 0 & \text{if } F < 0 \text{ or } \dot{F} < 0 \\ \\ \dot{f}\, \dfrac{\partial \phi}{\partial \sigma_i} & \text{if } F = \dot{F} = 0 \end{bmatrix} \tag{18}
$$

where $\dot{f}$ is a proportionality factor known as plastic multiplier
and $\phi = \phi(\sigma_i, W)$ is the plastic potential function at the interface. When yielding conditions exist, $\dot{F}$ can be approximated by

$$\frac{\partial F}{\partial \sigma_i} \dot{\sigma}_i + \frac{\partial F}{\partial W} \dot{W} = 0 \tag{19}$$

One can write the elasto-plastic constitutive relation at the interface as

$$\dot{\sigma}_i = K_{ij}^{(ep)} \dot{\Delta}_j \tag{20}$$

where $K_{ij}^{(ep)}$ are the elastic-plastic stiffness coefficients. From (15) to (20), one can show that

$$K_{ij}^{(ep)} = K_{ij}^{(e)} - \frac{1}{z} \frac{\partial F}{\partial \sigma_\ell} K_{\ell j}^{(e)} K_{im}^{(e)} \frac{\partial \phi}{\partial \sigma_m} \tag{21}$$

with

$$Z = \frac{\partial F}{\partial \sigma_\ell} K_{\ell m}^{(e)} \frac{\partial \phi}{\partial \sigma_m} - \frac{\partial F}{\partial W} \dot{W} \tag{22}$$

where $\dot{W}$ can be expressed in terms of the rate of the plastic work done at the interface location.

Therefore, the boundary condition at the interface S_3 and S_4 can be written as

$$\begin{bmatrix} \dot{P}_n^{(\alpha)} \\ \dot{P}_t^{(\alpha)} \end{bmatrix} = \begin{bmatrix} -K_{NN} & K_{NT} & -K_{NN} & K_{NT} \\ K_{TN} & K_{TT} & K_{TN} & -K_{TT} \end{bmatrix}^{(ep)} \begin{bmatrix} \dot{u}_n^{(\alpha)} \\ \dot{u}_t^{(\alpha)} \\ \dot{u}_n^{(\beta)} \\ \dot{u}_t^{(\beta)} \end{bmatrix}$$

$$= \begin{bmatrix} \dot{P}_n^{(\beta)} \\ \dot{P}_t^{(\beta)} \end{bmatrix} \tag{23}$$

which will be applied to the incremental matrix equation (14).

CONTACT PROBLEMS WITH DILATION

The dilation model proposed by Plesha and Belytschko[12] is considered in this section. This model consists of an interface with fully-mated asperity teeth defined by the asperity angle θ (Figure 3) which can be degraded as a result of the tangential plastic work done at the interface given by

$$\theta = \theta_o \exp[-cW_T^{(p)}] \tag{24}$$

where θ_o is the initial asperity angle at the interface, c is

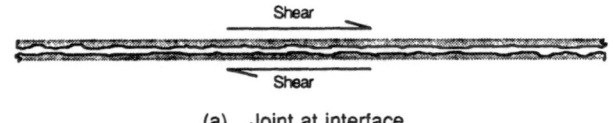

(a)   Joint at interface

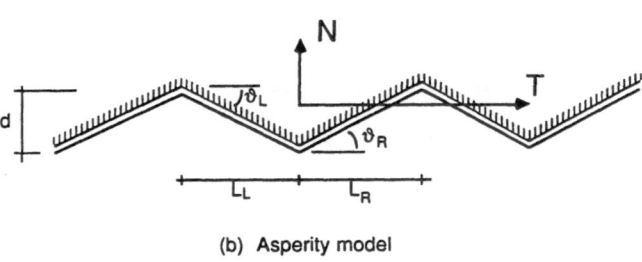

(b) Asperity model

Fig (3)   Dilation Model for Contact Problem

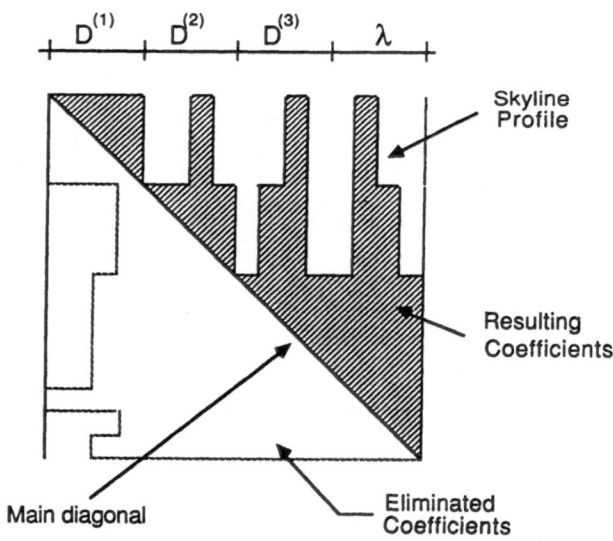

Fig (4)   The Profile System Matrix
Correspond to Fig(1)

the degradation constant defining the strength of the teeth and $W_T^{(p)}$ is the tangential plastic work done given incrementally as

$$\dot{W}_T^{(p)} = \sigma_T \dot{\Delta}_T^{(p)} \tag{25}$$

Therefore, the expression given by (19) can be replaced by

$$\frac{\partial F}{\partial \sigma_i} \dot{\sigma}_i + \frac{\partial F}{\partial \theta} \frac{\partial \theta}{\partial W_T^{(p)}} \dot{W}_T^{(p)} = 0 \tag{26}$$

The corresponding yield function is given by

$$F = [(\sigma_N \sin \theta + \sigma_T \cos \theta)^2]^{1/2}$$

$$+ \mu(\sigma_N \cos \theta - \sigma_T \sin \theta) \tag{27}$$

where μ is the coefficient of the friction at the asperity surface. The plastic potential function will be defined as

$$\phi = [(\sigma_N \sin \theta + \sigma_T \cos \theta)^2]^{1/2} + \text{constant} \tag{28}$$

where $F \neq \phi$, which is a non-associated plasticity.

The standard Coulomb friction surface condition can be approached by considering

$$\theta = 0 \; ; \quad K_{TN} = K_{NT} = 0$$

$$K_{TT} \rightarrow \infty \quad \text{and} \quad K_{NN} \rightarrow \infty \tag{29}$$

SYSTEM EQUATIONS

The assembly procedure given by Au[13] is implemented to formulate the system equation contributed from all subregions. For any degree of freedom within a subregion, a single row is used to store the coefficients resulting from the application of the boundary conditions S_1, S_2, S_3 and S_4 types. One can write a single row equation as

$$[-g_1^{(\alpha)} \quad h_2^{(\alpha)} \quad \pi_3^{(\alpha)}] \begin{bmatrix} P_1^{(\alpha)} \\ u_2^{(\alpha)} \\ u_3^{(\alpha)} \end{bmatrix} + [\pi_4^{(\alpha)}] [u_4^{(\beta)}] = b \tag{30}$$

where the subscripts (1) to (4) indicates the types of boundary; b is the resulting values from the prescribed boundary values,

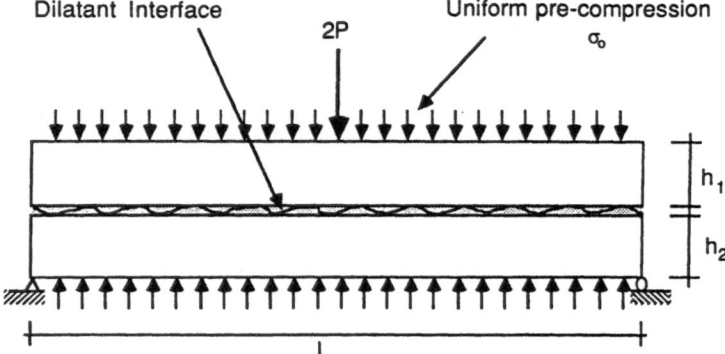

G_1 = Shear Modulus of the upper layer

G_2 = Shear Modulus of the lower layer

V_1 = Poisson ratio of the upper layer

V_2 = Poisson ratio of the lower layer

Fig (5)   Two elastic layers with a
             dilatant plane interface

i.e.,

$$b = [g_2^{(\alpha)}] \{\bar{P}_2^{(\alpha)}\} - [h_1^{(\alpha)}] \{\bar{u}_1^{(\alpha)}\} \tag{31}$$

and $[u_3^{(\alpha)}]$ are the displacements at S_3 which is in terms of element local system as

$$[u_3^{(\alpha)}] = \begin{bmatrix} u_n^{(\alpha)} \\ u_t^{(\alpha)} \end{bmatrix} \tag{32}$$

Similarly,

$$[u_4^{(\beta)}] = \begin{bmatrix} u_n^{(\beta)} \\ u_t^{(\beta)} \end{bmatrix} \tag{33}$$

The corresponding coefficient matrices are

$$[\pi_3^{(\alpha)}] = [[h_n^{(\alpha)} + g_n^{(\alpha)} K_{NN} - g_t^{(\alpha)} K_{TN}] ;$$

$$[h_t^{(\alpha)} - g_n^{(\alpha)} K_{NT} + g_t^{(\alpha)} K_{TT}]] \tag{34}$$

and

$$[\pi_4^{(\alpha)}] = [[g_n^{(\alpha)} K_{NN} - g_t^{(\alpha)} K_{TN}] ;$$

$$[-g_n^{(\alpha)} K_{NT} + g_t^{(\alpha)} K_{TT}]] \tag{35}$$

where all the K_{ij} in (34) and (35) can be obtained either from (15) or (20). For the single row equation in (30), the elimination of the coefficients can be applied based on the previously available system matrix coefficients. The resulting coefficients beyond the diagonal position will be stored in a one dimensional array as a profile system matrix as in Figure (4). The resulting system matrix will be stored in between the diagonal and the skyline. After considering all the degrees of freedom, back substitution phase is applied for the solution.

NUMERICAL RESULTS

The boundary element procedure discussed in the preceeding section has been applied to examine the behaviour of a composite two layer system with a dilatant interface. The problem under consideration consists of two identical elastic layers which are in contact with each other through a dilatant plane interface. The plane composite region is subjected to a uniform precompression σ_o and simply supported at the edges (Figure 5). The

interface region is modelled by a dilatant behaviour, the
fundamental response characteristics of which are shown in
Figure (6). The Figure (6) illustrates the manner in which
the normal and shear stress responses at the interface are
governed by the degradation parameters defined in equation (24).
The Figures (7) and (8) indicate the typical responses that
are obtained for the load vs central displacement of the
composite which is subjected to quasi-static load cycling
behaviour. These results also illustrate the manner in which
the non-linear dilatant responses can be influenced by the
degree of precompression imposed on the composite region. The
precompression stress is expressed as a proportion of the shear
modulus of the elastic regions.

REFERENCES

1. Gudehus, G., (Ed.) (1977). Finite Elements in Geomechanics,
    John Wiley, New York.
2. Desai, C.S. and Christian, J.T., (Eds.) (1977). Numerical
    Methods in Geotechnical Engineering, McGraw-Hill, New York.
3. Johnson, K.L. (1985). Contact Mechanics, Cambridge Univ-
    ersity Press., Cambridge.
4. Selvadurai, A.P.S. and Voyiadjis, G.Z., (Eds.) (1986),
    Mechanics of Material Interfaces, Studies in Applied
    Mechanics, Vol.II, Elsevier Scientific Publishing Co.,
    The Netherlands.
5. Selvadurai, A.P.S. and Au, M.C., (1986), Boundary Element
    Modelling of Interface Phenomena in Topics in Boundary
    Element Research, 5, (to appear).
6. Andersson, T. and Allan Persson, B.G., (1983). The boundary
    element method applied to two-dimensional contact problems,
    Ch.5 in Progress in Boundary Element Methods, (Brebbia,
    C.A., Ed.), CML Publications, U.K., 2, 136-157.
7. Paris, F. and Garrido, J.A. (1985), On the use of
    discontinuous element in two dimensional contact problems,
    Proc. of 7th Int. Conf. on Boundary Elements, Lake Como,
    Italy, Sept. 1985, 13:27-13:39.
8. Brebbia, C.A., (1978), The Boundary Element Method for
    Engineers, Pentech Press, London.
9. Mustoe, G.G.W., (1979), Coupling of boundary solution
    procedures and finite elements for continuum problems,
    Ph.D. Thesis, University of Wales, University College,
    Swansea.
10. Goodman, R.E. and St. John, C., (1977), Finite Element
    Analysis for Discontinuous Rocks in Numerical Methods in
    Geotechnical Engineering, Desai, C.S. and Christian, T.
    Eds., McGraw-Hill Book Co., New York, N.Y.
11. Heuze, F.E. and Barbour, T.G., (1982), New Models for Rock
    Joints and Interfaces, J. Geotechnical Engineering Division,
    ASCE, 108, 757-776.
12. Plesha, M. and Belytschko, T., (1986), On the modelling
    of contact problems with dilation, Studies in Applied
    Mechanics, Vol.11, Elsevier Scientific Publishing Co., The
    Netherlands.

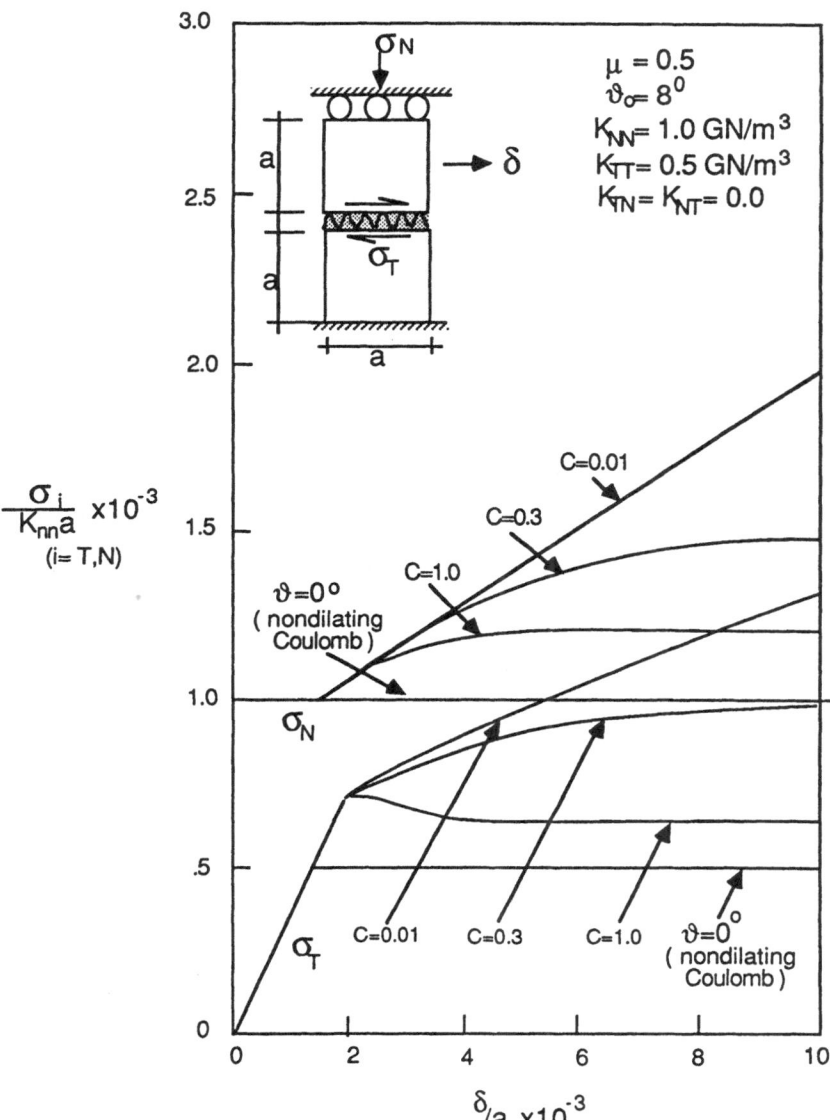

Fig (6)   The Tangential and Normal Stress due to the Tangential displacement (δ)

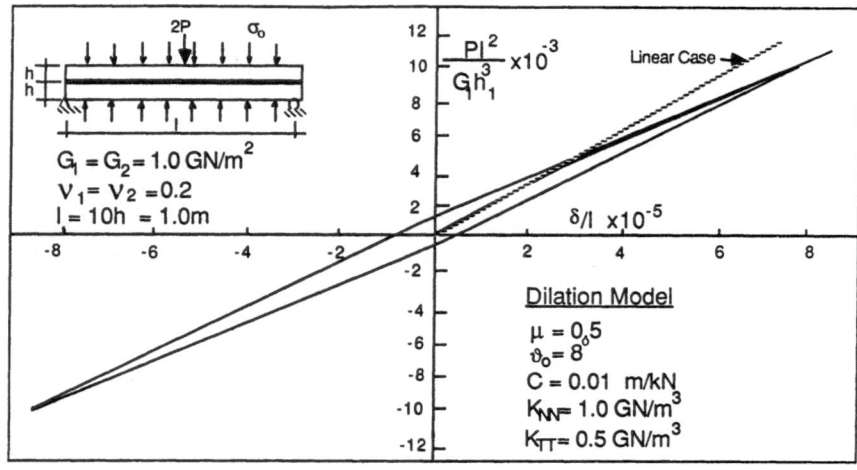

Fig(7)   Load-Dispacement Relationship for $\sigma_o/G_1 = 1.0 \times 10^{-5}$

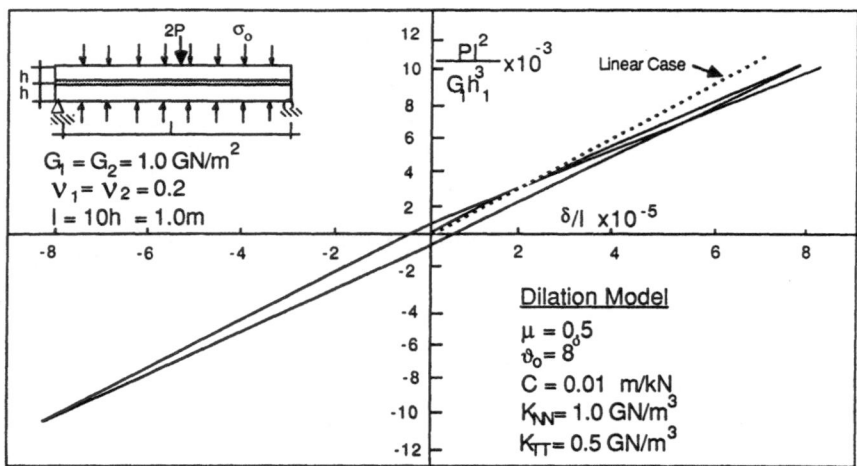

Fig (8)   Load-Displacement Relationship for $\sigma_o/G_1 = 2.0 \times 10^{-5}$

13.   Au, M.C., (1984), Boundary Element Direct Elimination
      Procedure in Multi-region Problems, Proc. 6th Int. Conf.
      on Boundary Element (Brebbia. C.A., Ed.), Springer, Berlin.

Optimum Positions for the Nodes in Discontinuous Boundary Elements

J.M. Xu, C.A. Brebbia
Computational Mechanics Institute, Southampton, U.K.

1. INTRODUCTION

This paper studies the optimum position of the nodes in discontinuous linear boundary elements, to obtain the most accurate results. Previous work on discontinuous elements has
demonstrated their advantages in representing edges and corners
and modelling discontinuous loads. Past researchers however
did not investigate the best position for the nodes but simply
put them in such a manner that for elements of equal size the
nodes were equidistant of each other.

Numerical investigations carried out by the present authors
demonstrate that the optimum positions for the nodes are those
corresponding to the Gauss integration points. Although further
studies are required to understand on a more mathematical basis
why this behaviour occurs , it is expected that the conclusions
obtained in the present paper for linear elements can also be
applied to higher order elements.

2. DIFFERENT TYPES OF ELEMENTS

In the early stages of BEM applications most codes used continuous elements such as the linear and quadratic elements
shown in figure 1. In practice however many problems present
discontinuities of tractions or fluxes due to corners or
different types of loadings such as occurred in the Points A
and B shown in figure 2. These discontinuities can be dealt
with, using several techniques.

The simplest way to treat a corner discontinuity is to
round off all edges or corners but this is not a satisfactory
solution as it fundamentally alters the character of the
problem.

Other approaches are the single-node representation [1],
[2],[3], multiple independent node [4] and the multiple

* Visiting Fellow from Congqing Institute of Architecture and
Engineering, China.

node concept with auxiliary relationships. The single-node
relationship assumes that the tractions as well as the displacements are the same for two adjacent elements, which may not be
the case at a corner. Hence the method can give very inaccurate
results. The multiple independent node consists of assuming two
nodes at the same point. The resulting equations are independent due to the different type of boundary conditions on one and
the other side. The idea of using corner nodes with auxiliary
relationships [5] can be useful in some problems but care should
be taken when working with the matrix system of equations as
auxiliary relationships may alter the reference system under
consideration as demonstrated by Georgiou [6]. This change in
reference system may be important when trying to combine
boundary elements with finite elements, joining together different boundary element regions, etc.

   While continuous elements are necessary in finite elements
to obtain positive definite matrices, they are not required in
BE theory. Indeed, the first - constant - boundary elements
demonstrated that continuity was not essential. These considerations motivated Brebbia and others [7][8] to develop a family
of higher order discontinuous elements which were successfully
applied to solve a series of engineering problems.

   The geometry of discontinuous elements is described in the
same way as for continuous ones but the nodes on which the
unknowns are taken and the singularities are positioned inside
the element as shown in figure 1. This allows,for interelement
discontinuities to occur and consequently do not present any
special difficulties when trying to model discontinuous tractions or fluxes.

   The main advantages of discontinuous elements is that they
completely solve the problem of representing edges or corners
and that they can model discontinuous loads. An added advantage of discontinuous elements is that they provide a simple
indication of the accuracy of the solution by looking at the
values of stresses, fluxes etc. in adjacent   elements. In
the cases in which these values ought to be the same, their
difference gives a simple and clear indication of the accuracy
of the solution. If their difference is larger than desired
the analyst should refine the mesh to increase the accuracy of
the results. In practice the most important advantage of discontinuous elements is that when using them, meshes can be
altered much more easily than if continuous elements are used.
Because of these advantages discontinuous boundary elements
should be incorporated in any well written BE code as otherwise
the full potential of the new method will not be fully exploited.

3. OPTIMUM POSITION FOR THE NODES IN DISCONTINUOUS ELEMENTS

Once the advantages of discontinuous boundary elements have
been accepted, the immediate question is where the nodes should

be placed. Previous work on linear and quadratic elements
established their position in such a manner that for elements
of equal size the nodes were equidistant of each other.
Patterson and Sheikh [9] and the BEASY code [10] adopted this
idea. In practice however, equidistant nodes are the exception
rather than the rule due to changes in element size and this
idea may not produce the most accurate results.

The present authors decided to study the best possible
position for the nodes inside the element in an effort to find
out a simple rule which yields high accuracy in all cases.
Results were compared with analytical solution for several
stress analysis problems.

4. BASIC THEORY

The boundary integral equation relating the displacements at a
source point 'S' on the boundary Γ to displacements and tractions at other points 'Q' over the same boundary and body forces
over the domain Ω, can be written as follows[11],

$$c_i U_i(S) + \int_\Gamma P^*(S,Q)U(Q)d\Gamma(Q) = \int_\Gamma U^*(S,Q)P(Q)d\Gamma(Q)$$

$$+ \int_\Omega U^*(s,q)B(q)d\Omega(q) \quad (1)$$

where P^* and U^* are the matrices of the fundamental solutions
for tractions and displacements respectively.

For two-dimensional isotropic plane strain case,

$$P^* = [p^*_{\ell k}] = -\frac{1}{4\pi(1-\nu)r}\left\{\frac{\partial r}{\partial n}[(1-2\nu)\Delta_{k\ell} + 2\frac{\partial r}{\partial x_k}\frac{\partial r}{\partial x_\ell}]\right.$$

$$\left. - (1-2\nu)\left[\frac{\partial r}{\partial x_\ell}n_k - \frac{\partial r}{\partial x_k}n_\ell\right]\right\} \quad (2a)$$

$$U^* = [u^*_{\ell k}] = \frac{1}{8\pi G(1-\nu)}\left[(3-4\nu)\ln(\frac{1}{r})\Delta_{\ell k} + \frac{\partial r}{\partial x_\ell}\frac{\partial r}{\partial x_k}\right] \quad (2b)$$

$$(\ell,k = 1,2 \quad)$$

in which   n   is the normal to the surface of the body

$\Delta_{\ell k}$ is the Kronecker delta

r   is the distance from the source point to the
point under consideration

n_j is the direction cosine.

U, P and B are the vectors of the displacement, tractions and body forces respectively.

Body forces B are always assumed to be prescribed but for simplicity they are ignored in this paper.

c_i is a coefficient. For a smooth boundary $c_i = \frac{1}{2}$; for a non-smooth boundary c_i can be obtained from the rigid body conditions [11]. In this paper discontinuous elements are used throughout, nodes are always inside the elements, i.e. the boundary at the nodes is always smooth, $c_i = \frac{1}{2}$.

Thus equation (1) can be written as

$$\frac{1}{2}U_i(S) + \int_\Gamma P^*(S,Q)U(Q)d\Gamma(Q) = \int_\Gamma U^*(S,Q)P(Q)d\Gamma(Q) \qquad (3)$$

The integral equation (3) is difficult to solve analytically, so a numerical approach to its solution is needed, in practice.

First the boundary of the object is discretized into N 'boundary' elements by mesh points. Inside each element there are two points, called nodes, as shown in Figure 3. The two nodes are positioned symmetrically to the middle of a boundary element and the distance between the middle and a node is called α, (Figure 4). Let us use the local dimensionless coordinate ξ, varying between -1 and $+1$. After discretization equation (3) becomes

$$\frac{1}{2}U_i + \sum_{j=1}^{N} \int_{\Gamma_j} P^*_{ij} U_j d\Gamma = \sum_{j=1}^{N} \int_{\Gamma_j} U^*_{ij} P_j d\Gamma \qquad (4)$$

Displacements and tractions are any points of an element are approximated from the nodal values in the element through interpolation functions.

For linear elements,

$$U_j(\xi) = [N_1 \ N_2] \begin{Bmatrix} U^1_j \\ U^2_j \end{Bmatrix} \qquad (5)$$

where
$$N_1 = \begin{bmatrix} \phi_1 & 0 \\ 0 & \phi_1 \end{bmatrix} = \phi_1[I] \qquad (6a)$$

$$N_2 = \begin{bmatrix} \phi_2 & 0 \\ 0 & \phi_2 \end{bmatrix} = \phi_2 [I] \tag{6b}$$

and

$$\phi_1 = \tfrac{1}{2}(1 - \frac{1}{\alpha} \xi) \tag{6c}$$

$$\phi_2 = \tfrac{1}{2}(1 + \frac{1}{\alpha} \xi) \tag{bd}$$

Similarly

$$P_j(\xi) = [N_1 \ N_2] \begin{Bmatrix} P_j^1 \\ P_j^2 \end{Bmatrix} \tag{7}$$

For expressions (5) and (7) $U_j^1(U_j^2)$ and $P_j^1(P_j^2)$ represent displacements and tractions of Node 1 (Node 2) in the jth element respectively.

Substituting equations (5), (6) and (7) into (4) yields

$$\sum_{k=1}^{NN} H_{ik} U_k = \sum_{k=1}^{NN} G_{ik} P_k \tag{8}$$

where NN is the total number of the nodes on the boundary of an object and NN = 2N.

U_k and P_k are vectors of displacements and tractions at Node k respectively and

$$U_k = U_{(k+1)/2}^1 \quad ; \quad P_k = P_{(k+1)/2}^1 \quad \text{(when k is an odd number)}$$

$$U_k = U_{k/2}^2 \quad ; \quad P_k = P_{k/2}^2 \quad \text{(when k is an even number)} \tag{9}$$

H_{ik} and G_{ik} are the components of influence matrices of BE, for which expressions see reference [12].

Applying equation (8) at each node in turn, i.e. i = 1,2,. ... ,NN, results in a system of linear algebraic equations relating all nodal displacements and tractions.

After introducing boundary conditions (either known nodal displacements or known nodal tractions) one can solve equation (8) and obtain all unknown nodal values.

The key to solve equation (8) is to calculate the components H_{ik} and G_{ik} of influences $\underset{\sim}{H}$ and $\underset{\sim}{G}$.

If the node i is on the integrated element Γ_j the H_{ik} and G_{ik} can be calculated analytically; if not, they are implemented numerically. In this paper numerical integration is calculated using Gaussian integration scheme.

Stresses and Displacements at Internal Points

Once the boundary values are known, stresses and displacements at any internal point m can be calculated by using the formulae below [12], i.e.

$$U_m = \sum_{k=1}^{NN} (-H_{mk} U_k + G_{mk} P_k) \tag{10}$$

$$\sigma_{ij} = \sum_{T=1}^{NN} \bar{D}_{ijT} P_T + \bar{S}_{ijT} U_T \tag{11}$$

where; (1) If T is an odd number

$$\bar{D}_{ijT} = \int_{\Gamma_{(T+1)/2}} \phi_1 D_{ij} d\Gamma \tag{12a}$$

$$\bar{S}_{ijT} = \int_{\Gamma_{(T+1)/2}} \phi_1 S_{ij} d\Gamma \tag{12b}$$

(2) If T is an even number

$$\bar{D}_{ijT} = \int_{\Gamma_{T/2}} \phi_2 D_{ij} d\Gamma \tag{12c}$$

$$\bar{S}_{ijT} = \int_{\Gamma_{T/2}} \phi_2 S_{ij} d\Gamma \tag{12d}$$

For expressions of D_{ij} and S_{ij} see [11].

Stresses on the Boundary

In most practical problems stresses at the boundary are needed.

Brebbia, Telles and Wrobel [13] outlined a simple procedure to obtain the stresses on the boundary in a local coordinate system. Their approach is followed in this paper.

The procedure for two-dimensional discontinuous linear elements in a global coordinate system is described in detail in reference [12].

5. NUMERICAL EXPERIMENTS TO OBTAIN THE OPTIMUM POSITION OF
    THE NODES

Once the code had been completed it was decided to carry out a
series of numerical experiments to find the optimum position
of the internal nodes for the linear elements. The examples
studied were representative of the different types of loadings
on two dimensional stress analysis, i.e.

i)   Square plate under uniaxial tension (Figure 5).
ii)   Simply supported beam with bending moments at the
      ends (Figure 6).
iii) Simply supported beam under uniform load (Figure 7).
iv)   Cantilever beam with end loading (Figure 8).

i) Square Plate under Uniaxial Tension
This example presents two axes of symmetry and consequently
only one quarter of the plate needs to be discretized as shown
in figure 5. The analytical solution for this problem is
simply

$$\sigma_x^e = 2 \quad ; \qquad \text{for point B;} \qquad \sigma_x^e = 2.0000$$

$$\sigma_y^e = 0.6$$

$$\tau_{xy}^e = 0$$

$$u^e = 0.364x \quad ; \quad \text{for point A;} \quad u^e = 0.7280$$

$$v^e = 0$$

The position of the nodes was moved in function of α and the
accuracy of the results was plotted in figure 9. The results
indicate that the accuracy of the numerical solution is very
high within a wide range of α ($\alpha = 0.0001$ to 0.90). For all
these α the results are within less than 0.05% of the exact
solution for the displacements and 0.01% for stresses. (See
Table 1).

ii) Simply Supported Beam with Bending Moments at the Ends
This application represents a case of pure bending, with one
axis of symmetry and another of antisymmetry as shown in
figure 6. The analytical solution is

$$\sigma_x^e = \frac{M}{I} y = y \quad ; \quad \text{for Point B ;} \quad \sigma_x^e = 1.8000$$

$$\sigma_y^e = \tau_{xy}^e = 0$$

$$u^e = \frac{M}{EI} \left(x - \frac{\ell}{2}\right) y = \frac{1}{E}\left(x - \frac{\ell}{2}\right) y$$

$$v^e = \frac{M}{2EI} (\ell-x)x - \frac{\nu M}{2EI} y^2 = \frac{1}{2E} [x(\ell-x) - \nu y^2] \quad ;$$

$$\text{for Point A ;} \quad v^e = 0.3771$$

Numerical and analytical solutions are shown in table 2 and the accuracy of the former is plotted in figure 10. For a range of α from 0.01 to 0.90 this solution was within 2% error for vertical displacements at the centre and less than 3% in plane stresses. As in the previous example no particular position was detected for the nodes to yield the most accurate solution.

iii) Simply Supported Beam under Uniform Load
In this case there is only one axis of symmetry as shown in figure 7. The analytical solution is

$$\sigma_x^e = -\frac{6q}{h^3} \left(\frac{\ell^2}{4} - x^2\right)y - q\frac{y}{h}\left(4\frac{y^2}{h^2} - \frac{3}{5}\right);$$

$$\text{for Point A;} \quad \sigma_x^e = 0.40364$$

$$\sigma_y^e = -\frac{q}{2}\left(1 - \frac{y}{h}\right)\left(1 + \frac{2y}{h}\right)^2;$$

$$\text{for Point B;} \quad \sigma_y^e = -0.5000$$

$$\tau_{xy}^e = \frac{6q}{h^3}x\left(\frac{h^2}{4} - y^2\right)$$

Table 3 and figure 11 show the accuracy of the numerical results in function of α. In this case the best position for the nodes is for $\alpha = 0.7746$ which is the position of a 3 point Gauss integration scheme. At that position the error in σ_x, σ_y stresses is smaller than for any other position (2.26% and 0.04% respectively).

iv) Cantilever Beam with End Loading
This beam has an axis of antisymmetry as shown in figure 8.

The beam theory gives

$$v_A^e = \frac{ph\ell^3}{3EI} = 5.0000$$

$$\sigma_{xB}^e = \frac{M_B y_B}{I} = 225$$

The accuracy of the solution varies with the position of the nodes as shown in Table 4 and figure 12. Notice that in this case a substantial number of linear elements are required to model the variation of curvature along the beam.

From the results shown in figure 12 it is also evident that the 2 point Gauss position at $\alpha = 0.5774$ gives a local minimum for the error and that this is also the case for the 3 point Gauss position with $\alpha = 0.7746$.

6. CONCLUSIONS

The work reported in this paper aims to improve the performance of discontinuous elements by finding the best positions for their nodes. These positions were defined as those for which the results were more accurate when compared against analytical solutions.

In general when the variation of stresses and displacements is not significant over the domain (such as is the case in examples i) and ii)) the position of the nodes is not important, except if they are set very near the middle or the extremities of the elements. When the variables vary significantly within the domain however, the position of the nodes is of fundamental importance to optimize the accuracy of the numerical solution.

It was found through numerical experimentation that the best positions were those corresponding to the Gauss integration points. Notice that in practice, the position used in many codes for linear elements is $\alpha = 0.5$ which produces a substantial error in many cases.

The above conclusion has recently been reinforced by numerical studies carried out by Chang [14] for elastodynamics problems. He also found that the optimum position for discontinuous 3-D boundary elements coincides with the Gauss integration points.

Further studies are required to understand on a more mathematical basis why this behaviour occurs and extend the present studies to quadratric and higher order elements.

REFERENCES

1. LACHAT, J.C.   Further Development of Boundary Integral Technique for Elastostatics, Ph.D. thesis, Southampton University, 1975.

2. LACHAT, J.C. and WATSON, J.O.   A Second Generation Boundary Integral Equation Program for Three-dimensional Stress Analysis, in T.A. Cruse and F.J. Rizzo (eds), Proc. ASME Conf. on Boundary Integral Equations, AMD-11, ASME, New York, 1975.

3. LACHAT, J.C. and WATSON, J.O.   Effective Numerical Treatment of Boundary Integral Equation, Int. J. Num. Meth. in Engng., 10, 991-1005, 1976.

4. RICCARDELLA, P.   An Implementation of the Boundary Integral Techniques for Plane Problems in Elasticity and Elasto-plasticity, Ph.D. Thesis, Carnegie Mellon University, Pittsburgh, 1973.

5. CHAUDONNERET, M.,   On the Discontinuity of the Stress
    Vector in the Boundary Integral Equation Method for Elastic
    Analysis,   In Recent Advances in Boundary Element Methods
    (C.A. Brebbia, Ed.), Pentech Press, London, 1978.

6. GEORGIOU, P. The Coupling of the Direct Boundary Element
    Method with the Finite Element Displacement Technique in
    Elastostatics. Ph.D. Thesis, Southampton University, 1981.

7. BREBBIA, C.A. The Boundary Element Method in Engineering
    Practice. International Journal of Engineering Analysis,
    Vol.1, No.1, January 1984.

8. BEASY Boundary Element Analysis System,
    Computational Mechanics Centre, MANUAL, Southampton, UK,
    1986.

9. PATTERSON, C. and SHEIKH, M.A. Interelement Continuity in
    the Boundary Element Method, In Topics in Boundary Element
    Research, Volume 1 (C.A. Brebbia, Ed.).

10. BREBBIA, C.A., D. DANSON & J. BAYNHAM, BEASY Boundary
    Element Analysis System,   in Finite Element Systems Hand-
    book, Computational Mechanics Publications and Springer
    Verlag, Southampton, NY, 1985.

11. BREBBIA, C.A. The Boundary Element Method for Engineers,
    Pentech Press, London, Comp. Mechanics, Boston, 1978,1985.

12. XU, J.M., C.A. BREBBIA AND D. NARDINI, A Two Dimensional
    Elastostatics Program using Discontinuous Boundary Elements,
    Submitted for Publication in Microsoftware for Engineers.

13. BREBBIA, C.A., L.C. WROBEL and J.C.F. TELLES, Boundary
    Element Method: Theory and Applications in Engineering,
    Springer-Verlag, Berlin, 1984.

14. CHANG, O.V. Boundary Elements Applied to Three Dimensional
    Elastodynamics Problems, submitted for presentation as
    Ph.D. thesis, Southampton University, 1986.

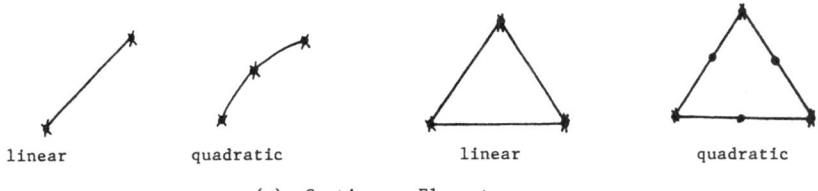

linear   quadratic   linear   quadratic

(a)   Continuous Elements

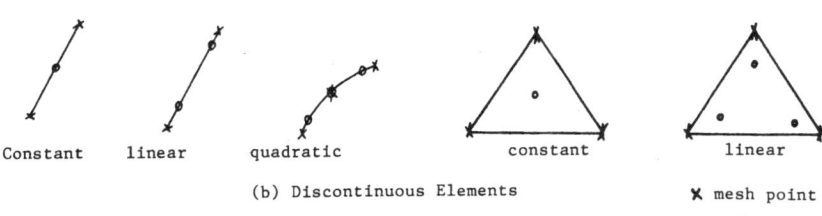

Constant   linear   quadratic   constant   linear

(b) Discontinuous Elements   ✗ mesh point

Fig. 1   Continuous and Discontinuous Elements   ○ node

Fig. 2   Problem about Edges   and Corners

Fig. 3 Discretized Boundary using Discontinuous
       Elements

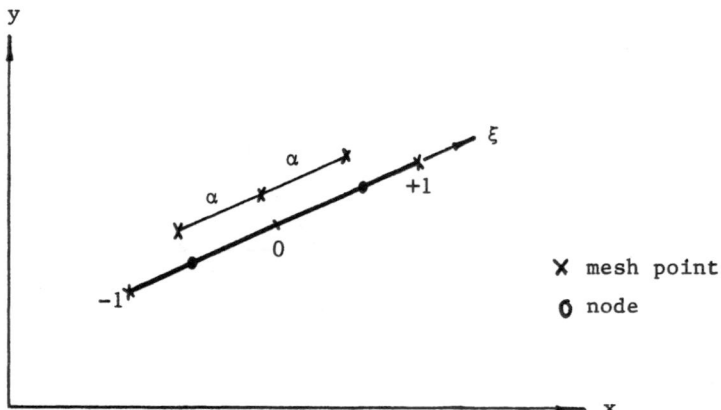

Fig. 4 Discontinuous Linear Element

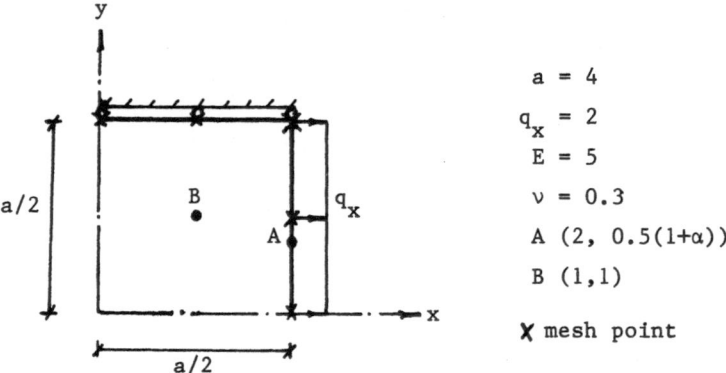

Fig. 5 Mesh for Square Plate under Uniaxial
       Tension

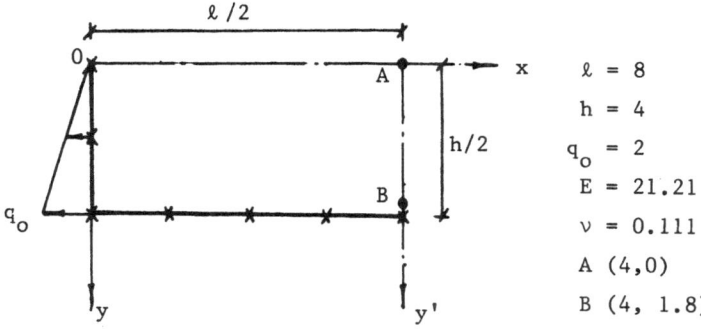

$\ell = 8$
$h = 4$
$q_o = 2$
$E = 21.21$
$\nu = 0.111$
A $(4,0)$
B $(4, 1.8)$
✗ mesh point

Fig. 6   Mesh for Simply Supported Beam
with bending Moments at Ends

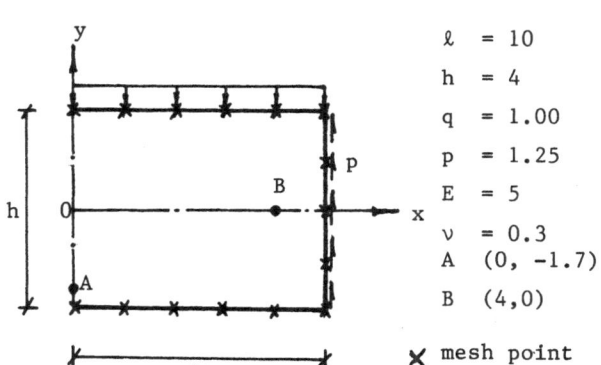

$\ell = 10$
$h = 4$
$q = 1.00$
$p = 1.25$
$E = 5$
$\nu = 0.3$
A $(0, -1.7)$
B $(4,0)$

✗ mesh point

Fig. 7   Mesh for Simply Supported Beam
under Uniform Load

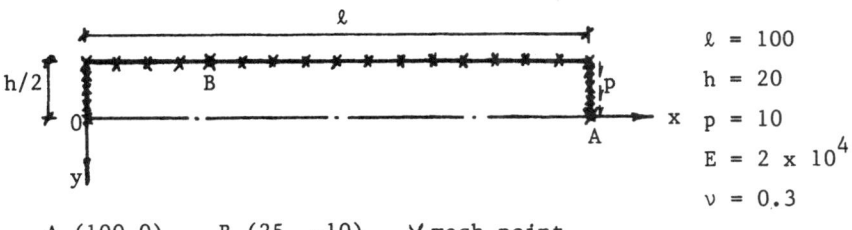

$\ell = 100$
$h = 20$
$p = 10$
$E = 2 \times 10^4$
$\nu = 0.3$

A (100.0),   B $(25, -10)$,   ✗ mesh point

Fig. 8   Mesh for Cantilever Beam with
End Loading

α	0.0001	0.0100	0.1000	0.3000	0.5000	0.5774	0.6000	0.7000	0.7746	0.8000	0.9000	0.9900	0.9990
u^e	0.7280												
u	0.7279	0.7279	0.7279	0.7279	0.7279	0.7279	0.7279	0.7279	0.7279	0.7279	0.7283	0.7270	0.6720
ε(%)	0.01	0.01	0.01	0.01	0.01	0.01	0.01	0.01	0.01	0.01	0.04	1.51	7.69
σ_x^e	2.0000												
σ_x	2.0000	2.0000	2.0000	2.0000	2.0000	2.0000	2.0000	2.0000	2.0000	2.0000	1.9999	1.9374	1.8728
ε(%)	0.00	0.00	0.00	0.00	0.00	0.00	0.00	0.00	0.00	0.00	0.01	3.13	6.36

Table 1   Influence of Node Positions on Results

< Square Plate under Uniaxial Tension >

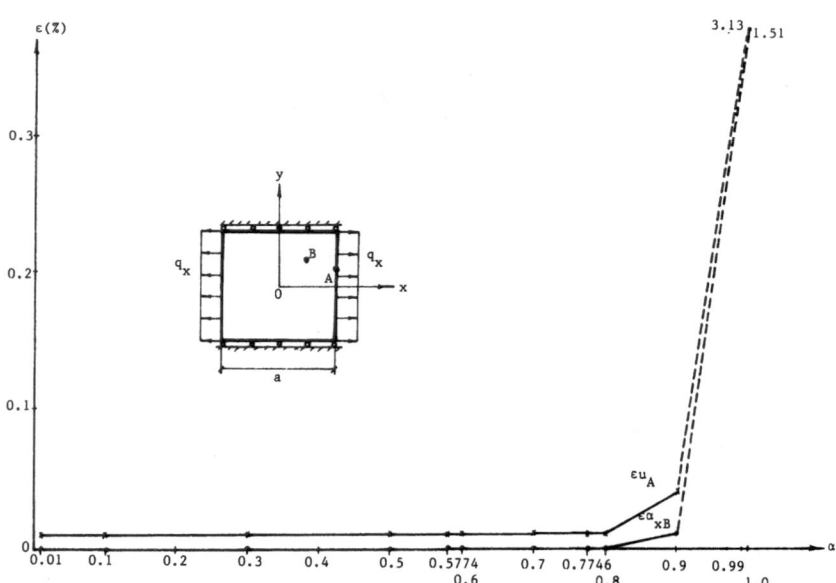

Fig.   9   Square Plate under Uniaxial Tension

α	0.0010	0.0100	0.1000	0.3000	0.5000	0.5774	0.6000	0.7000	0.7746	0.8000	0.9000	0.9500
v^e						0.3771						
v	0.2228	0.3740	0.3773	0.3773	0.3769	0.3767	0.3766	0.3761	0.3757	0.3754	0.3696	0.4221
$\varepsilon(\%)$	40.92	0.82	0.05	0.05	0.05	0.11	0.13	0.27	0.37	0.45	1.99	11.93
σ_x^e						1.8000						
σ_x	1.5742	1.7853	1.7899	1.7892	1.7864	1.7845	1.7838	1.7802	1.7770	1.7753	1.7482	1.9625
$\varepsilon(\%)$	12.54	0.82	0.56	0.60	0.76	0.86	0.90	1.10	1.28	1.37	2.88	9.03

Table 2   Influence of node Positions on Results

< Simply Supported Beam with Bending Moments at Ends >

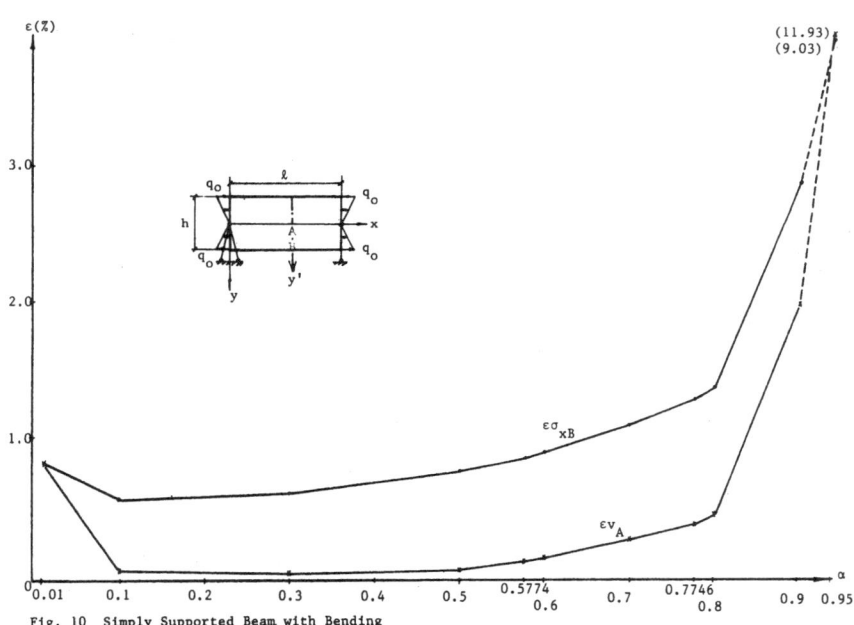

Fig. 10   Simply Supported Beam with Bending
         Moments at Ends

α	0.0500	0.1000	0.3000	0.5000	0.5774	0.6000	0.7000	0.7746	0.8000	0.9000	0.9500	0.9900
σ_x^e	4.0364											
σ_x	3.8899	3.8954	3.9136	3.9288	3.9340	3.9354	3.9411	3.9450	3.9451	3.8982	4.3541	2.6902
$\varepsilon(\%)$	3.63	3.49	3.04	2.67	2.54	2.50	2.36	2.26	2.26	3.43	7.87	33.35
σ_y^e	-0.5000											
σ_y	-0.6069	-0.5452	-0.5037	-0.4981	-0.4980	-0.4981	-0.4990	-0.5002	-0.5007	-0.5012	-0.5224	-0.4696
$\varepsilon(\%)$	21.38	9.04	0.74	0.38	0.38	0.38	0.20	0.04	0.14	0.24	4.48	6.08

Table 3   Influence of Node Positions on Results

< Simply Supported Beam under Uniform Load >

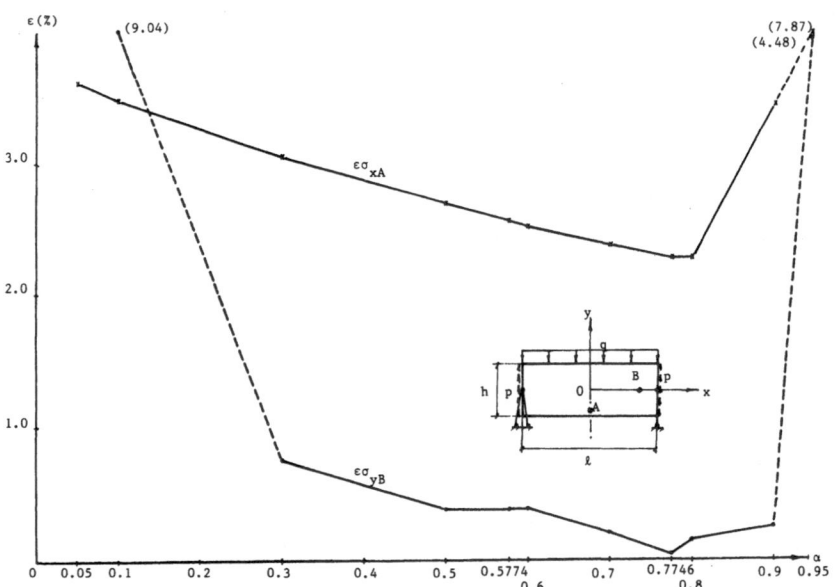

Fig. 11   Simply Supported Beam under Uniform Load

α	0.0001	0.0100	0.1000	0.3000	0.5000	0.5774	0.6000	0.7000	0.7746	0.8000	0.8100	0.8200
v^e	5.0000											
v	5.3975	5.3974	5.3875	5.3168	5.1788	5.0336	4.9661	4.5643	4.7898	5.5344	6.1390	7.1295
$\varepsilon(\%)$	7.95	7.95	7.75	6.34	3.58	0.67	0.68	8.71	4.20	10.69	22.78	42.59
σ_x^e	225.00											
σ_x	225.91	225.90	225.53	222.90	217.79	212.26	209.67	194.37	204.46	234.94	259.55	299.79
$\varepsilon(\%)$	0.40	0.40	0.24	0.93	3.20	5.66	6.81	13.61	9.13	4.42	15.36	33.24

Table 4   Influence of Node Positions on Results

< Cantilever Beam with End Loading >

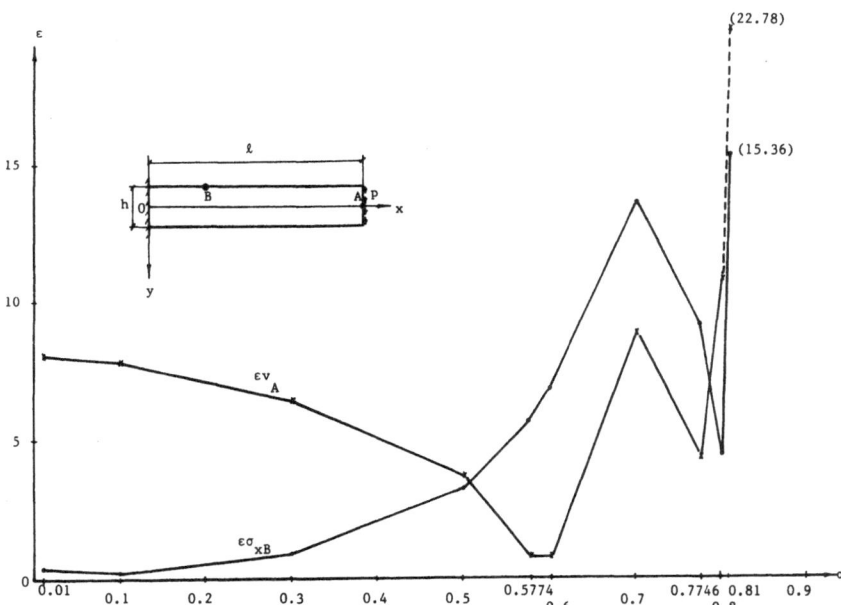

Fig. 12 Cantilever Beam with End Loading

SECTION XIV   FLUID MECHANICS

Three-Dimensional Analysis of Potential Flow in a Centrifugal Impeller by Panel Method

Y. Miyake, K. Bando and Y. Masuda
Department of Mechanical Engineering, Osaka University, 2-1 Yamada-oka, Suita, Osaka 565, Japan
S. Nagamatsu
Toyota Motor Company, 1 Toyota, Toyota, Aichi 471, Japan

INTRODUCTION

Widely used quasi three-dimensional method for the analysis of flow in turbomachine impellers gives satisfactorily accurate results at the design flow rate.[1,2] However, recently it is needed to develop a new method which is more accurate and applicable to wide range of flow rate to meet the demand for a high performance turbomachine impeller. For this purpose the fully three-dimensional analysis of flow in the impeller is necessary. Due to the rapid development of computers, reports on the fully three-dimensional analysis are more and more increasing recently, e.g., for inviscid flows the finite element method[3-5] and the time marching method.[6-9] Very few reports have been published using the singularity method[10,11] for axial-flow impellers. These methods[10,11] are the extention of a lifting line theory and capable of including the effect of blade thickness. However, they deal with only radial element blades in an annular flow passage of a free vortex type of flow. Panel method[12] widely used for the analysis of external flow about, e.g., aircraft can deal with arbitrary blade shape. However, if it is applied to the internal flow analysis with low-order singularities, an accurate solution can not be available due to the flow leakage between control points[13]. Because of this difficulty reports for internal flows based on panel methods have been limited to the analysis of simple internal flows, e.g. nacelles[14].

In this report it is intended to present a procedure of analysis to give the accurate results of three-dimensional, inviscid, incompressible flow in a centrifugal impeller by a panel method. The present method is an improved version of the panel method[15] presented by the authors for the axial-flow impeller, applying to more complicated flow in the centrifugal impeller.

In order to give calculation examples for this method, a centrifugal impeller is designed using the conventional procedure and
three calculation results, namely, three-dimensional solution
by the panel method, two-dimensional one by the blade-to-blade
method and one-dimensional one obtained from the design process
are compared to clarify the degree of approximation of each
flow and to demonstrate strong three-dimensionality of the
flow in this centrifugal impeller.

ANALYSIS BY THE PANEL METHOD

The equation governing a three-dimensional, inviscid, incompressible flow in an impeller is $\nabla^2\Phi=0$, where Φ is a velocity
potential. Let this velocity potential express as $\Phi=\phi+\phi'$,
where ϕ and ϕ' are harmonic functions, ϕ' being introduced to
make the numerical accuracy of Φ best. Applying Green's identity to the equation $\nabla^2\phi=0$ and taking into account the flow
periodicity give the following boundary integral equation.

$$2\pi\phi_i - \sum_{m=1}^{N}\!\!\iint_{A_m}\!\!\!\!\!' \frac{\partial}{\partial n}\left(\frac{1}{R}\right)\phi dS = \sum_{m=1}^{N}\iint_{W_m}\frac{\partial}{\partial n}\left(\frac{1}{R}\right)\Delta\phi_W dS$$
$$- \sum_{m=1}^{N}\!\!\iint_{A_m}\!\!\!\!\!' \frac{1}{R}\frac{\partial\phi}{\partial n} dS \qquad \cdots(1)$$

where the subscript i denotes a control point on the boundary,
N the number of blades, $\iint'$ Cauchy's principal value integral,
A_m and W_m the one-pitch surfaces of the impeller boundary and
that of the wake, respectively, $\partial/\partial n$ the normal derivative into
the fluid domain on the boundary, R the distance between the
control point and the dummy integration point and $\Delta\phi_W$ the
potential jump across the wake surface.

The inlet surface of the impeller is taken perpendicularly to
the rotating axis far upstream and the outlet one as a cylindrical surface far downstream. The boundary conditions are then
given as

$$V \equiv w\cdot n = \begin{cases} 0 & : \text{on blade, hub, shroud surfaces} \\ c_z(=\text{const.}) & : \text{on inlet surface} \\ -c_r & : \text{on outlet surface} \end{cases} \qquad \cdots(2)$$

where W denotes the relative velocity vector, and c_z and c_r the
axial and radial components of the absolute velocity vector C,
respectively. c_z is determined from the flow rate Q. c_r is
assumed to be constant in the circumferential direction and to
vary satisfying the following two equations[5].

$$\frac{\partial c_r^2}{\partial z} = \frac{\omega}{\pi}\frac{\partial\Gamma}{\partial z} - \frac{1}{(2\pi r)^2}\frac{\partial\Gamma^2}{\partial z} \qquad \cdots(3)$$

$$2\pi\int_{\text{Shroud}}^{\text{Hub}} rc_r dz = Q \qquad \cdots(4)$$

where ω denotes the rotational angular velocity and $\Gamma=2\pi rc_\theta$ the
circulation. The vectors C and W are dependent with each other
as

$$w = c - \omega r e_\theta \qquad \cdots (5)$$

where e_θ denotes the unit circumferential vector. Equations(2-5) give the value of $\partial\phi/\partial n$ on the boundary as

$$\frac{\partial\phi}{\partial n} = \omega r e_\theta \cdot n + V - \frac{\partial\phi'}{\partial n} \qquad \cdots (6)$$

The geometry of wake surface is assumed to be along the averaged downstream flow of the impeller blades expressed by

$$w_r = \frac{Q}{2\pi rB} , \quad w_\theta = \frac{K}{r} - \omega r , \quad w_z = 0 \qquad \cdots (7)$$

where B is the width of diffuser and $K=rc_\theta$ remains constant along a stream line, the value of which is calculated at the blade trailing edge such that

$$K = r(\omega r - w_r \cot\beta) \qquad \cdots (8)$$

where β is the angle of blade. On the wake surface $\partial\phi/\partial n$ vanishes and $\Delta\phi_w$ retaining a constant value along the stream line of Equation(7) is determined from the Kutta condition as

$$\Delta\phi_w = \Delta\phi_t \qquad \cdots (9)$$

where the subscript t denotes the blade trailing edge. Therefore, $\Delta\phi_w$ is expressed in terms of potentials on the blade surface.

The panel shape, the distribution of ϕ and $\partial\phi/\partial n$ on it and the calculation procedure are the same as those reported in our previous paper[15]. Solving a system of linear equations gives ϕ at the control points on the boundary and the velocity and the pressure are determined by

$$w = \nabla\Phi - \omega r e_\theta \qquad \cdots (10)$$

$$p = p_{st} + \frac{\rho}{2}(\omega^2 r^2 - w^2) \qquad \cdots (11)$$

where ρ is the density of fluid and the subscript st means the upstream stagnation point. The velocity at points interior to the boundary is calculated by

$$4\pi c = \sum_{m=1}^{N} \iint_{A_m} \left\{ \frac{n}{R^3} - \frac{3(n \cdot R)R}{R^5} \right\} \phi dS + \sum_{m=1}^{N} \iint_{W_m} \left\{ \frac{n}{R^3} - \frac{3(n \cdot R)R}{R^5} \right\} \Delta\phi_w dS$$

$$- \sum_{m=1}^{N} \iint_{A_m} \frac{R}{R^3} \frac{\partial\Phi}{\partial n} dS \qquad \cdots (12)$$

Equation(12) which is obtained by the analytical differentiation of Equation(1) gives more accurate solution than using numerical differentiation of Φ. In order to calculate Equation (12) numerically, one panel is divided into subpanels on each of which the singularity strength is constant. The velocity influence coefficient is then expressed analytically[12].

NUMERICAL EXAMPLE FOR TWO-DIMENSIONAL IMPELLER

The three-dimensional panel method mentioned above is applied to a virtually two-dimensional impeller and the results are compared with exact two-dimensional solutions in order to examine the computational accuracy. Figure 1 shows the impeller, the blade leading and trailing edge radii being .08m and .15m, respectively, the rotational velocity 2000r.p.m, the flow rate 0.2m^3/s, the logarithmic spiral blade center line angle 30.7

deg., the blade width 0.04m and
the number of panels 464. The
calculated results are shown in
Figures 2 and 3. Figure 2 shows
the pressure coefficient distribution along the blade element defined by

$$c_p = (r/r_t)^2 - (w/\omega r_t)^2 . \quad (13)$$

H in Figure 2 is a distance
from the leading edge along the
blade surface. The two-dimensional solution obtained by the
singularity method is almost
exact because sufficient number
of control points are assigned
on the blade. The three-dimensional solution obtained by the
panel method hardly varies in
the blade spanwise direction
and agrees well with the two-
dimensional one. Figure 3 shows
the calculated flow rate across
the axi-symmetric cross sections located between the blades to compare with the prescribed flow rate. LE and TE
in Figure 3 are the leading and
trailing edges, respectively.
According to these results, it
is confirmed that the sufficiently accurate solution is obtained for the two-dimensional
centrifugal impeller and that
the present three-dimensional
panel method is valid.

NUMERICAL EXAMPLES FOR A THREE-
DIMENSIONAL IMPELLER

Design of the centrifugal
impeller
A centrifugal impeller is designed as follows and it is
used as a model impeller to
compare the predictions of flow
by the present 3D, approximated
2D and 1D methods. The impeller
meridional configuration is
prescribed and this is divided
into many part-impellers. Let
the pressure difference across
the blade be expressed as

Figure 1. Two-dimensional
impeller shape

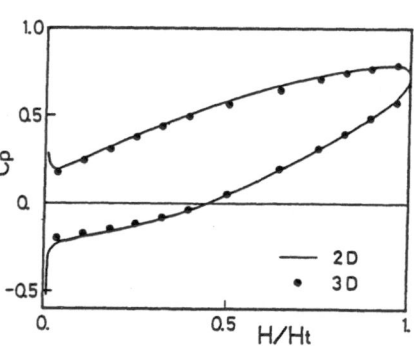

Figure 2. Pressure distribution
along blade element

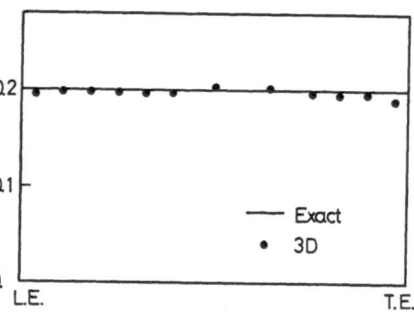

Figure 3. Flow rate between
blades

$$\Delta p = k_p f_p \quad (0 \leq f_p \leq 1) \qquad \cdots (14)$$

where k_p is constant determined to satisfy a required head
and f_p is a mode function of blade loading selected as in Figure
4, in which s denotes a distance along the meridional line.
By equating the force moment of the impeller and the angular
momentum flow rate one obtains the averaged circumferential
absolute velocity $\overline{c}_\theta$ as the flow rate Q, the head H, the number
of blades N, the rotational angular velocity ω, the pre-rotation
$\lambda = rc_\theta$ and f_p in Equation(14) are prescribed. The averaged meridional velocity $\overline{c}_s$ is calculated assuming that the blade thickness is small and that the averaged flow direction coincides
with the blade centerline. The distribution of blade angle from
leading to trailing edges is determined from the $\overline{c}_\theta$ and $\overline{c}_s$ thus
obtained. Then the impeller configuration is determined. The
blade thickness t_n and the blade angle β are shown in Figures 5
and 6, respectively. The $\overline{c}_\theta$, $\overline{c}_s$, β, Δp and Bernoulli's equation
for the relative flow give the magnitudes of flow velocity and
pressure on both sides of the blade.

Figure 7 shows the prescribed meridional configuration of impeller. The impeller specifications are as follows: the blade
trailing edge radius is .15m, the flow rate 0.201m^3/s, the head
594.6Pa being constant in the blade spanwise direction, the
number of blades 8, the rotational velocity 2000r.p.m, the pre-
rotation zero and the trailing edge rake angle zero.

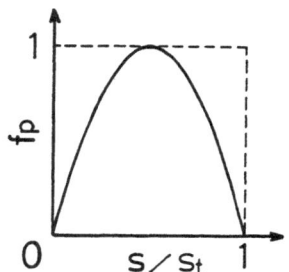

Figure 4. Mode function of
blade loading

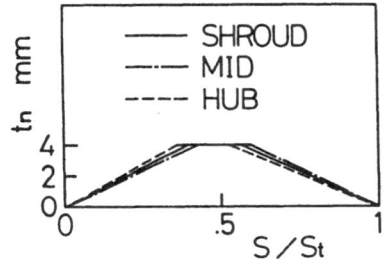

Figure 5. Blade thickness
distribution

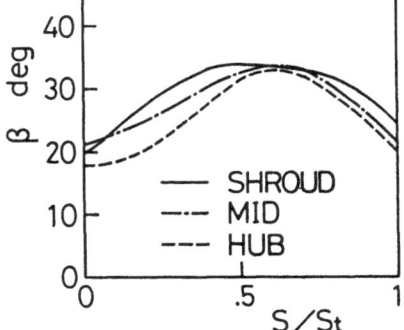

Figure 6. Blade angle distribution

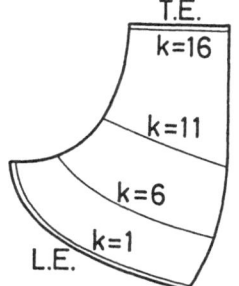

Figure 7. Impeller meridional
configuration and cross
sections between blades

Figure 8 shows the panel network system of the impeller designed
above. The numbers in the figure mean: [1]inlet surface, [2]out-
let surface, [3]hub surface before blades, [4]shroud surface
before blades, [5]hub surface between blades, [6]shroud surface
between blades, [7]hub surface after blades, [8]shroud surface
after blades, [9]pressure surface of blade, [10]suction surface
of blade, [11]wake surface. The total number of panels is 765.

Analysis by panel method
The flow calculations are carried out at the 100% and 75% of
designing flow rate.

Figure 9 shows the calculated flow rates across the axi-symmetric cross sections between the blades located in Figure 7.
It is shown that the accurate solutions satisfying the continuity condition are obtained.

Figures 10 and 11 show the pressure distributions along the
blade elements on hub, shroud
and mid stream surfaces of axisymmetric potential flow. The
chain curve denotes the one-
dimensional solution obtained
from the impeller design process, the solid curve denotes
the two-dimensional solution
obtained from the blade-to-
blade calculation on the axisymmetric stream surface given
by the potential hub-to-shroud
flow and the solid circle denotes the three-dimensional
solution obtained by the present panel method. The designed
pressure difference across the
blade increases from hub to
shroud, but the resulting head
is constant in the blade spanwise direction. This is because
that the meridional length of

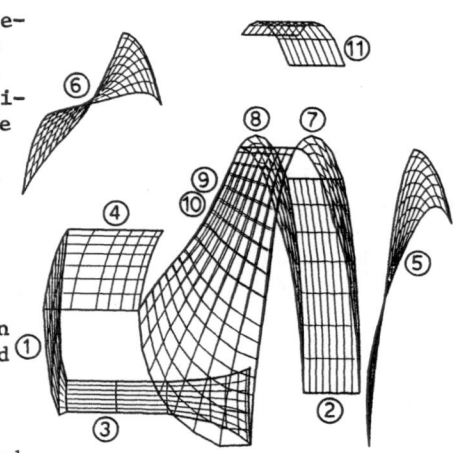

Figure 8. Panel network system
of the centrifugal
impeller one-pitch
surface

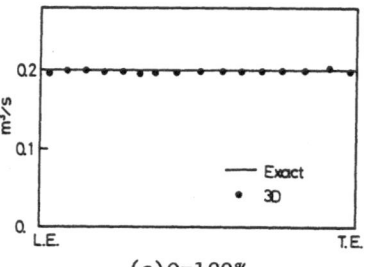

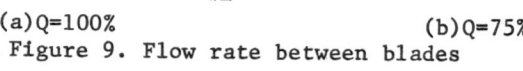

(a)Q=100%   (b)Q=75%
Figure 9. Flow rate between blades

blade element becomes smaller and the velocity becomes higher
from hub to shroud. The two- and three-dimensional solutions
agree quite well with one another on the hub surface. This is
because the configuration of part-impeller on the hub employed
for the calculation of three-dimensional flow and that of axisymmetric potential flow are almost identical. This configuration scarcely changes when a small flow attack takes place at
leading edge, as shown in Figure 11(a) for the 75% flow rate
case. The difference between the solid and chain curves in Figure 10(a) is due to the flow slip increasing towards the trailing edge, since in designing the impeller the fluid is assumed

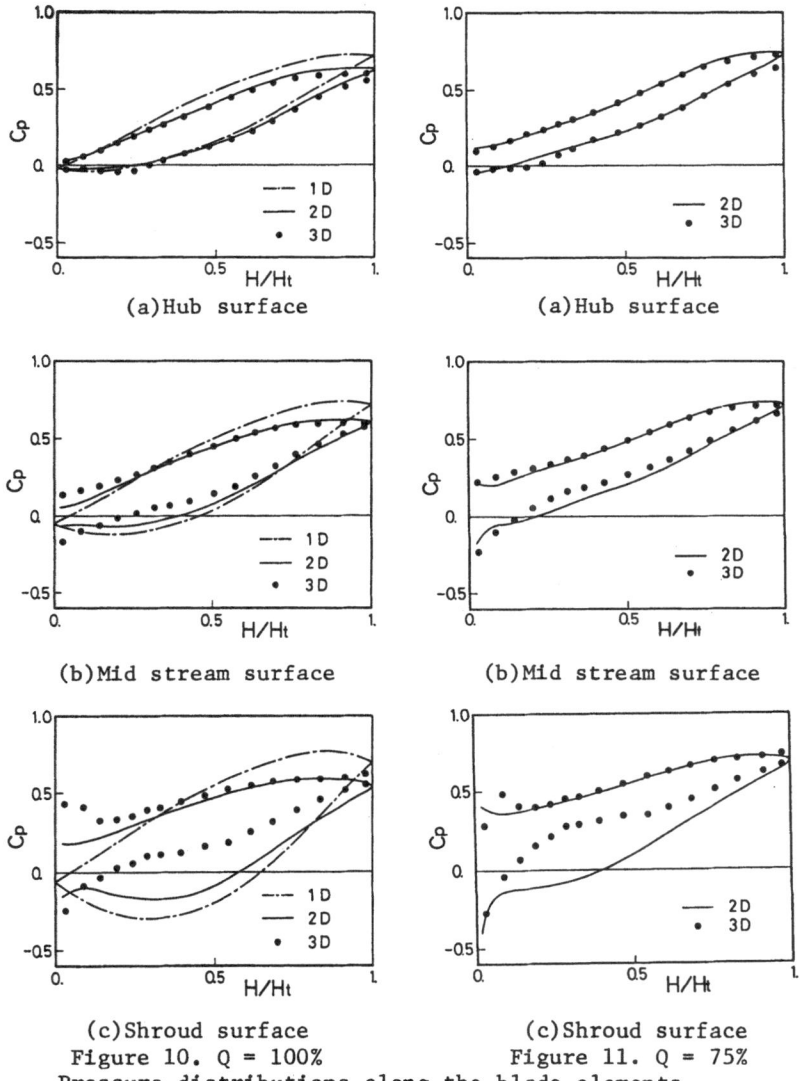

(a)Hub surface   (a)Hub surface

(b)Mid stream surface   (b)Mid stream surface

(c)Shroud surface   (c)Shroud surface
Figure 10. Q = 100%   Figure 11. Q = 75%
Pressure distributions along the blade elements

to flow along the blade centerline. At the designing flow rate there exists the flow attack on the mid and shroud surfaces due to a large leading edge blade angle near the shroud. The attack is small on the mid surface so that the two- and three-dimensional solutions agree well. On the shroud surface the flow attack is larger than that on the mid surface. Therefore, the two- and three-dimensional solutions differ there due to strong reverse flow accompanying spanwise flow that occurs near the leading edge in the three-dimensional solution. However, the attack is a local phenomenon so that both solutions yield slight difference with respect to the blade pressure surface except near the leading edge.

Figures 12 through 15 show the pressure coefficient contours and the relative velocity vectors on the blade pressure and suction surfaces. These figures indicate the flow attack, the reverse flow and the spanwise flow on the blade pressure surface near the shroud. It is recognized that the centrifugal force has the dominant effect on the pressure rise as shown in Figures 12 and 13.

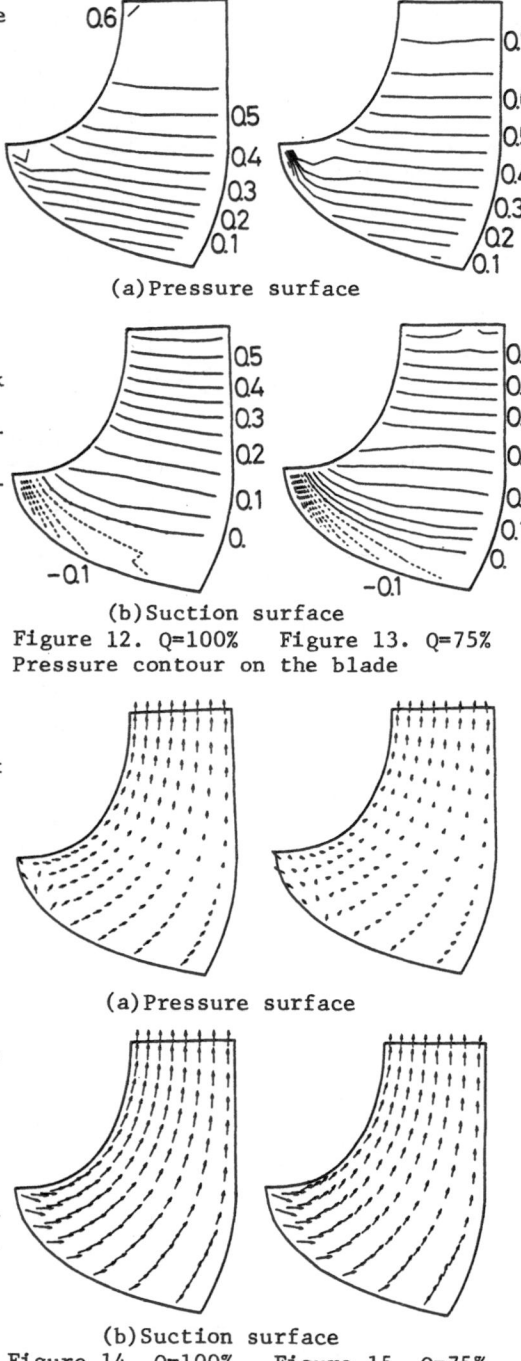

(a) Pressure surface

(b) Suction surface

Figure 12. Q=100%   Figure 13. Q=75%
Pressure contour on the blade

(a) Pressure surface

(b) Suction surface

Figure 14. Q=100%   Figure 15. Q=75%
Relative velocity vector on the blade

Figure 16 shows the contour of the normal relative velocity to
the axi-symmetric cross section located just after the leading
edge(k=1) in Figure 7. The right, left, upper and lower side
lines correspond to the pressure, suction, shroud and hub surfaces, respectively. The broken curve denotes the reverse flow
in the meridional plane. The reverse flow on the pressure side
due to the flow attack is shown. However, the reverse flow region shown in Figure 16 is rather exaggerated. That is, near
the stagnation point on the blade surface, even the normal
streamlines locally direct to the impeller inlet and these local
pseudo-reverse region is also included as the reverse region in
Figure 16. Therefore, the net reverse flow region is smaller
than that shown in Figure 16.

Figure 17 shows the pressure coefficient contour on the hub and
shroud surfaces at the 100% flow rate.

Figure 18 shows the spanwise local head distribution in the
blade outlet surface. T in the figure denotes the spanwise distance from the shroud. ψ is the head coefficient defined by

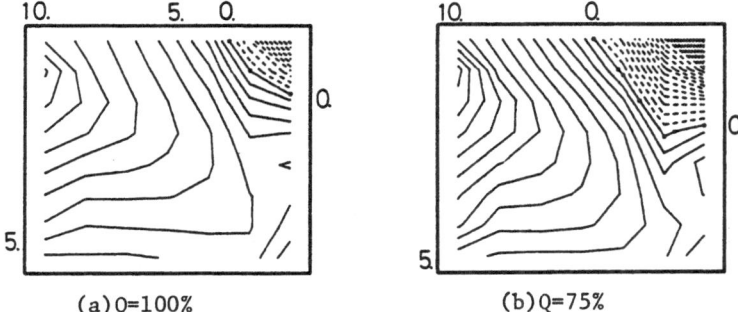

(a)Q=100%   (b)Q=75%

Figure 16. Contour of normal relative velocity to the axisymmetric cross section located just after the
leading edge

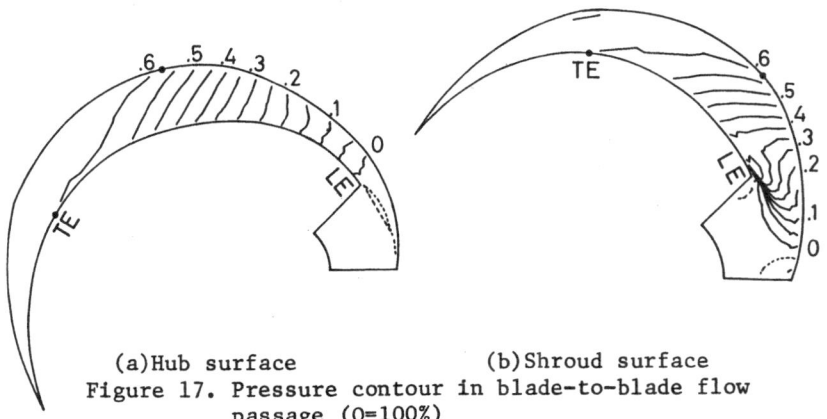

(a)Hub surface   (b)Shroud surface
Figure 17. Pressure contour in blade-to-blade flow
passage (Q=100%)

$$\psi = 2gh/(\omega r_t)^2 \qquad\qquad \cdots (15)$$

where g is the acceleration of gravity and h is the local head
in the blade outlet surface calculated by

$$h = \frac{\omega r_t}{g} \frac{\overline{c_r c_\theta}}{\overline{c_r}} \qquad\qquad \cdots (16)$$

The chain curve, the solid curve and the solid circle are the
heads obtained by the design, the two-dimensional method and
the panel method, respectively. The difference between the
chain and solid curves indicates the flow slip increasing from
hub to shroud. The head distributions obtained by the three-
dimensional solutions have characteristics which the two-dimensional head distributions are relaxed.

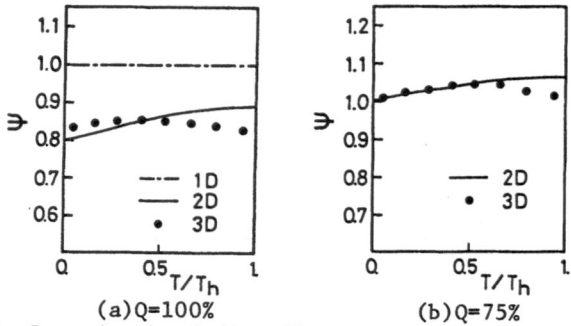

(a) Q=100%   (b) Q=75%

Figure 18. Spanwise head distribution in the blade outlet
surface

CONCLUSION

A panel method to analyse accurately a three-dimensional, inviscid, incompressible flow in a centrifugal impeller is presented. This method can be applied to axial- and mixed-flow impellers by slight modification of boundary conditions. The results for the centrifugal impeller show that the three-dimensional analysis is necessary in the case where the strong reverse flow occurs near the blade leading edge although the two-
dimensional calculation is valid to some extent because of the
localization of the reverse flow. The blade spanwise head distribution obtained by the three-dimensional method has the
characteristic that the two-dimensional head distribution is
relaxed.

REFERENCES

1. Adler D. and Krimerman Y. (1979), Comparison between the Calculated Subsonic Inviscid Three-Dimensional Flow in a Centrifugal Impeller and Measurements, Performance Prediction of
Centrifugal Pumps and Compressors, (Ed. Gopalakrishnan S.,
et al.), pp.19-26, ASME.
2. Jennions I.K. and Stow P. (1985), A Quasi-Three-Dimensional
Turbomachinery Blade Design System: Part II – Computerized
System, ASME Journal of Engineering for Gas Turbines and
Power, Vol.107, No.2, pp.308-316.

3. Laskaris T.E. (1978), Finite-Element Analysis of Three-Dimensional Potential Flow in Turbomachines, AIAA Journal, Vol.16, No.7, pp.717-722.
4. Nagafuji T. and Morii H. (1980), A Flow Study in Francis Turbine Runner, pp.583-594, Proceedings of IAHR 10th Symp., Tokyo, Japan, 1980.
5. Daiguji H. (1983), Finite Element Analysis for 3-D Compressible Potential Flow in Turbomachinery, pp.455-462, Proceedings of International Gas Turbine Congress, Tokyo, Japan, 1983.
6. Sarathy K.P. (1982), Computation of Three-Dimensional Flow Fields through Rotating Blade Rows and Comparison with Experiment, ASME Journal of Engineering for Power, Vol.104, No.2, pp.394-402.
7. Singh U.K. (1982), A Computation and Comparison with Measurements of Transonic Flow in an Axial Compressor Stage with Shock and Boundary Layer Interaction, ASME Journal of Engineering for Power, Vol.4, No.2, pp.510-515.
8. Hove W.V. (1984), Calculation of Three-Dimensional, Inviscid, Rotational Flow in Axial Turbine Blade Rows, ASME Journal of Engineering for Gas Turbines and Power, Vol.106, No.2, pp.430-436.
9. Holmes D.G. and Tong S.S. (1985), A Three-Dimensional Euler Solver for Turbomachinery Blade Rows, ASME Journal of Engineering for Gas Turbines and Power, Vol.107, No.2, pp.258-264.
10. Tamura A. and Lakshminarayana B. (1976), Assessment of Three-Dimensional Inviscid Effects in Turbomachinery Using Simple Models, ASME Journal of Fluids Engineering, Vol.98, No.2, pp.163-172.
11. Howells R. and Lakshminarayana B. (1977), Three-Dimensional Potential Flow and Effects of Blade Dihedral in Axial Flow Propeller Pumps, ASME Journal of Fluids Engineering, Vol.99, No.1, pp.167-175.
12. Johnson F.T. (1980), A General Panel Method for the Analysis and Design of Arbitrary Configurations in Incompressible Flows, NASA Contractor Report 3079.
13. Bristow D.R. and Grose G.G. (1978), Modification of the Douglas Neumann Program to Improve the Efficiency of Predicting Component Interference and High Lift Characteristics, NASA Contractor Report 3020.
14. Clark D.R., Maskew B. and Dvorak F.A. (1984), The Application of a Second Generation Low-Order Panel Method - Program "VSAERO" - to Powerplant Installation Studies, AIAA paper 84-0122.
15. Miyake Y., et al. (1983), Purely Three-Dimensional Analysis of a Flow in an Axial Rotor by Panel Method, pp.447-454, Proceedings of International Gas Turbine Congress, Tokyo, Japan, 1983.
16. Murata S., et al. (1980), Quasi Three-Dimensional Analysis of Compressible Flow in Centrifugal Impeller, Technology Reports of the Osaka Univ., Vol.30, No.1575, No.1575, pp.519-526.

Free Surface Stokes Flow Over an Obstacle

E.B. Hansen

Laboratory of Applied Mathematical Physics, Technical University of Denmark, DK-2800 Lyngby, Denmark

SUMMARY

The Stokes flow of a viscous fluid layer over a cylindrical obstacle on a tilted plane is determined by means of an integral
equation method. The method is based on an integral formula for
the stream function expressed in terms of two field quantities
on the surface of the obstacle and the fluid velocity in the
free surface. From the formula three integral equations are derived. The location of the surface and the field quantities are
found by means of an iterative procedure. Numerical examples
with fluids with or without surface tension are presented.

INTRODUCTION

Free boundary or free surface problems occur in many branches of
fluid mechanics such as the theory of surface waves, jets, flows
in porous media or Hele Shaw cells, and flows associated with
coating processes. Apart from the case when the surface shape
only deviates slightly from a known surface, so that the boundary
conditions in the surface may be linearized, whereby the problem becomes a classical boundary value problem, free surface
problems are difficult to solve. Consequently, a large variety
of solution methods have been suggested and used. A review of
methods relevant to water waves has been given by Yeung[1]. Another review, of which the emphasis is more relevant to the problem considered in this paper, is contained in the recent book by
Ingham and Kelmanson[2].

Since the field quantities in the surface and, in particular, the
location of the surface itself are often of foremost interest in
free surface problems, a boundary integral equation method seems
to be an obvious choice for solving them. Nevertheless, as far as
non-linear free surface problems are concerned, the use of these
methods have only recently become widespread, and most of the applications so far have been to problems governed by Laplace's

equation, that is to potential or porous flow problems. In fact, to the author's knowledge, until now the only applications of boundary integral equations to viscous free surface problems are those by Kelmanson[2,3] and Hansen[4,5], which treat the flow in the Stokes approximation. The problem studied in [4] and [5] is that of a fluid film flowing down a plane wall into an infinite pool. In this paper we use a method similar to the one used in [4] and [5] to determine the Stokes flow of a fluid layer over a cylindrical obstacle placed on a tilted plane. However, since not all of the fixed boundary is plane in the present case, some modifications become necessary.

In classical viscous flow problems two conditions are valid on the boundary. In free surface problems a third condition is added, compensating for the fact that the location of the surface is unknown. Irrespective of the method used, in all previous work on viscous free surface flows other than that in [4] and [5] only two of the three conditions are applied directly into the numerical scheme, while the third one is used as a criterion for the free surface location. In [4] and [5] and in the present investigation all the three boundary conditions are introduced right from the beginning. This fact gives rise to an advantage which we point out below in connection with the detailed description of the method.

THE PROBLEM

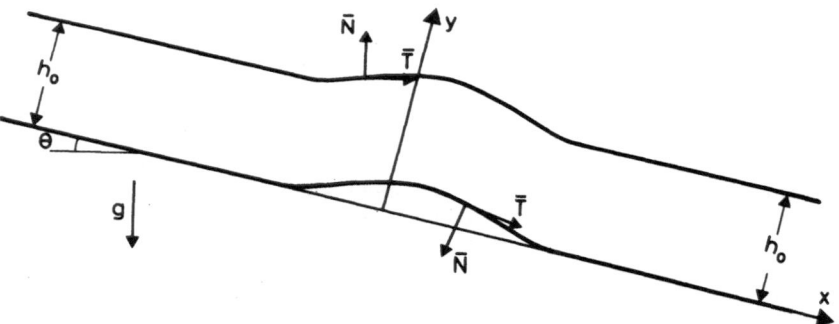

Figure 1. A cylindrical obstacle is placed on a tilted plane surface and a viscous fluid is flowing down the plane and over the obstacle.

Figure 1 shows an inclined plane which forms an angle θ with the horizon. A cylindrical obstacle is placed on the plane and a fluid is flowing down the plane and over the obstacle. Far from the obstacle, in the upstream as well as in the downstream directions, the thickness of the fluid layer approaches the value h_0. The fluid is assumed to be incompressible and the velocity is expressed in terms of a stream function $\psi = \psi(x,y)$ as

$\vec{v} = (\psi_y, -\psi_x)$. It is assumed that the Stokes approximation is applicable to the flow. Then the equations of motion are

$$\frac{\partial p}{\partial x} = \mu\Delta\frac{\partial\psi}{\partial y} + \rho g \sin\theta \ , \ \frac{\partial p}{\partial y} = -\mu\Delta\frac{\partial\psi}{\partial x} - \rho g \cos\theta \ , \tag{1}$$

where p is the pressure, μ the viscosity, ρ the density, and g the acceleration of gravity. On the bottom the stream function assumes a constant value, which we denote by $-\psi_0$, and its normal derivative vanishes. In the surface the stream function also assumes a constant value, which we put equal to zero. Furthermore, the shear stress vanishes and the normal stress balances the surface tension force . Since $\psi = 0$, these two boundary conditions may be written as

$$\Delta\psi + 2\kappa\frac{\partial\psi}{\partial N} = 0 \quad \text{and} \quad p + 2\mu\frac{\partial}{\partial s}\left(\frac{\partial\psi}{\partial N}\right) + \sigma\kappa = 0 \ , \tag{2}$$

respectively, where $\partial/\partial N$ denotes differentiation in the direction of the outward normal vector and κ is the curvature of the intersection between the fluid surface and the xy-plane, κ being positive if the centre of curvature is over the fluid surface.

Far upstream and far downstream the flow is a parallel flow so that ψ only depends on y there. It therefore follows from the equations of motion and the boundary conditions that the velocity approaches the value

$$v_\infty \equiv 3\frac{\psi_0 y}{h_0^2}\left(1 - \frac{1}{2}\frac{y}{h_0}\right) \ , \tag{3}$$

and that

$$\psi_0 = \frac{h_0^3 \rho g \sin\theta}{3\mu} \ . \tag{4}$$

We introduce non-dimensional variables by measuring all lengths in units of h_0, the stream function in units of ψ_0, and the pressure in units of the quantity $p_0 = \mu\psi_0/h_0^2$. In terms of the non-dimensional variables $x_1 = x/h_0$, $\psi_1 = \psi/\psi_0$, $p_1 = p/p_0$ etc. the equations of motion are

$$\frac{\partial p}{\partial x} = \Delta\frac{\partial\psi}{\partial y} + 3 \quad , \ \frac{\partial p}{\partial y} = -\Delta\frac{\partial\psi}{\partial x} - 3\cot\theta \ , \tag{5}$$

the boundary conditions on the bottom are

$$\psi = -1 \quad \text{and} \quad \frac{\partial \psi}{\partial N} = 0 , \tag{6}$$

and the boundary conditions in the surface are

$$\psi = 0 , \quad \Delta\psi + 2\kappa \frac{\partial \psi}{\partial N} = 0 , \quad \text{and} \quad p + 2 \frac{\partial}{\partial s}\left(\frac{\partial \psi}{\partial N}\right) + \beta\kappa = 0 , \tag{7}$$

where

$$\beta = \frac{3\sigma}{\rho g h_o^2 \sin\theta} . \tag{8}$$

In equations (5), (6), and (7) and in the sequel the subscript $_1$ on the non-dimensional variables is omitted.

THE METHOD OF SOLUTION

It follows from the equations of motion (5) that the stream function is biharmonic. Consequently, if Ω is a region in the xy-plane located entirely in the fluid and $\vec{r}_o$ is an interior point in Ω, $\psi(\vec{r}_o)$ may be expressed as

$$\psi(\vec{r}_o) = \int_{\partial\Omega}\left[\frac{\partial \Delta\psi}{\partial N} G - \Delta\psi \frac{\partial G}{\partial N} + \frac{\partial \psi}{\partial N} \Delta G - \psi \frac{\partial \Delta G}{\partial N} \right] ds . \tag{9}$$

Here $\partial/\partial N$ denotes differentiation in the direction of the outward normal to the boundary $\partial\Omega$ of Ω, and G is a fundamental solution to the biharmonic equation in Ω. For G we choose the function

$$G(\vec{r},\vec{r}_o) = \frac{1}{16\pi}\left[(x-x_o)^2+(y-y_o)^2 \right] \ell n \frac{(x-x_o)^2+(y+y_o)^2}{(x-x_o)^2+(y-y_o)^2} - \frac{yy_o}{4\pi} \tag{10}$$

with the properties that $G = \partial G/\partial y = 0$ on $y = 0$.

Let $\partial\Omega$ be the closed curve consisting of parts of the intersections, S and B, between the xy-plane and the free surface and the bottom, respectively, and straight line sections on $x = \pm x_\infty$. If $x_\infty \to \infty$, the contributions to the integral from the line sections vanish. Since $\partial\psi/\partial N = 0$ and $\psi = -1$ on B, the second last term in the integrand vanishes and the last one is known on B. By integrating along a closed curve consisting of a part of B and a curve in the half plane $y > 0$ and letting this curve ex-

pand towards infinity we find the value of the integral along B
of the last term to be −1. On the part of B which coincides with
the x-axis the two first terms vanish. Hence, the integral along
B is equal to −1 plus the integral of the first two terms taken
along the part, B_0, of B which is on the obstacle.

In the integral along S we use the equations of motion to rewrite $\partial\Delta\psi/\partial N$ as

$$\frac{\partial\Delta\psi}{\partial N} = -\frac{\partial p}{\partial s} + 3(T_x - T_y \cot\theta) \, , \tag{11}$$

where (T_x, T_y) is the unit tangent vector shown in Figure 1. Using
the last boundary condition in (7) and integrating by parts twice
with respect to the arc length we may therefore express the first
term of the integrand in terms of known quantities and $\partial\psi/\partial N$. By
means of the middle boundary condition in (7) the same can be
done with the second term in the integrand. Thus, since $\psi = 0$ in
the surface, the integral along S can be expressed in terms of
one unknown, $\partial\psi/\partial N$, and formula (9) becomes

$$\psi(\vec{r}_0) = -1 + f(\vec{r}_0)$$

$$+ \int_S \left[\frac{\partial^2 G}{\partial N^2} - \frac{\partial^2 G}{\partial T^2} \right] v(s) \, ds - \int_{B_0} \left[G \frac{\partial p}{\partial s} + \frac{\partial G}{\partial N} \Delta\psi \right] ds \, . \tag{12}$$

Here we have written v(s) for $\partial\psi/\partial N$, which is equal to the fluid
velocity in the surface, and used formula (11) to introduce
$\partial p/\partial s$ in the integral over B_0. The function $f(\vec{r}_0)$ is

$$f(\vec{r}_0) = \int_S \left[\beta\kappa \frac{\partial G}{\partial s} - wG \right] ds + \int_{B_0} wG \, ds \, , \tag{13}$$

where $w = 3(T_x - T_y \cot\theta)$.

From formula (12) we derive a system of three integral equations.
The first one is obtained by differentiating in the direction of
the normal $\bar{N}'$ at an arbitrary point $\vec{r}' \in S$ and passing to the limit
$\vec{r}_0 \to \vec{r}'$. This equation is

$$\frac{1}{2} v(s') = \frac{\partial f}{\partial N'} (\vec{r}')$$

$$+ \int_S \frac{\partial}{\partial N'} \left[\frac{\partial^2 G}{\partial N^2} - \frac{\partial^2 G}{\partial T^2} \right] v(s) \, ds - \int_{B_0} \left[\frac{\partial G}{\partial N'} \frac{\partial p}{\partial s} + \frac{\partial^2 G}{\partial N'\partial N} \Delta\psi \right] ds , \tag{14}$$

where s' is the value of the arc length parameter s at $\vec{r}'$. The second equation is obtained by letting $\vec{r}_0$ approach a point $\vec{r}'$ on B_0 and using the boundary condition $\psi = -1$. The third one is obtained by operation on both sides of formula (12) with Laplace's operator and then letting $\vec{r}_0$ approach $\vec{r}' \in B$. These two equations are

$$-f(\vec{r}') = \int_S \left[\frac{\partial^2 G}{\partial N^2} - \frac{\partial^2 G}{\partial T^2} \right] v(s)ds - \int_{B_0} \left[G \frac{\partial p}{\partial s} + \frac{\partial G}{\partial N} \Delta\psi \right] ds \quad (15)$$

and

$$\frac{1}{2} \Delta'\psi(\vec{r}') = \Delta'f(\vec{r}')$$

$$+ \int_S \Delta' \left[\frac{\partial^2 G}{\partial N^2} - \frac{\partial^2 G}{\partial T^2} \right] v(s)ds - \int_{B_0} \left[\Delta'G \frac{\partial p}{\partial s} + \Delta'\frac{\partial G}{\partial N} \Delta\psi \right] ds. \quad (16)$$

If the three integral equations are solved and the solutions inserted in formula (12), the right hand side is the stream function on the surface and, consequently, it is equal to zero. If, on the other hand, S in the integral equations and formula (12) is replaced by some other curve, C, and the solution functions are inserted in formula (12), the right hand side will not, in general, vanish. Since we do not know the location of the surface, we shall therefore solve the problem numerically by using an iterative procedure starting from some approximation, C_0, to S. At each step of the procedure we solve the system of equations and adjust the curve, C, which approximates S, until the absolute values of the right hand side of formula (12) at a set of collocation points become smaller than some prescribed number.

Before we describe the method in more detail we shall point out two features which we believe distinguish the present method in comparison with other methods which might have been chosen. The one of these is that by using the fundamental solution given by formula (10) we limit the interval of integration on the bottom to the surface of the obstacle, thereby reducing the number of collocation points on the bottom, which are needed in order to obtain a certain accuracy. The other is that if we had used only two of the boundary conditions, which hold in the surface, and introduced the third one later as a criterion for the location of the surface, we would have obtained a system of four integral equations to be solved in each iteration step. Since, in the present method, all the three boundary conditions are used when setting up the integral formula, the number of equations is reduced to three. Instead we must compute the stream function in the surface. Nevertheless, with a given total number

of collocation points the number of matrix elements to be computed in each iteration step is reduced.

The location of the free surface, S, was found numerically by
means of an optimization method developed by Madsen[6] and implemented in the subroutine VG02AD of the Harwell subroutine library. In this method a set of N collocation points defining S was
shifted along normals to an initial curve until the absolute
values of the stream function at all the collocation points were
less than 10^{-5}. At each iteration step the set of integral equations had to be solved. This was done by means of a collocation
method using the N points on S and M points on B_0. In most cases
N was 27 and M was 20 leading to a system of 67 linear algebraic
equations with 67 unknowns, which was solved by means of the
subroutine LEQT2F from the IMSL library. In the integral equations S was approximated by the asymptote y = 1 and v by its
asymptotic value, v_∞ = 1.5, except for an interval, usually extending from −20 to +10, in which the collocation points were
located. Between the collocation points S was approximated by
parabolas and v by first degree polynomials. In all cases the
obstacle extended over the interval [−1,1] and was given by a
function $y = \frac{1}{2}a(1 + \cos \pi x)$, where a was a positive constant less
than unity. On B_0 the unknowns were approximated by piecewise
constant functions. Since B_0 is fixed, those matrix elements,
for which the field point as well as the integration point are
on B_0, do not change during the iteration process. Therefore they
only had to be computed once. At each step the solution func-
tions for v, $\partial p/\partial s$, and $\Delta \psi$ were inserted in formula (12) and the
stream function computed at the collocation points.

The computations were started with a = 0.25, β = 0 and y = 1 as
the initial curve. By using steps in a of about 0.20 and steps
in β of 10 or more, and applying the final approximation curve
to S for each set of values (a,β) as the next starting curve, we
were able to find the free surface and v, $\partial p/\partial s$ and $\Delta \psi$ for parameter values in the intervals 0.1 to 0.75 of a and 0 to 50 of β.
the largest number of iterations needed was 14 but usually 7 or
8 were enough. The computations were carried out on the IBM 3033
computer of the UNI·C computing centre in Lyngby. A typical iteration process took about 90 cpu seconds.

RESULTS

Some examples of the results obtained are shown in Figures 2, 3,
and 4. They all apply to a plane forming an angle θ = 20° with
the horizon. The plots in Figure 2 show that when there is no
surface tension the surface raises considerably only in a rather
short interval near the obstacle, while the elevation is extended
to a larger interval in the case of a fluid with surface tension.
This is to be expected since the surface tension tends to smooth
out large deflections of the surface. Based upon the following
material parameters for glycerine, $\rho = 1.2613 \mathrm{g/cm^3}$, $\sigma = 63.4$

Dynes/cm and μ = 14.9 g cm^{-1}sec^{-1}, given in Weast[7] the value
β = 25 for θ = 20° is found to correspond to a layer thickness
h_O = 1.3 mm. In this case Reynold's number defined as $\rho h_O v_\infty (h_O)/\mu$
is .003. The corresponding surface velocities as functions of the
abscissa x are shown in Figure 3. The variation of the surface
velocity reflects the variation of the depth of the fluid layer.

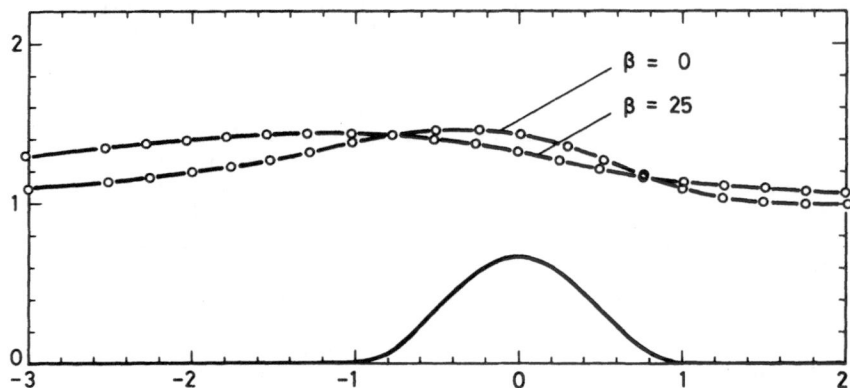

Figure 2.   The free surface over an obstacle with profile
$y = \frac{1}{2}a(1 + \cos \pi x)$ for $a = 2/3$, $\theta = 20^{\circ}$, and $\beta = 0$ or 25.

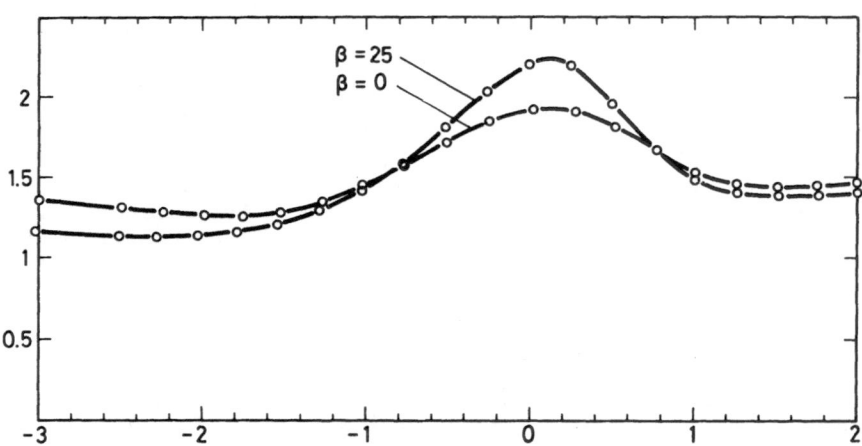

Figure 3.   The surface velocity in the free surfaces shown in
Figure 2.

Since ψ is constant, and its normal derivative vanishes on the bottom, $\Delta\psi$ is equal to the tangential stress on the bottom surface. Figure 4 shows $\Delta\psi$ on B_0 as a function of the x-coordinate for obstacles with a = 1/20, 1/4, and 2/3 in the case of a fluid without surface tension. If there were no obstacle present the non-dimensional tangential stress at the bottom would be equal to 3. The curve for a = 1/20 confirms that this value is approached as the obstacle becomes small. For a = 1/4 the tangential stress is positive along the whole surface of the obstacle. For a = 2/3, a region with negative shear stress exists on the front as well as on the rear side, showing that in this case there are separated flow regions on both sides of the obstacle.

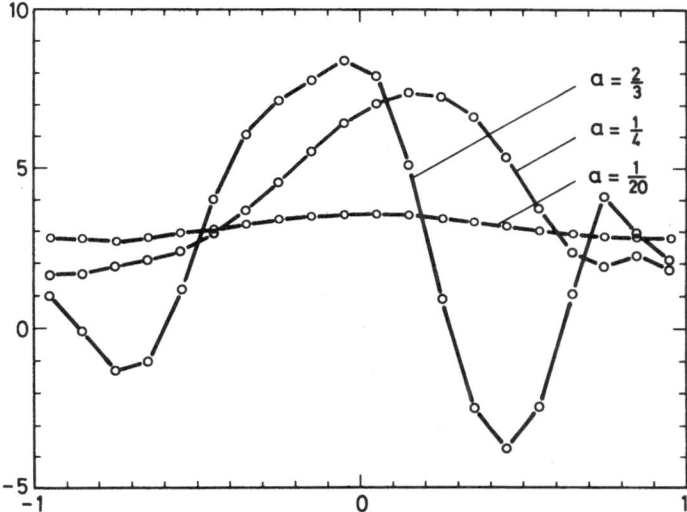

Figure 4. The tangential stress on the surfaces of obstacles with a = 1/20, 1/4, and 2/3. In all cases θ = 20° and β = 0. The abscissa is the x-coordinate.

REFERENCES

1. Yeung R.W. (1982), Numerical Methods in Free Surface Flows, Annual Review of Fluid Mechanics, (Eds. v.Dyke M., Wehausen J.V. and Lumley J.L.), Vol.14, pp. 395-442, Annual Review Inc., Palo Alto.

2. Ingham D.B. and Kelmanson M.A. (1984), Boundary Integral Equation Analysis of Singular, Potential, and Biharmonic Problems, Lecture Notes in Engineering, 7, Springer-Verlag, Berlin and New York.

3. Kelmanson M.A. (1983), Boundary Integral Equation Solution of
    Viscous Flows with Free Surfaces, Journal of Engineering Mathe-
    matics, Vol.17, pp.329-343.

4. Hansen E.B. (1985), Stokes Flow into a Large Pool, Free Boundary
    Problems: Applications and Theory, (Eds. Bossavit A., Damlamian
    A. and Fremont M.), Vol.4, pp.391-394, Pitman, London.

5. Hansen E.B. (1986), Stokes Flow Down a Wall Into an Infinite
    Pool, Journal of Fluid Mechanics, submitted.

6. Madsen K. (1975), Minimax Solution of Non-Linear Equations
    Without Calculating Derivatives, Mathematical Programming Study,
    Vol.3, pp.110-126.

7. Weast R.C. (Ed.) (1981). CRC Handbook of Chemistry and Physics,
    62nd Edition, pp. C 312, F 38, and F 46.

Characteristics Incorporated Boundary Element for One-Dimensional Convective Diffusion Problem

M. Sakakihara

The Graduate School of Science, Okayama University of Science, Ridai-cho 1-1, Okayama 700, Japan

ABSTRACT

The aims of this paper are to present characteristics incorporated a boundary method for solving unsteady convective diffusion problems, and to discuss on a error estimate of the present method. The error estimate is similar to Thomee's theorem[1] which is established to a parabolic equation. The present method is also able to be applied to a nonlinear convective diffusion problem, though this is a boundary method. Numerical results for linear and nonlinear convective diffusion problems is shown.

1. INTRODUCTION

The boundary element method is a method which gives us the stable and accurate numerical solution for convective diffusion problems. For steady and unsteady convective diffusion problems in a piston flow, the boundary element methods were presented[2,3]. In order to formulate the boundary element method, it is necessary to obtain the fundamental solution with respect to the adjoint problem. It is, however, difficult to obtain the fundamental solution for the convective diffusion problem under the variable flow. In order to avoid the difficulty, a integral equation method with boundary integral terms was proposed[4]. If we apply the integral equation method to those problems, we have to solve the large and dense linear or nonlinear algebraic systems. In this paper the author presents a method of characteristic incorporated a boundary method for solving unsteady linear and nonlinear convective diffusion problems. A time truncation error estimate is shown. Moreover, we show numerical examples for linear and nonlinear problems in which the nonlinear problem is the boundary-initial value problem of Burgers' equation.

2. CONVECTIVE DIFFUSION PROBLEM

Let us consider an unsteady convective diffusion problem of the form

(1) $\dfrac{\partial u}{\partial t} + v(x)\dfrac{\partial u}{\partial x} - \nu\dfrac{\partial^2 u}{\partial x^2} = f(x,t)$ $in\ (a,b)\times(0,T]$,

(1a) $u(x,0) = g(x)$ $in\ (a,b)$,

(1b) $u(a,t) = 0$, $u(b,t) = 0$,

where $v(x)$, $f(x,t)$ and $g(x)$ are given functions, and ν and T are positive constants. Throughout this paper we assume that

(A1) $g(x) \in L^{\infty}(I)$ where $I = (a,b)$,

(A2) there exists a positive constant K such that

$$| v(x) | + | dv(x)/dx | \le K$$

and

(A3) the solution u of the problem (1) satisfies

(a) $u \in L^{\infty}(0,T;W^{q,2}(I))$,

(b) $\partial u/\partial t \in L^2(0,T;W^{q-1+\theta,2}(I))$,

$\theta = 1$ if $q = 2$ and $\theta = 0$ if $q > 2$,

and

(c) $\partial^2 u/\partial x^2 \in L^2(0,T;L^2(I))$,

for some $q \geq 2$. Here $W^{p,q}(I)$ denotes the usual Sobolev space in I. Characteristics direction with respect to

$$\partial u/\partial t + v(x)\partial u/\partial x$$

is denoted by $\tau=\tau(x)$. Then derivation to the characteristic direction is expressed as

(2) $\dfrac{\partial}{\partial\tau(x)} = \dfrac{1}{\psi(x)}\left(\dfrac{\partial}{\partial t} + v(x)\dfrac{\partial}{\partial x} \right)$

where $\psi(x)=[1+v(x)^2]^{1/2}.$ Applying Eq.(2) to Eq.(1) we obtain

(3) $\psi(x)\dfrac{\partial u}{\partial\tau(x)} - \nu\dfrac{\partial^2 u}{\partial x^2} = f(x,t)$ $in\ (a,b)\times(0,T]$.

The problem (3), (1a) and (1b) is equivalent to the problem (1) since Jacobian for the transformation is 1. The finite difference approximation for the characteristic derivative is expressed as

(4) $\psi(x)\dfrac{\partial u}{\partial\tau} \simeq \psi(x)\dfrac{u(x,t^n)-u(\bar{x},t^{n-1})}{[(x-\bar{x})^2+(\Delta t)^2]^{1/2}} = \dfrac{u(x,t^n)-u(\bar{x},t^{n-1})}{\Delta t}$

where Δt is a time step size, $t^n = n\Delta t$ and $\bar{x}$ is defined by

(5) $\bar{x} = x - v(x)\Delta t$.

Applying Eq.(4) to Eq.(3) we obtain the semi-discrete system of the form

(6) $\dfrac{\hat{u}(x,t^n) - \hat{u}(\bar{x},t^{n-1})}{\Delta t} - \nu\dfrac{\partial^2}{\partial x^2} u(x,t^n) = f(x,t^n).$

Eq.(6) is rewritten into

(7) $-\nu\dfrac{\partial^2}{\partial x^2}\hat{u}(x,t^n) + \dfrac{1}{\Delta t}\hat{u}(x,t^n) = f(x,t^n) + \dfrac{1}{\Delta t}\hat{u}(\bar{x},t^{n-1})$

which is the elliptic type equation with respect to $u(x,t^n)$. For the nonlinear convective diffusion problem such as

(8) $\dfrac{\partial u}{\partial t} + v(x,u)\dfrac{\partial u}{\partial x} - \nu\dfrac{\partial^2 u}{\partial x^2} = f(x,t)$ $in\ (a,b)\times(0,T]$

subject to (1a) and (1b) we obtain a similar discrete equation to the linear problem. Note that if $v(x,u)=u$ then Eq.(8) is Burgers' equation. The semi-discrete equation for Eq.(8) is Eq.(7) in which $\bar{x}$ is defined by

(9) $\bar{x} = x - v(x,\hat{u}^{n-1}(x))$

where $\hat{u}^{n-1}(x)=\hat{u}(x,t^{n-1})$.

3. BOUNDARY METHOD

In order to solve Eq.(7) we present a boundary method which is formulated with the fundamental solution for the differential operator : $-d^2/dx^2 + \lambda^2$.
The fundamental solution

(10) $\phi(x,y) = \dfrac{1}{2\lambda}exp[\ -\lambda|x-y|\ ]$

satisfies the equation such as

(11) $-d^2\phi/dx^2 + \lambda^2\phi = \delta(|x-y|)$ $in\ x,y\epsilon R$

where $\delta(|x-y|)$ is the Dirac's delta function and

(12) $\lambda = \sqrt{1/(\nu\Delta t)}$.

With the integral operator, in which the kernel of integral operator is the fundamental solution (10), we obtain

(12) $\int_I \phi(x,y)[-\dfrac{\partial^2\hat{u}}{\partial y^2} + \lambda^2\hat{u}]dy = \dfrac{1}{\nu}\int_I \phi(x,y)[f^n(y) + \dfrac{1}{\Delta t}\hat{u}^{n-1}(\bar{y})]dy$

where $f^n(y)=f(y,n\Delta t)$. Applying the integration by parts to

Eq.(12) the equation of the form

$$(13) \quad \hat{u}^n(x) + [\frac{\partial}{\partial y} \phi(x,y)\hat{u}^n(y)]_a^b - [\phi(x,y)\frac{\partial}{\partial y}\hat{u}^n(y)]_a^b$$

$$= \frac{1}{\nu} \int_I \phi(x,y)\{f^n(y) + \frac{1}{\Delta t}\hat{u}^{n-1}(y)\}dy$$

is given where $x\epsilon(a,b)$. If x tends to the boundary points a and b then we obtain the boundary equation such as

$$(14) \quad \frac{1}{2} \hat{u}^n(X_i) + [\frac{\partial}{\partial y} \phi(X_i,y)\hat{u}^n(y)]_a^b - [\phi(X_i,y)\frac{\partial}{\partial y} \hat{u}^n(y)]_a^b$$

$$= \frac{1}{\nu} \int_I \phi(X_i,y)\{f^n(y) + \frac{1}{\Delta t} \hat{u}^{n-1}(\bar{y})\}dy, \quad (i=1,2),$$

where $X_1=a$ and $X_2=b$. From the Dirichlet boundary condition (1a) Eq.(14) is written into the linear equation of the form

$$(15) \quad
\begin{bmatrix}
\frac{1}{2\lambda} & \frac{1}{2\lambda} exp[-\lambda|a-b|] \\
\\
\frac{1}{2\lambda} exp[-\lambda|a-b|] & \frac{1}{2\lambda}
\end{bmatrix}
\begin{Bmatrix}
\hat{q}_a^n \\
\\
\hat{q}_b^n
\end{Bmatrix}$$

$$=
\begin{Bmatrix}
\frac{1}{\nu} \int_I \phi(a,y)[f^n(y) + \frac{1}{\Delta t} \hat{u}^{n-1}(\bar{y})]dy \\
\\
\frac{1}{\nu} \int_I \phi(b,y)[f^n(y) + \frac{1}{\Delta t} \hat{u}^{n-1}(y)]dy
\end{Bmatrix}$$

where

$$\hat{q}_a^n = \partial\hat{u}^n(y)/\partial y \mid_{y=a} \quad \text{and} \quad \hat{q}_b^n = \partial\hat{u}(y)/\partial y \mid_{y=b} .$$

4. ERROR ESTIMATE

In this section we prove a error estimate for the present method. Douglas and Russel developed some mathematical considerations for the method of characteristic. The following theorem is one of those results.

THOREM 1.

$$\| \psi \frac{\partial u^n}{\partial \tau} - \frac{u^n - u^{n-1}}{\Delta t} \|^2 \leqslant 2\| \psi^4 \|_\infty \| \frac{\partial^2 u}{\partial \tau^2} \|_{L^2((a,b)\times[t^{n-1}, t^n])}$$

where $\bar{u}^{n-1}=u^{n-1}(\bar{x})$ and $\| \cdot \|$ denotes L^2-norm.

With theorem 1 we prove the following theorem :

TEOREM 2.

The present method satisfies the error estimate such as

$$\| u^n - \hat{u}^n \| \leq C\Delta t + C^{\curlyvee} \| u^0 - \hat{u}^0 \|$$

where C and $C^{\curlyvee}$ are generic constants.

Proof : From

$$(16) \quad -\nu \frac{\partial^2}{\partial x^2}(u^n - u^n) = -[\psi \frac{\partial u}{\partial \tau} \Big|_{t^n} - (\frac{\hat{u}^n - U^{n-1}}{\Delta t})] \quad ,$$

where $U^n = \hat{u}^n(\bar{x})$, we obtain

$$(17) \quad -\nu \frac{\partial^2}{\partial x^2}(u^n - \hat{u}^n)$$

$$= -[\psi \frac{\partial u}{\partial \tau} \Big|_{t^n} - (\frac{u^n - \bar{u}^{n-1}}{\Delta t}) + (\frac{u^n - \bar{u}^{n-1}}{\Delta t}) - (\frac{\hat{u}^n - U^{n-1}}{\Delta t})].$$

Then

$$(18) \quad (1 - \nu \Delta t \frac{\partial^2}{\partial x^2})(u^n - \hat{u}^n)$$

$$= -[(u^{n-1} - u^n) + \Delta t \psi \frac{\partial u}{\partial \tau} \Big|_{t^n}] + (\hat{u}^{n-1} - U^{n-1})$$

is given. We therefore obtain the inequality such as

$$(19) \quad \begin{aligned} & \| u^n - \hat{u}^n \|^2 + \Delta t \| \nu \frac{\partial^2}{\partial x^2} (u^n - \hat{u}^n) \|^2 \\ & \leq [C(\Delta t)^2 + \| \hat{u}^{n-1} - U^{n-1} \|] \| u^n - \hat{u}^n \| . \end{aligned}$$

From theorem 1 and Eq.(17) the estimate of the form

$$(20) \quad \| u^{n-1} - u^n + \Delta t \psi \frac{\partial u^n}{\partial \tau} \| \leq C(\Delta t)^2$$

is given, then

$$(21) \quad \| u^n - \hat{u}^n \| \leq C(\Delta t)^2 + \| \hat{u}^{n-1} - U^{n-1} \| .$$

Since the determinant of the transformation is $1 + O(\Delta t)$, we obtain

$$(21) \quad \| u^n - \hat{u}^n \| \leq C(\Delta t)^2 + (1 + C^{\curlyvee}\Delta t) \| u^{n-1} - \hat{u}^{n-1} \| .$$

Therefore iterating,

$$(22) \quad \| u^n - \hat{u}^n \| \leq nC(\Delta t)^2 + nC^{\curlyvee}\Delta t \| u^0 - \hat{u}^0 \|$$

$$+ \| u^0 - \hat{u}^0 \| .$$

Hence we proved the theorem 2.

5. NUMERICAL EXAMPLES

Applying the trapezoidal rule to Eq.(15) the discretized equation is expressed as

(23)

$$
\begin{bmatrix}
\dfrac{1}{2\lambda} & \dfrac{1}{2\lambda}exp[\ -\lambda|a-b|\ ] \\[3mm]
\dfrac{1}{2\lambda}exp[\ -\lambda|a-b|\ ] & \dfrac{1}{2\lambda}
\end{bmatrix}
\begin{Bmatrix}
\hat{q}_a^n \\[3mm]
\hat{q}_b^n
\end{Bmatrix}
$$

$$
= \dfrac{1}{\nu}
\begin{Bmatrix}
\Sigma_{i=1}^{N} \dfrac{1}{2}\{\ F_a^n(y_i) + F_a^n(y_{i+1})\ \}h_i \\[4mm]
\Sigma_{i=1}^{N} \dfrac{1}{2}\{\ F_b^n(y_i) + F_b^n(y_{i+1})\ \}h_i
\end{Bmatrix}
$$

where

$$
F_a^n(y_i) = \phi(a,y_i)\{\ f^n(y_i) + \frac{1}{t}\ \hat{u}^{n-1}(\bar{y}_i)\ \}\quad,
$$

$$
F_b^n(y_i) = \phi(b,y_i)\{\ f^n(y_i) + \frac{1}{\Delta t}\ \hat{u}^{n-1}(\bar{y}_i)\ \}
$$

and $h_i = y_{i+1} - y_i$. From solutions for Eq.(23) and u^{n-1}, internal values is evaluated by

$$
(24)\quad u^n(x) = [\phi(x,y)\hat{q}^n(y)]_b^a + \frac{1}{\nu}\Sigma_{i=1}^{N}\frac{1}{2}\{\ F_x^n(y_i) + F_x^n(y_{i+1})\ \}h_i
$$

In this paper we show numerical examples for following problems such as :

[Problem 1] Equation : $\dfrac{\partial u}{\partial t} + v\dfrac{\partial u}{\partial x} = \dfrac{\partial^2 u}{\partial x^2}$ $\quad in\ (0,1)\times(0,T]$,

P1-1 | Initial condition : $u(x,0) = \dfrac{1}{2}sin(\pi x)$

  | Boundary condition: $u(0) = u(1) = 0$

P1-2 | Initial condition : $u(x,0) = 0$

  | Boundary condition: $u(0) = 1,\ u(1) = 1$

[Problem 2] Equation : $\dfrac{\partial u}{\partial t} + u\dfrac{\partial u}{\partial x} = \dfrac{\partial^2 u}{\partial x^2}$ $\quad in\ (0,1)\times(0,T]$

P2 | Initial condition : $u(x,0) = A\ sin(\pi x)$

  | Boundary condition: $u(0) = u(1) = 0$

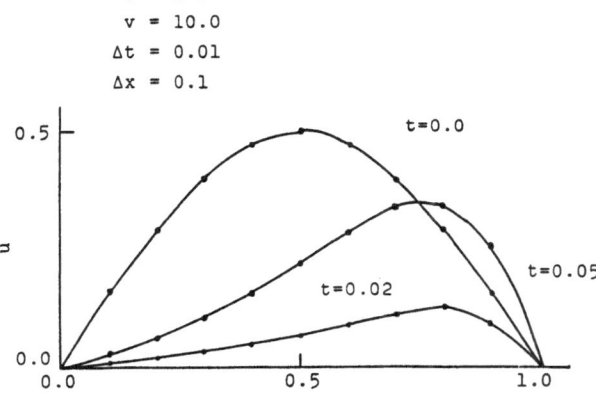

Fig. 1.   Numerical solutions of Pl-1.

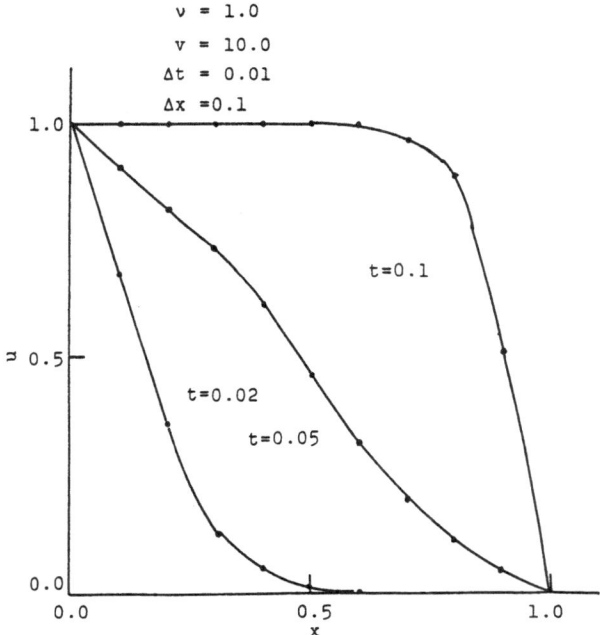

Fig. 2. Numerical solutions of Pl-2.

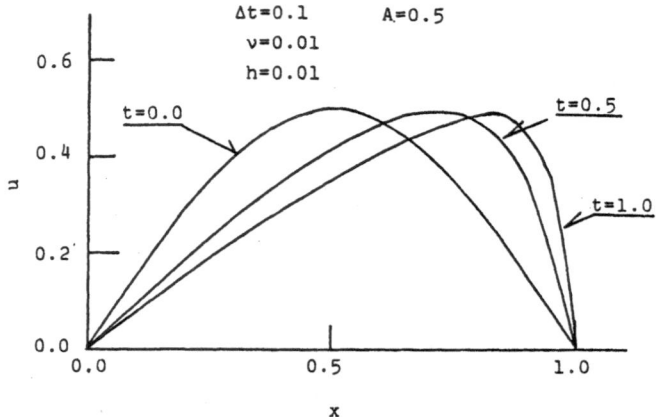

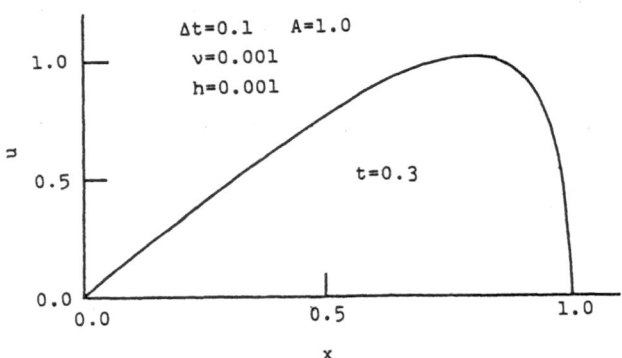

Fig. 3. Numerical solution of P2.

	Problem	Diffusion number	Courant number
	P1-1	1.0	1.0
	P1-2	1.0	1.0
P2	$A=0.5$	10.0	5.0
	$A=1.0$	100.0	100.0

Table 1. Table of Diffusion and Courant numbers.

6. CONCLUSION

The author presented a method of characteristic incorporated a boundary method for solving one-dimensional unsteady convective diffusion problems. The present method has a merit, that is, it is possible to apply the present method to the convective diffusion problem in the variable flow and nonlinear convective diffusion problems as Burgers' equations. From numerical results P1-1, P1-2 and P2 it is shown that the stable solutions for high diffusion numbers and high Courant numbers are given by the present method.

REFERENCE :

1. Bramble J.H. and Thomee V. (1972), Semi-discrete Last Squares Methods for a Parabolic Boundary Value Problem, Math. Comput. vol.26, pp.663-648.

2. Ikeuchi M., Sakakihara M. and Onishi K. (1983), Constant Boundary Element Solution for Steady-State Convective Diffusion Equation in Three Dimensions, Trans. IECE Japan, vol.E66, pp.373-376.

3. Ikeuchi M. (1985), Transient Solution of Convective Diffusion Problem by Boundary Element Method, Trans. IECE Japan, vol.E68, pp.435-440.

4. Skerget P. and Brebbia C. (1983), The Solution of Convective Problem in Laminar Flow, Boundary Elements, Proceedings of 5th International Conference, eds. Brebbia C.A., Futagami T. and Tanaka M., pp.251-274.

5. Douglas, Jr J. and Russell T.F. (1982), Numerical Methods for Convective-Dominated Diffusion Problems Based on Combining the Method of Characteristics with Finite Element or Finite Diference Procedures, SIAM Numer. Anal., vol.19, pp.871-885.

6. Kesavan S. and Vasudevamurthy A.S. (1985), On Some Boundary Element Methods for the Heat Equation, Numer. Math., vol.46, pp.101-120.

7. Sakakihara M and Ikeuchi M. (1985), Characteristics Integration Method for One-Dimensional Initial Value Problem of Convective Diffusion Equation, Numerical Methods in Laminar and Tublent Flow, eds. Taylor C., Olson M.D., Gresho P.M. and Habashi W.G., Pineridg Press, pp.1003-1010.

Integral Equation Analysis of Laminar Natural Convection Problems

N. Tosaka, N. Fukushima
Department of Mathematical Engineering, College of Industrial Technology, Nihon University, Narashino-shi, Chiba 275, Japan

INTRODUCTION

The coupled convective and conductive heat transfer phenomena due to the buoyancy-driven flow under heating of incompressible viscous fluids are commonly encountered in many important natural science and engineering problems. These include many practical problems, which are known as the natural convection problems, such as thermal insulation of buildings, cooling of electronic equipment and general circulation of planetary atomospheres.

The differential equations governing the phenomena, the Navier-Stokes equations and the equation of energy, are a highly nonlinear coupled set. Because of difficulties in obtaining analytical solutions of the problem, several numerical procedures based on the finite difference method[1] and the finite element one[2,3] have been developed with the advance of electronic computers. However, a limited attempt[4~8] has been only done to solve the two-dimensional natural convection problems by using the integral equation method.

In this paper, we wish to examine possibilities on the integral equation analysis of laminar natural convection problems. The integral equation formulation of the problems is given in terms of velocity vector, pressure and temperature known as the primitive variable formulation[8]. With this formulation, we can treat easily three-dimensional problems as well as two-dimensional ones. But, the scope of this study is limited to steady-state two-dimensional laminar convection problems. Following the proposed methodology[6,7] we derive the nonlinear integral equation set by means of the weighted residual form and the fundamental solution tensor for linear operators of the system of governing differential equations. Fundamental

solutions which are indispensable for construction of our
fundamental solution tensor are shown explicitly concerning the
problems. Numerical solution procedures based on a discretized
scheme by using of both boundary elements and interior cells
are presented. The derived nonlinear system of equations in
terms of the unknown nodal vectors is solved with aid of the
Newton-Raphson method which is efficient in analyzing of
problems for the higher Rayleigh number. Sample results for
two-dimensional natural convection problems in a square, and
some nonrectangular enclosures are presented in order to
illustrate the accuracy and efficiency of the newly proposed
method.

The summation convention is used for the repeated index. A
comma "," is used to denote partial differentiation with
respect to a space variable (i.e. $,i \equiv D_i \equiv \partial/\partial x_i$) and
$\Delta \equiv \partial^2/\partial x_i^2$ denotes the Laplacian operator.

STATEMENT OF PROBLEM

Let us consider the steady problem of an incompressible viscous
heat-conducting Newtonian fluid. The region of the flow
contained in Euclidean space R^2 is denoted by Ω with a
piecewise smooth boundary Γ possessing the unit outward
normal component n_i . We are concerned with the coupled
convection and conduction heat transfer problems based on the
familiar Boussinesq approximation. The governing differential
equations simplified by use of the above flow conditions can be
expressed in the non-dimensional form as follows :
The equations of momentum

$$Re\, u_j\, u_{i,j} = \tau_{ij,j} + \frac{Gr}{Re}\, \theta\delta_{i2} \qquad \text{in } \Omega \qquad (1)$$

The equation of continuity

$$u_{i,i} = 0 \qquad \text{in } \Omega \qquad (2)$$

The equation of energy

$$Pr\, u_j\, \theta_{,j} = \frac{1}{Re}\, \theta_{,jj} \qquad \text{in } \Omega \qquad (3)$$

where u_i is the component of velocity vector, p is the
pressure, θ is the temperature and τ_{ij} is the component
of stress tensor given by the following constitutive equations

$$\tau_{ij} = -\,Re\, p\, \delta_{ij} + u_{i,j} + u_{j,i} \qquad \text{in } \Omega \qquad (4)$$

We introduce the Reynolds number Re , the Grashof number Gr ,
the Prandtl number Pr and the Rayleigh number $Ra = Gr \cdot Pr$ as
the several dimensionless parameters in problems being
currently investigated. Equations (1)-(4) must be solved in
conjunction with the following appropriate boundary conditions
on each boundary Γ_u , Γ_τ , Γ_θ and Γ_q :

$$u_i = \hat{u}_i \quad \text{on } \Gamma_u, \quad \tau_i \equiv \tau_{ij} n_j = \hat{\tau}_1 \quad \text{on } \Gamma_\tau \qquad (5)$$

$$\theta = \hat{\theta} \quad \text{on } \Gamma_\theta, \quad q \equiv \theta_{,j} n_j = \hat{q} \quad \text{on } \Gamma_q \qquad (6)$$

For derivation of the integral equation formulation, it is convenient at this stage to rewrite the above set (1)-(4) to the following matrix form :

$$L_{IJ} U_J = B_I \qquad (I,J=1,2,3,4) \qquad (7)$$

or we can express concretely as

$$
\begin{bmatrix}
\Delta + D_1^2 & D_1 D_2 & 0 & -ReD_1 \\
D_2 D_1 & \Delta + D_2^2 & \dfrac{Gr}{Re} & -ReD_2 \\
0 & 0 & \dfrac{1}{Re}\Delta & 0 \\
D_1 & D_2 & 0 & 0
\end{bmatrix}
\begin{bmatrix}
u_1 \\
u_2 \\
\theta \\
p
\end{bmatrix}
=
\begin{bmatrix}
Re u_j u_{1,j} \\
Re u_j u_{2,j} \\
Pr u_j \theta_{,j} \\
0
\end{bmatrix}
\qquad (8)
$$

INTEGRAL EQUATION FORMULATION

In order to transform the govening differential equation set (7) into the integral equation one, let us follow the methodology proposed in Refs. [6,7,8]. We start with the following weighted residual statement of equation (7) over the domain and for the weighting tensor V_{IL}^* :

$$\int_\Omega (L_{IJ} U_J - B_I) V_{IL}^* \, d\Omega = 0 \qquad (9)$$

Integrating by parts and taking the differential operator L_{IJ} into consideration, we arrive at a set of integral equations of the form

$$U_I = \int_\Gamma (u_i \Sigma_{iI}^* - \tau_i V_{iI}^*) \, d\Gamma$$

$$+ \int_\Gamma \frac{1}{Re} (\theta V_{3I,n}^* - q V_{3I}^*) \, d\Gamma$$

$$+ \int_\Omega B_J V_{JI}^* \, d\Omega \qquad (10)$$

$$(i = 1,2 ; I,J = 1,2,3,4)$$

where V_{JL}^* must be determined as the fundamental solution tensor which satisfies the differential equation for the adjoint operator L_{IJ} of L_{IJ} such that

$$L_{IJ} \, V_{JL}^* = \delta_{IL} \, \delta(x-y) \tag{11}$$

and Σ_{iL}^* , which corresponds to the traction vector τ_i , is defined by

$$\Sigma_{iL}^* = (\, -V_{4L}^* \delta_{ij} + V_{iL,j}^* + V_{jL,i}^* \,) \, n_j \tag{12}$$

Let us consider the derivation of our fundamental solution tensor V_{JL}^* according to the methodology developed in Refs. [6,7]. The fundamental solution tensor can be expressed as

$$V_{JL}^* = M_{JL} \, \phi^* \tag{13}$$

where $[\, M_{JL} \,]$ is the transposed co-factor matrix of $[\, L_{IJ} \,]$ given by

$$[\, M_{JL} \,] = \begin{bmatrix} D_2^2 \Delta & -D_1 D_2 \Delta & 0 & \frac{1}{Re} D_1 \Delta^2 \\[2mm] -D_1 D_2 \Delta & D_1^2 \Delta & 0 & \frac{1}{Re} D_2 \Delta^2 \\[2mm] D_1 D_2 Gr & -D_1^2 Gr & \Delta^2 Re & -\frac{Gr}{Re} D_2 \Delta \\[2mm] -D_1 \Delta^2 & -D_2 \Delta^2 & 0 & 2\frac{1}{Re} \Delta^3 \end{bmatrix} \tag{14}$$

and ϕ^* is the fundamental solution for the differential operator $L \equiv det\, [\, L_{IJ} \,]$ which satisfies

$$L \, \phi^* = \delta(\, x - y \,) \tag{15}$$

For the problem discussed in this paper, L is given explicitly as the tri-harmonic differential operator, i.e. $L \equiv \Delta^3$. Consequently, we can determine the fundamental solution ϕ^* for the tri-harmonic differential operator as

$$\phi^* = \frac{1}{128\pi} \, r^4 \, log \, r \tag{16}$$

where $r = \| x - y \|$ denotes the distance between x and y . Therefore, the fundamental solution tensor can be given completely by the relation (13) with the aid of (14) and (16).

DISCRETIZATION

In order to get approximate solutions of the integral equation (10), let the boundary Γ and the domain Ω divide into n

elements and m cells, respectively. The unknown functions involved in equation (10) may be represented approximately in terms of the shape functions and nodal vectors as follows :
On the boundary element,

$$u_i = \phi_N(\ \pmb{x}\ )\ u_{Ni} \qquad , \qquad \tau_i = \phi_N(\ \pmb{x}\ )\ \tau_{Ni}$$

$$\theta = \phi_N(\ \pmb{x}\ )\ \theta_N \qquad , \qquad q = \phi_N(\ \pmb{x}\ )\ q_N \qquad\qquad (17)$$

In the interior cell,

$$u_i = \psi_M(\ \pmb{x}\ )\ u_{Mi} \qquad , \qquad \theta = \psi_M(\ \pmb{x}\ )\ \theta_M \qquad\qquad (18)$$

in which $\phi_N(\pmb{x})$ and $\psi_N(\pmb{x})$ denote the shape fanctions defined on the boundary element and in the interior cell, respectively. Substitution of the discretizations of unknown functions (17) and (18) into equation (10) and arrangement of the resulting equations for all nodes result in the following pair of element equations :

$$\mathsf{H}_u\ \mathsf{u} = \mathsf{G}_u\ \tau - \mathsf{T}_u\ \theta + \mathsf{S}_u\ q - \mathsf{N}_u(\mathsf{u})\ \mathsf{u} - \mathsf{N}_u(\theta)\ \mathsf{u} \quad (19)$$

$$\mathsf{T}_\theta\ \theta = \mathsf{S}_\theta\ q - \mathsf{H}_\theta\ \mathsf{u} + \mathsf{G}_\theta\ \tau - \mathsf{N}_\theta(\mathsf{u})\ \mathsf{u} - \mathsf{N}_\theta(\theta)\ \mathsf{u} \quad (20)$$

where u is the nodal velocity vector, τ is the nodal traction vector, θ is the nodal temperature vector, q is the nodal heat flax vector, and H and G are the so-called influence matrixes, and N represents a nonlinear mapping defined with the convective terms. Equations (19) and (20) which are nonlinear with respect to u and θ are solved effectively with the Newton-Raphson method.

NUMERICAL EXAMPLES

In this chapter we present numerical results for a number of natural convection problems in two-dimensional rectangular and non-rectangular enclosures subjected to different side heating. All the results are obtained by using both the linear segment and the linear triangular cell.

A square enclosure

Let us consider the buoyancy-driven flow in a square closed enclosure induced by thermal gradients between the hot and cold walls. Fig.1 shows the geometry, boundary conditions and the element mesh.

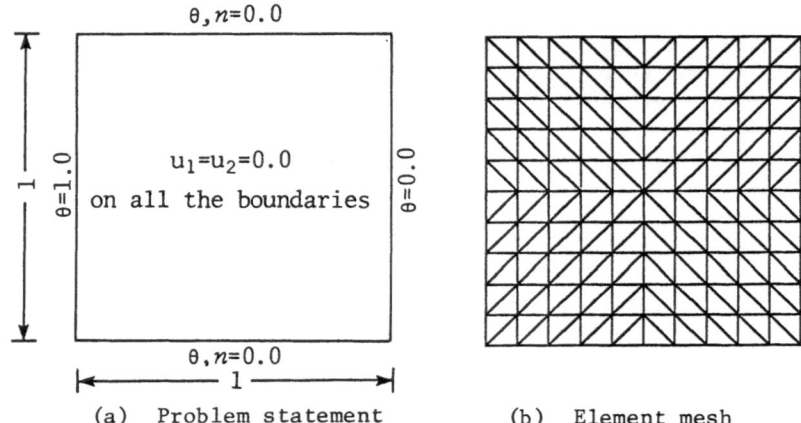

(a) Problem statement   (b) Element mesh

Figure 1. A square enclosure

Fig.2 shows the velocity vector fields and isotherms for Rayleigh number simulations of 10^3 , 10^4 and 10^5 (for fixed Prandtl number, $Pr=1.0$), respectivily.

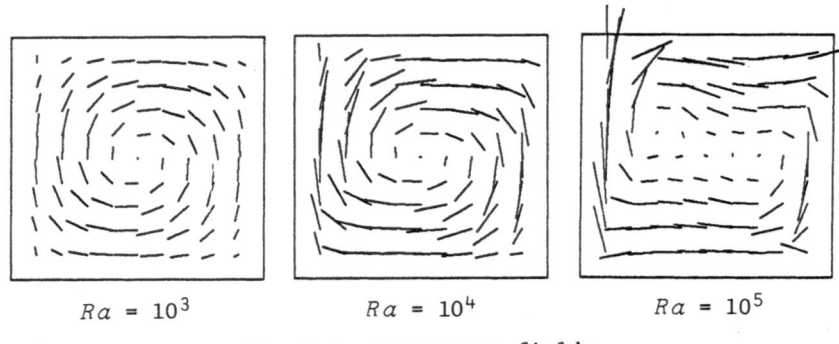

$Ra = 10^3$ $Ra = 10^4$ $Ra = 10^5$

(a) Velocity vector fields

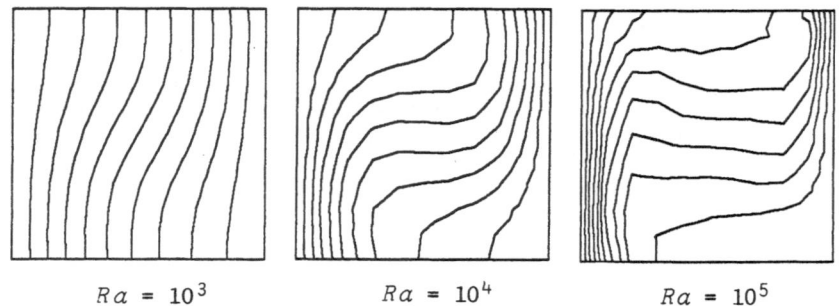

$Ra = 10^3$ $Ra = 10^4$ $Ra = 10^5$

(b) Temperature contours

Figure 2. Numerical results ($Pr=1.0$)

A plot of the average Nusselt number – Rayleigh number is shown in Fig.3 through the comparison with another results obtained by different methods.

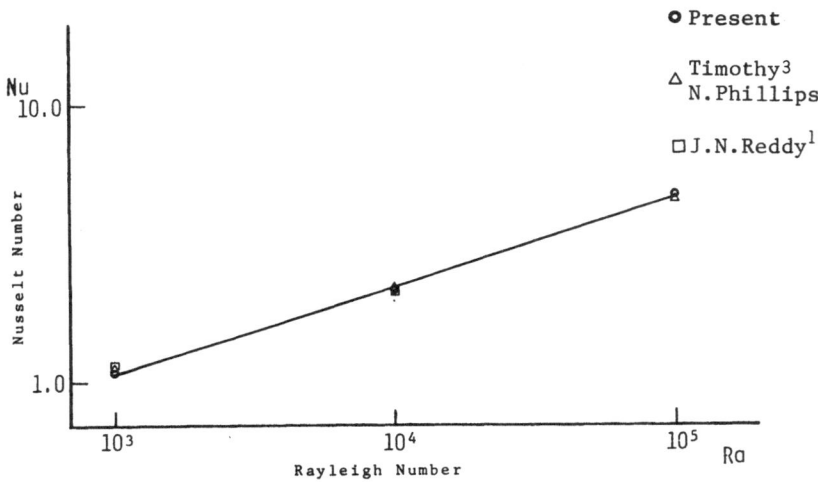

Figure 3. Average Nusselt number

A nonrectangular enclosure

Next, let us present numerical results for a nonrectangular enclosure shown in Fig.4.

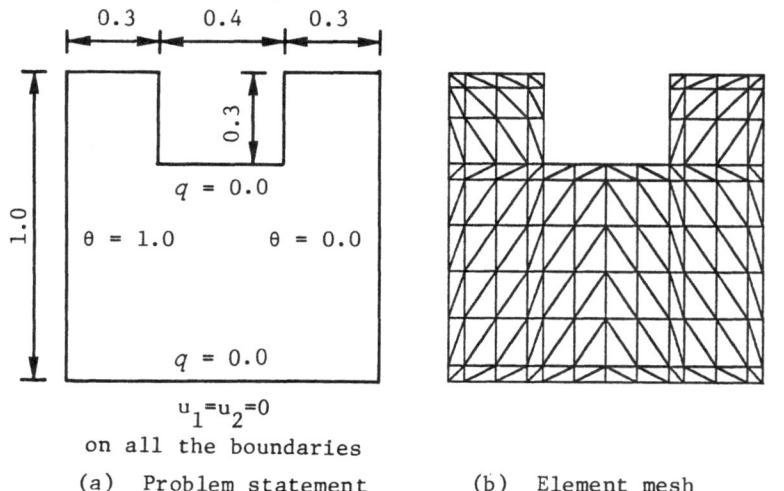

(a) Problem statement   (b) Element mesh

Figure 4. A nonrectangular enclosure

Fig.5 shows velocity vectors and isotherms for $Ra=10^4$ and 10^5 ($Pr=1.0$), respectively.

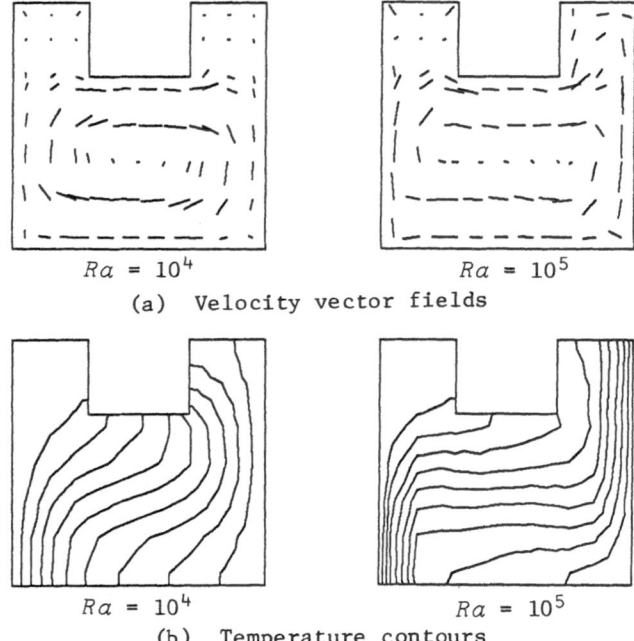

(a)   Velocity vector fields

(b)   Temperature contours

Figure 5.   Numerical results $(Pr=1.0)$

A circular annulus

As an example of the analysis of natural convection in arbitrary enclosures we consider the results for a circular annulus problem. We show the natural convection problem given in an annulus between two concentric circular cylinders in Fig.6.

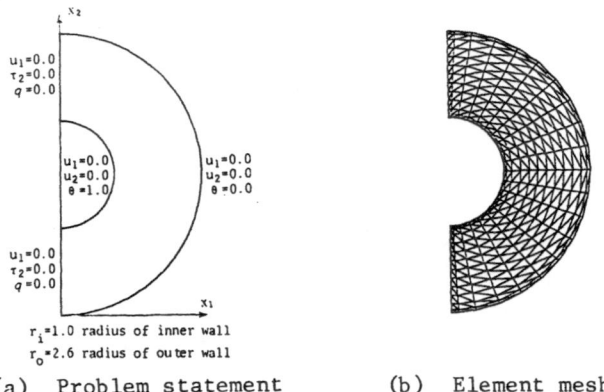

(a)   Problem statement   (b)   Element mesh

Figure 6.   A circular annulus

Fig.7 shows the obtained results for two Rayleigh numbers, $Ra = 5.0 \times 10^3$ and 10^4, with $Pr = 0.706$.

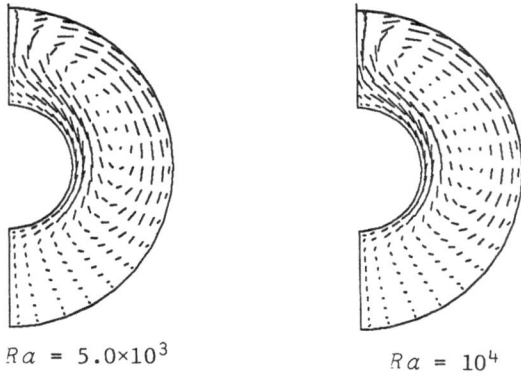

$Ra = 5.0 \times 10^3$ $Ra = 10^4$

(a)   Velocity vector fields

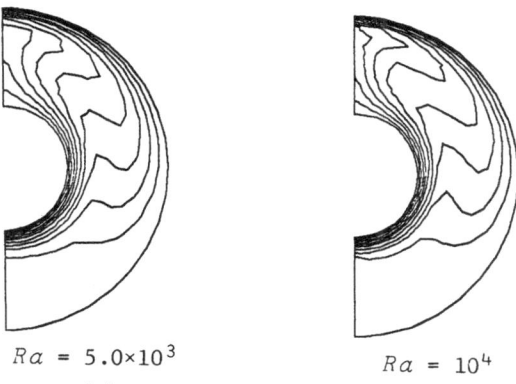

$Ra = 5.0 \times 10^3$ $Ra = 10^4$

(b)   Temperature contours

Figure 7.   Numerical results $(Pr=0.706)$

CONCLUSIONS

Numerical analyses based on the new integral equation formulation and its solution procedure were presented for the coupled convective and conductive heat transfer problems under heating of an incompressible viscous fluid. Steady-state numerical solutions were presented for two-dimensional natural convection in rectangular enclosure, a non-rectangular one and a circular annulus, subjected to thermal gradients. Despite relatively coarse meshes with lower order elements used in this study, the obtained numerical results were in agreement with results available in the literature. Efficiency, accuracy and simplicity of the present methodology make its use in three-dimensional simulations effectively.

REFERENCES

[1] Philips,T.N. (1984) ; Natural Convection in an Enclosed
    Cavity, J.Comput. Phys., 54, pp.365-381.

[2] Zienkiewicz,O.C., Gallagher,R.H. and Hood,P. (1976) ;
    Newtonian and Non-Newtonian Viscous Incompressible Flow.
    Temperature Induced Flows. Finite Element Solutions. 20,
    The Mathematics of Finite Elements and Applications 2,
    (Ed. Whiteman J.R.), pp.235-267, Academic Press.

[3] Reddy,J.N. (1983) ; Penalty-Finite Element Methods
    in Conduction and Convection Heat Transfer. Chapter 6,
    Numerical Methods in Heat Transfer, (R.W.Lewis, K.Morgan,
    and B.A.Schrefler, Eds), Vol.2, pp.145-178,
    John Wiley & Sons.

[4] Kuroki,T., Onishi,k. and Tosaka,N. (1985) ; Thermal Fluid
    Flow with Velocity Evaluation Using Boundary Elements and
    Penalty Function Method, in Boundary Elements VII, Vol 2,
    (Eds. C.A.Brebbia and G.Maier), pp.2-107 to 2-114,
    Springer-Verlag.

[5] Onishi,K. (1986) ; Boundary Element Analysis of Thermal
    Fluid Flow Using Ψ-ω, in Innovatibe Numerical Methods
    in Engineering, (Eds. Shaw,R.P., Periaux,J., Chaudouet,A.,
    Wu,J., Marino,C. and Brebbia,C.A.), pp.269-274,
    Springer-Verlag.

[6] Tosaka,N. (1985) ; Integral Equation Formulations for
    Viscous Flow Problems, Proc. 2nd Japan National Symposium
    on Boundary Element Method, Tokyo (1985), pp.155-160.

[7] Tosaka,N. (1986) ; Numerical Methods for Viscous Flow
    Problems Using an Integral Equation, Proc. Third Int. Symp.
    on River Sedimentation, Mississippi, USA, pp.1514-1525.

[8] Tosaka, N. and Onishi.K. (1986) ; Integral Equation Method
    for Thermal Fluid Flow Problems, Proc. Int. Con.
    on Computational Mechanics, Tokyo, Japan, 1986.

Numerical Simulations for Incompressible Viscous Flow Problems Using the Integral Equation Methods

N. Tosaka, K. Kakuda

Department of Mathematical Engineering, College of Industrial Technology, Nihon University, 1-2-1 Izumi-cho, Narashino-shi, Chiba 275, Japan

INTRODUCTION

The integral equation method or the boundary element method has been used increasingly for nonlinear problems of fluid mechanics. In particular, the general integral equation formulation for the flow problems of an incompressible viscous fluid has attracted interesting attention in many researchers. Several papers [1~4] for the problems can be available.

Recently, an innovative integral equation formulation for the flow problems of an incompressible viscous fluid in terms of the well known primitive-variable was presented systematically by Tosaka[5~8] . The typical numerical examples for two-dimensional steady-state flow fields[9] were effectively computed by the use of an incremental Newton-Raphson strategy , in which the Reynolds number is incremented, instead of the simple iterative scheme [4]. And, the numerical simulations for unsteady-state flow problems [10] were also implemented by applying the standard Newton-Raphson scheme to solve the equation set of each time step. More recently, the methodology was also extended to natural convection problems[11] and sample results, illustrating the effectiveness of the method, were presented and compared with the ones obtained by the other methods.

In this paper, in order to steady the accuracy and versatility of the above methodology we demonstrate the numerical simulations for the steady-state, and the unsteady-state flow problems of an incompressible viscous fluid. In the case of steady-state flow problems, the obtained nonlinear integral equation set is discretized spatially by using both the boundary elements and the internal cells. On the other hand, in the case of unsteady-flow problems we use both the spatial discretization and the time one. Making use of the usual time-marching scheme we apply also the Newton-Raphson scheme to the obtained equation set on each time step. As the numerical examples, the two-dimensional steady solutions and unsteady ones of flow past a step and flow past double steps are demonstrated.

Throughout this paper, we employ the summation convention on repeated indices. A comma is used to denote partial differentiation with respect to a space variable.

FUNDAMENTAL EQUATIONS

Let us consider the two-dimensional flow problems of an incompressible viscous fluid governed by the Navier-Stokes equations. Let Ω be a bounded domain in R^2 , and have a piecewise smooth boundary Γ . Let Γ_u and Γ_τ be subsets of Γ satisfying $\bar{\Gamma}_u \cup \bar{\Gamma}_\tau = \Gamma$ and $\Gamma_u \cap \Gamma_\tau = \phi$. The unit outward normal vector component to Γ is denoted by n_i . Also let $\mathcal{T}$ be a closed time interval $[0,T]$.

As the basic differential equations governing flow problems of an incompressible viscous fluid, we can write compactly the equation set into the following matrix form [5,6] :

$$[L_{IJ}]\{U_J\} = \{B_I\} \qquad\qquad in\ \Omega \times \mathcal{T}\ , \qquad (1)$$

wrere

$$[L_{IJ}] \equiv \begin{bmatrix} (-Re D_t + \Delta)\delta_{ij} + D_i D_j & -Re D_i \\ \\ D_j & 0 \end{bmatrix} , \qquad (2)$$

$$\{U_J\} \equiv \{\ u_j \quad p\ \}^\mathsf{T} \qquad , \qquad (3)$$

$$\{B_I\} \equiv \{\ Re\ u_j u_{i,j} - f_i \quad 0\ \}^\mathsf{T} \qquad , \qquad (4)$$

$$(i,j = 1,2)$$

in which u_i denotes the velocity vector component, Re is the Reynolds number, f_i is the external force vector component, p is the pressure, δ_{ij} is the Kronecker's delta, $D_t \equiv \partial/\partial t$, $D_i \equiv \partial/\partial x_i$, and $\Delta = D_i D_i$ denotes the Laplacian.

In addition to the above equation set, the boundary and initial conditions are imposed as follows :

$$u_i = \hat{u}_i \qquad\qquad\qquad on \ \Gamma_u \times \Upsilon , \qquad (5)$$

$$\tau_i \equiv \tau_{ij} n_j = \hat{\tau}_i \qquad\qquad\qquad on \ \Gamma_\tau \times \Upsilon , \qquad (6)$$

$$u_i(\boldsymbol{x}, 0) = u_{0i} \qquad\qquad\qquad in \ \Omega , \qquad (7)$$

where τ_{ij} is the Cauchy stress tensor, $\hat{u}_i$ represents the velocity vector component prescribed on the velocity boundary Γ_u , $\hat{\tau}_i$ designates the traction vector component prescribed on the traction boundary Γ_τ , and u_{0i} is the initial velocity.

If we do not take account of the terms with respect to time in Eqs.(1) to (7), then the above problem are reduced to an steady-state flow one.

INTEGRAL EQUATION FORMULATIONS

We intend to derive the integral equation set for the preceding unsteady-state flow problem. Moreover, the same set for the steady-state flow problem may be given lastly in this chapter. First of all, we start with the weighted residual statement of the basic nonlinear equation set (1). The weighted residual statement of (1) can be given as follows :

$$\iint_\Upsilon \int_\Omega (L_{IJ} U_J - B_I) \ U_{IL}^* \ d\Omega \, dt = 0 \qquad , \qquad (8)$$

$$(I, J, L = 1, 2, 3)$$

where U_{IL}^* is the weighting time-dependent tensor function.

Integrating by parts over the domain and the time interval, and after some manipulations, we can obtain the following nonlinear integral equation set in terms of the primitive variables :

$$C_{LK}(\boldsymbol{y}) U_K(\boldsymbol{y}, s) = \iint_\Upsilon \int_\Gamma u_i(\boldsymbol{x}, t) \Sigma_{iL}^*(\boldsymbol{x}, \boldsymbol{y}; t, s) \ d\Gamma(\boldsymbol{x}) \, dt$$

$$- \iint_\Upsilon \int_\Gamma \tau_i(\boldsymbol{x}, t) U_{iL}^*(\boldsymbol{x}, \boldsymbol{y}; t, s) \ d\Gamma(\boldsymbol{x}) \, dt$$

$$+ \int_\Omega [Re u_i(\boldsymbol{x}, t) U_{iL}^*(\boldsymbol{x}, \boldsymbol{y}; t, s)]_0 \ d\Omega(\boldsymbol{x})$$

$$+ \iint_\Upsilon \int_\Omega B_I(\boldsymbol{x}, t) U_{IL}^*(\boldsymbol{x}, \boldsymbol{y}; t, s) \ d\Omega(\boldsymbol{x}) \, dt \quad , \quad (9)$$

where the coefficient matrix $C_{LK}(y)$ generally depend on both the location of a boundary node x and the local geometry at the source point y .

In the following stage, let us consider the steady-state problem. The integral equation set corresponding to Eq.(9) can be easily derived as follows [7] :

$$C_{LK}(y)U_K(y) = \int_\Gamma u_i(x)\Sigma_{iL}^*(x,y) \; d\Gamma(x)$$

$$- \int_\Gamma \tau_i(x)U_{iL}^*(x,y) \; d\Gamma(x)$$

$$+ \int_\Omega B_I(x)U_{IL}^*(x,y) \; d\Omega(x) \quad . \tag{10}$$

In Eqs.(9) and (10), U_{IL}^* is the fundamental solution tensor which is given concretely in Refs.[7,8]. And, Σ_{iL}^* is the traction tensor defined in terms of the fundamental solution tensor.

In addition, the pressure can be explicitly determined from the integral equation (9) or (10) in which we set the subscript K to 3.

DISCRETIZATIONS AND SOLUTION PROCEDURES

We consider the discretizations and solution procedures of the obtained integral equations (9) and (10). Let us assume that the boundary Γ is divided into n elements and the domain Ω is discretized into m cells. In the case of Eq.(9) the time interval $\mathcal{T}$ is also discretized into f time steps. The unknowns of the integral equation set (9) or (10) can be approximated by using the interpolation functions and the nodal vectors [9,10] . Consequently, we can obtain the following discrete forms of (9) and (10) :
The discrete form of (9)(i.e.,the equation of unsteady-flow),

$$\Lambda_{lkRM} u_{kRM} + G_{liNM} \tau_{iNM} = H_{liNM} u_{iNM} - C_{liQM} u_{iQM}$$

$$+ N_{liQMjAE} u_{iQM} u_{jAE} - F_{liQM} f_{iQM} \quad . \tag{11}$$

The discrete form of (10)(i.e., the equation of steady-flow),

$$\Lambda_{lkR} u_{kR} + G_{liN} \tau_{iN} = H_{liN} u_{iN}$$

$$+ N_{liQjA} u_{iQ} u_{jA} - F_{liQ} f_{iQ} \quad . \tag{12}$$

Each coefficient matrix in the above is given in Ref.[10].

Setting Eq.(11) or (12) for all the boundary nodes and the internal ones, we can obtain easily the following global matrix forms :
  For the unsteady-state,

$$\mathbf{H}\,\mathbf{u}_f - \mathbf{G}\boldsymbol{\tau}_f - \mathbf{C}\,\mathbf{u}_{f-1} + \mathbf{N}(\mathbf{u}_f)\mathbf{u}_f - \mathbf{F}_f = \mathbf{0} \quad , \tag{13}$$

 For the steady-state,

$$\mathbf{H}\,\mathbf{u} - \mathbf{G}\boldsymbol{\tau} + \mathbf{N}(\mathbf{u})\mathbf{u} - \mathbf{F} = \mathbf{0} \quad , \tag{14}$$

where the subscript f implies the f-th time steps, $\mathbf{u}$ is the nodal velocity vector, $\boldsymbol{\tau}$ is the nodal traction vector, $\mathbf{H}$ and $\mathbf{G}$ are the so-called influence matrices, $\mathbf{N}(\mathbf{u})$ is the nonlinear convective force matrix, and $\mathbf{F}$ is the generalized force vector.

Since Eqs.(13) and (14) are nonlinear, we must make use of some iterative scheme to solve the equations. We had found in Ref.[9] that it is effective to use the Newton-Raphson scheme instead of the simple iterative scheme. In the case of unsteady-state flow problems, the standard time-marching scheme can be also employed in solving Eq.(13). The values of velocities at the end of each time interval are then used as initial values for the next time step. And, the usual Newton-Raphson scheme is applied to solve exactly the discretized equation set for each time step.

The convergence criterion employed herein is given by

$$\|\,\mathbf{R}(\mathbf{u},\boldsymbol{\tau})\,\| \leqq \varepsilon \quad , \tag{15}$$

where $\mathbf{R}(\mathbf{u},\boldsymbol{\tau})$ is the vector equivalent to the left-hand side of Eq.(13) or (14), $\|\ \|$ denotes the Euclidean norm , and ε is very small positive number.

NUMERICAL EXAMPLES

In the following we present a sample of problems which demonstrate the versatility and effectiveness of the present method. Throughout in this paper, we use linear boundary segments with eight Gaussian points and linear triangular cells with Hammer's quintic quadrature rule. The convergence criterion value, ε , is set to 10^{-6} . In the case of unsteady-flow problem, we assume the constant element on a time variable within each time step. Moreover, we assume that the initial velocities are zero everywhere in the interior domain.

Flow past a step

The problem statement and element mesh are depicted in Fig.1. The Reynolds number, Re, determined based on the velocity and length at the inlet is 100 . A fixed time step, Δt, employed in calculations of unsteady-state flow is equal to 1 .

In Fig.2 we present results of steady-state velocity fields. This calculation at $Re=100$ was made using of the obtained solution at $Re=50$ as initial values.

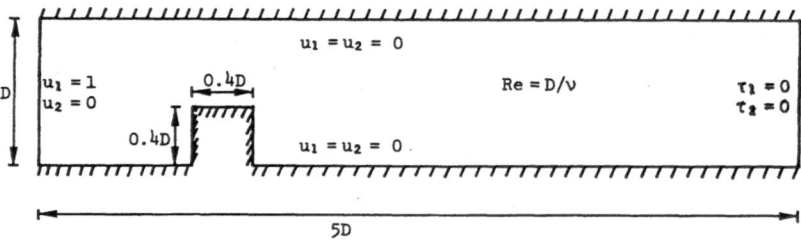

(a) Problem statement

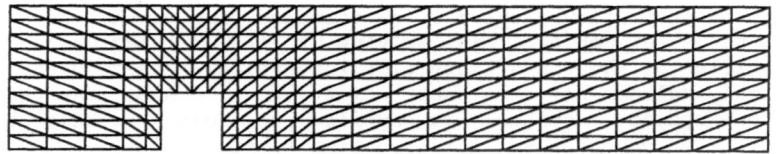

(b) Element mesh

Figure 1. Flow past a step

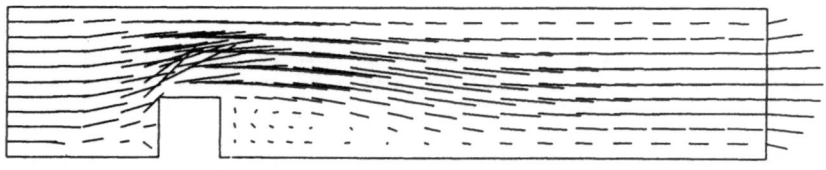

Figure 2. Steady-state velocity vector ($Re = 100$)

Numerical results for the flow pattern of unsteady-state velocity fields are shown in Fig.3. As can be seen from Fig.3, a steady-flow is attained after approximately 10 time steps, corresponding to 10 time units. The flow patterns at $t=10$ are in very good agreement with the steady-state solutions in Fig.2. The pressure contours at $t=10$ are illustrated in Fig.4.

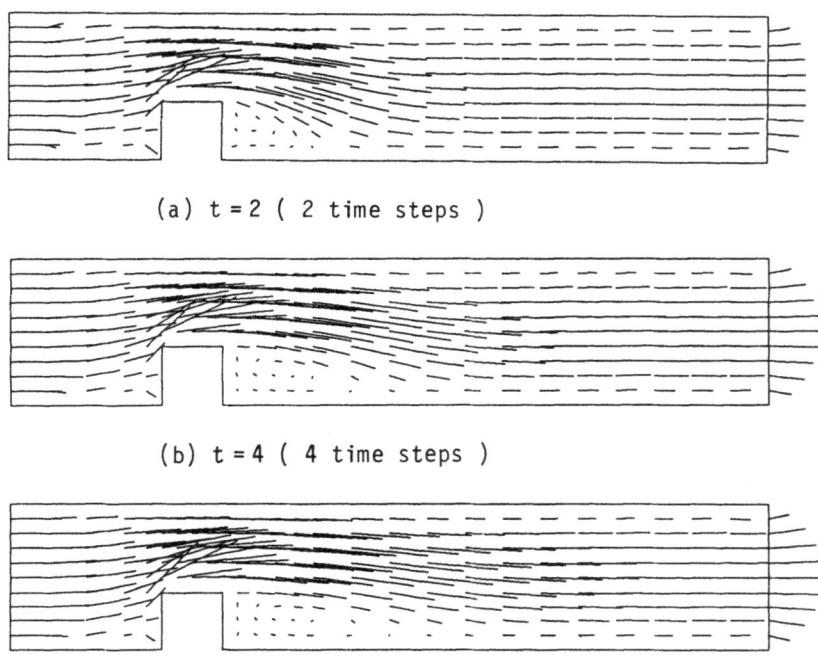

(a) t = 2 (2 time steps)

(b) t = 4 (4 time steps)

(c) t = 10 (10 time steps)

Figure 3.   Unsteady-state velocity vectors at each time step $(Re = 100, \Delta t = 1)$

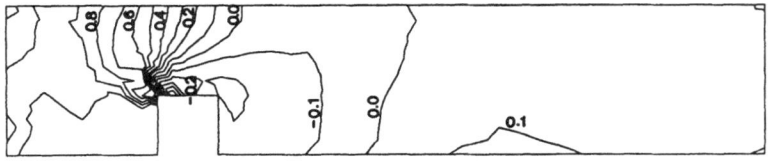

Figure 4.   Pressure contours at t = 10 (10 time steps)

Flow past double steps

Next, let us consider the flow past double steps. A problem description is shown in Fig.5.

Results for steady-flow and unsteady-flow at $Re=100$ are presented in Figs.6 and 7, respectively. In the case of unsteady-flow, the time step is constant at 1.

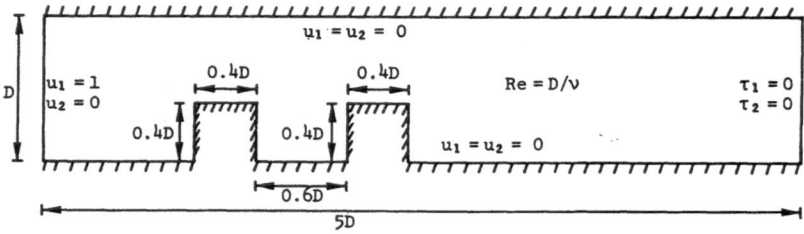

(a) Problem statement

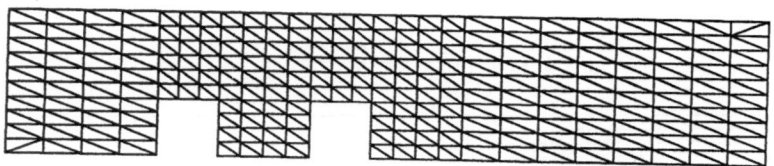

(b) Element mesh

Figure 5. Flow past double steps

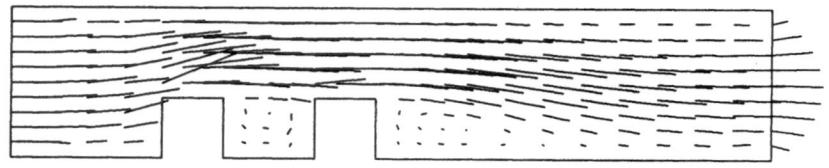

Figure 6. Steady-state velocity vector $(Re = 100)$

It took approximately *10* time steps for the flow to become steady-flow solutions in Fig.6. In Fig.8 we present the pressure contours at *t=10* (*10* time steps).

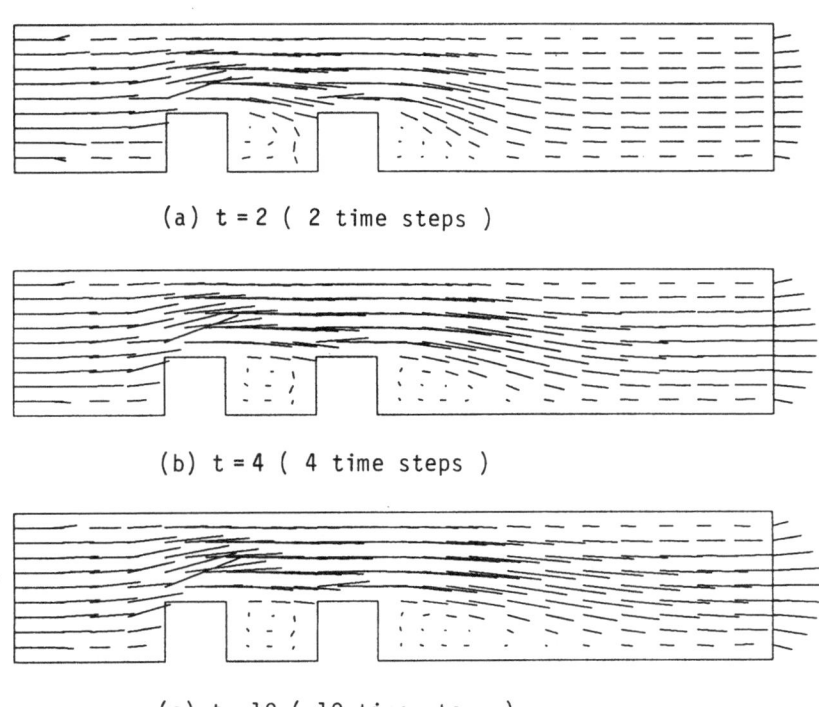

(a) t = 2 (2 time steps)

(b) t = 4 (4 time steps)

(c) t = 10 (10 time steps)

Figure 7. Unsteady-state velocity vectors at each time step
($Re = 100, \Delta t = 1$)

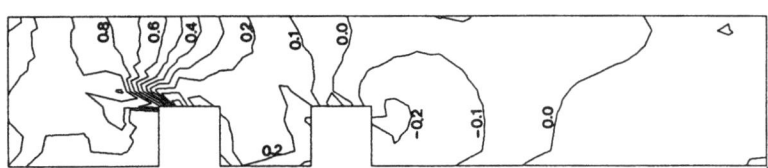

Figure 8. Pressure contours at t = 10 (10 time steps)

CONCUSIONS

In this paper, we have presented results of the numerical simulations for the steady-state, and the unsteady-state incompressible viscous flow problems by using the integral equation formulation. Numerical examples for two-dimensional problems of flow past a step and flow past double steps are demonstrated. In both problems, the flow patterns emerging from unsteady-state solutions at several time steps agree quite with the case of steady-state solutions. Moreover, the obtained results for unsteady-state solutions are extremely stable even for large time steps. Consequently, we conclude that the present method is very effective to numerical analysis of incompressible viscous flow problems.

REFERENCES

1. Wu,J.C. (1982), Problems of General Viscous Flow, Developments in Boundary Element Methods-2,(Eds.,P.K.Banerjee and R.P.Shaw), Applied Science Publishers,pp.69-109
2. Onishi,K.,T.Kuroki and M.Tanaka (1984), An Application of Boundary Element Method to Incompressible Laminar Viscous Flows, Engineering Analysis 1(3),pp.122
3. Skerget,P.,A.Alujevic and C.A.Brebbia (1984), The Solution of Navier-Stokes Equations in terms of Vorticity-Velocity Variables by Boundary Elements, Boundary Elements, (Ed.,C.A.Brebbia),pp.4/41, Springer-Verlag
4. Tosaka,N.,K.Kakuda and K.Onishi (1985.9), Boundary Element Analysis of Steady Viscous Flows Based on P-U-V Formulatin, pp.9/71-9/80,Boundary Elements VII Vol.2,(Eds.,C.A.Brebbia and G.Maier), Springer-Verlag
5. Tosaka,N. and K.Onishi (1985), Boundary Integral Equation Formulations for Steady Navier-Stokes Equations Using the Stokes Fundamental Solutions,Engng. Analy.2(3),pp.128-132
6. Tosaka,N. and K.Onishi, Boundary Integral Equation Formulations for Unsteady Incompressible Viscous Fluid Flow by Time-Differencing, Engineering Analysis (in press)
7. Tosaka,N. (1985),Integral Equation Formulations for Viscous Flow Problems (in Japanese), Proc. Second Japan National Symp. BEM, JASCOME, Tokyo, pp.155-160
8. Tosaka,N.(1986),Numerical Methods for Viscous Flow Problems Using an Integral Equation, Proc. third Int. Symp. on River Segmentation, Mississippi, pp.1514-1525
9. Tosaka,N. and K.Kakuda (1986),Numerical Solutions of Steady Incompressible Viscous Flow Problems by the Integral Equation Method, Proc. 4th Int. Symp.(Eds.,R.P.Shaw et al.) Georgia Institute of Technology,USA, pp.211-222
10. Tosaka,N. and K.Kakuda (1986), Numerical Simulations of the Unsteady-State Incompressible Viscous Flows Using an Integal Equation,Int. Con. BE,Tsinghua University, China
11. Tosaka,N. and K.Onishi (1986), Integral Equation Method for Termal Fluid Flow Problems,Proc. Int. Con. on Computational Mechanics, Tokyo

The Boundary Element Spectral Method and Applications in 3-D Viscous Hydrodynamics

W. Borchers, F.K. Hebeker
Fb. 17, Universität, D-4790 Paderborn, West Germany

INTRODUCTION

Recently, boundary element methods have found dramatically increasing interest for the numerical treatment of a large variety
of problems arising in the engineering sciences (e.g. Brebbia
et al.[3]). They serve as highly efficient but slightly specialized numerical methods. Hence a strong line of research is to
combine boundary element methods with other well experienced
methods in order to optimize their specific (often complementary) advantages. This has been successfully done mixing boundary elements and finite elements (Zienkiewicz, Kelly and
Bettess[13]; Johnson and Nedelec[9]; Wendland [12];Hsiao and Porter[8]).
A different combination which seems to be even more advantageous under certain circumstances is that of boundary elements
and fast spectral methods (Hebeker[6]; Borchers, Hebeker and Rautmann[2]). This hybrid approach, called boundary element spectral
method, is briefly described in the present note.

To be concrete, let us restrict ourselves to the problem of
viscous incompressible fluid flow, but most of the proposed
methods carry over easily to other problems of the applications.
In Hebeker[6] we considered the stationary Stokes flow, here we
extend the method to the nonstationary Stokes problem (cf. Borchers, Hebeker and Rautmann[2], too).

THE NONSTATIONARY STOKES FLOW

We consider the nonstationary Stokes differential system

$$v_t - \nabla^2 v + \nabla \rho = f \quad , \quad \text{div } v = 0 \text{ in } \Omega \tag{1}$$

for the velocity field $v = (v_1, v_2, v_3)$ and the (scalar) pressure function ρ of a homogeneous and viscous fluid contained
in a (smoothly bounded) 3-D cavity Ω. ∇^2 denotes the Laplacean, and the problem is normalized so that the (constant)
viscosity $\nu = 1$. Here we neglect the nonlinear convective

terms of Navier Stokes flow: this problem is commonly reduced
to the linear one by means of iterative or time-stepping pro-
dures.

As boundary conditions we choose the no-slip condition on the
boundary $\partial\Omega$, and we assume additionally that the fluid flow
is known at an initial time $t = 0$:

$$v\big|_{\partial\Omega} = 0 \quad , \quad v\big|_{t=o} = v_o \quad . \tag{2}$$

For the numerical treatment of (1,2) the time grid $t_k = k \cdot \tau$
is introduced, with a time step-size $\tau > 0$ and $k = o,1,2,\ldots$
By $v_k,\ldots$ we denote the values of $v,\ldots$ at time $t = t_k$. As
time-stepping procedure we choose a fully implicit scheme which
results in a stationary Stokes-like problem to be solved at
each time step:

$$\lambda v_k - \nabla^2 v_k + \nabla\rho_k = \lambda v_{k-1} + f_k \equiv g$$

$$\text{div} \quad v_k = 0 \qquad \text{in} \quad \Omega \tag{3}$$

$$v_k\big|_{\partial\Omega} = 0 \quad . \quad v\big|_{t=o} = v_o \quad .$$

Here $\lambda = \dfrac{1}{\tau}$ is positive, hence the numerical properties of
(3) are expected to be even better than in case of $\lambda = o$.

THE BOUNDARY ELEMENT SPECTRAL METHOD

Due to the complex geometry of many flow regions appearing in
practice, and due to the "simple" structure of problem (3) the
use of boundary elements to numerically solve (3) seems to be
the best way in general, but one is afraid of the volume potentials to treat the right-hand terms g in the differential
equations. Recently, Varnhorn[10] has shown in his thesis that
this approach is applicable (even in the nonlinear case) but
is extremely expensive due to the frequent evaluations of the
time-consuming volume potentials. Hence to circumvent this
task the way of using fast spectral solvers has been introduced by the present authors. The basic idea is that the volume potential just serves as a particular solution of the
nonhomogeneous systems (3): but this can often be achieved in
a more efficient way by different methods, under certain circumstances best by means of fast spectral methods.

We split the solution of (3) into two parts:

$$v_k = u_1 + u_2 \quad , \quad \rho_k = q_1 + q_2 \quad , \quad (k = 1,2,\ldots). \tag{4}$$

Here (u_1, q_1) is any particular solution of the differential equations (arbitrary boundary data), but (u_2, q_2) solves (3) with zero right-hand data $(g = 0)$, and boundary data

$$u_2|_{\partial\Omega} = - u_1|_{\partial\Omega} \cdot \qquad (5)$$

The first subproblem will be solved with a spectral method (next section) whereas the boundary element method (next but one section) is well suited to solve the second subproblem.

THE SPECTRAL PART

Extend g of (3) smoothly to a cube C (of side-length a) so that $g|_{\partial C} = 0$. Consider the (vector) series

$$u_1 = \frac{g_0}{\lambda} + \sum_{k \in \mathbb{Z}^3 \setminus \{0\}} (\lambda + |\frac{2\pi}{a}k|^2)^{-1} (g_k - \frac{g_k \cdot k}{|k|^2}k) \, e^{\frac{2\pi}{a}ik \cdot x},$$

$$q_1 = -\frac{ia}{2\pi} \sum_{k \in \mathbb{Z}^3 \setminus \{0\}} |k|^2 (g_k \cdot k) e^{\frac{2\pi}{a}ik \cdot x}, \qquad (6)$$

where the vector coefficients g_k are given as Fourier integrals

$$g_k = \frac{1}{a^3} \int_C g(x) e^{-\frac{2\pi}{a}ik \cdot x} \, dx \cdot \qquad (7)$$

Here k are vectors with integer components, $g_k \cdot k$ or $|k|$ denotes the scalar product or the euclidean norm resp. It has been proved in Borchers[1] that the series (6) are the periodical solution of the differential system

$$\lambda u_1 - \nabla^2 u_1 + \nabla q_1 = g \quad , \quad \text{div } u_1 = 0 \quad \text{in} \quad C. \qquad (8)$$

Hence it play the same role as the volume potential to treat (3) but has important numerical advantages: first, the integrals (7) are computed quickly and accurately using the trapezoidal rule; secondly, the series are efficiently evaluated using the FFT.

THE BOUNDARY ELEMENT PART

Consider the second subproblem

$$\lambda u_2 - \nabla^2 u_2 + \nabla q_2 = 0 \quad , \quad \text{div } u_2 = 0 \quad \text{in} \quad \Omega , \qquad (9)$$

with (5) as boundary condition. The fundamental matrix

$$\Gamma_{ij}(x) = -\frac{\delta_{ij}}{4\pi|x|}e^{-\sqrt{\lambda}|x|} + \frac{1}{4\pi\lambda}\left[(\frac{\delta_{ij}}{|x|^3} - 3\frac{x_i x_j}{|x|^5})(1-e^{-\sqrt{\lambda}|x|}\right.$$
$$\left. -\sqrt{\lambda}|x|\ e^{-\sqrt{\lambda}|x|}) + \frac{\lambda x_i x_j}{|x|^3}e^{-\sqrt{\lambda}|x|}\right] \qquad (10)$$

is well known, and the ith column $\Gamma_{(i)}$ of Γ_{ij} solves (9) with pressure function

$$P_i(x) = -x_i / (4\tilde{\pi}|x|^3) + \text{const.} \qquad (11)$$

Hence we can introduce the potentials of the simple and of the double layer as usual.

Let us look for a solution of (9) in terms of a double layer potential, shortly written as

$$u_2 = W\psi \quad , \quad q_2 \text{ analogously.} \qquad (12)$$

The jump relations (Varnhorn[10]) lead to an integral equation of the second kind for the unknown vector surface source ψ :

$$(I + 2 W + 2 N)\psi = -2u_1 \quad \text{on the boundary} \quad \partial\Omega. \qquad (13)$$

Here I denotes the unit matrix and N the boundary operator (n the exterior normal vector on $\partial\Omega$):

$$N\psi(x) = n(x) \int_{\partial\Omega} n(z)\cdot\psi(z)do_z. \qquad (14)$$

It has been shown in Borchers[1] that (13) has a unique solution and consequently the corresponding potential (12) solves the problem (9).

An efficient boundary element method to solve (13) numerically has been developed in Hebeker [4]. It consists, roughly speaking, of a collocation procedure using piecewise bilinear but globally continuous trial functions on the parameter space of the boundary. The resulting algebraic systems is then solved using a fast multigrid iteration procedure. For details concerning the numerical quadrature or the rigorous convergence analysis we refer to Hebeker [4,5,6], and concerning the mathematical treatment of the flow problem in cavities with corners and edges cf. Hebeker[7].

ON COMPUTATIONAL RESULTS

The boundary element method (described shortly above) has been tested numerically in 3-D by the second author (loc.cit.). The tests include the viscous drag of a sphere (test against Stokes' analytical formula), Oseen's flow, the slow viscous flow past an irregular shaped body, to mention just a few examples. For an application of the indicated methods and programs to the non-linear Navier Stokes problem see Varnhorn[10]. Concerning the given spectral method several numerical tests in 3-D have been carried out by Borchers [1] (cf. also Borchers, Hebeker and Rautmann[2]).

A first test of the combined boundary element spectral method has shown the following encouraging results. Here we compute the stationary tokes flow where f is chosen in this way so that we are given an analytic solution for reasons of comparison. Our flow region is a ball, but our computations are fully 3-D in order to have results to be considered typical also for more complicated flow regions.

The spectral series (6) has been evaluated using 16^3 vector coefficients g_k (7). The boundary element method has been carried out by substituting the unknown vector field ψ of (13) by an approximate vector field, where 867 unknowns are to be determined by solving a linear algebraic system with dense matrix. Then a mean relative error of

about 0,3 %

only turns out. Several hours of CPU time have been taken on our PRIME 750 computer.

CONCLUSION

A hybrid boundary element spectral method has been proposed and discussed to compute numerically the nonstationary Stokes flow in a 3-D cavity. A first numerical test computation has shown that the given methods work quite satisfactory and should be further developed for more complicated flow models.

REFERENCES

1. Borchers, W. (1985), Eine Fourier-Spektralmethode für das Stokes Resolventenproblem. Preprint U. of Paderborn, 13 pp., submitted.

2. Borchers, W., Hebeker, F.K., and Rautmann, R. (1985), A Boundary Element Spectral Method for Nonstationary Viscous Flows in Three Dimensions. In: Finite Approximationen in der Strömungsmechanik (Ed. Hirschel E.H.), pp. 14-28, Vieweg, Braunschweig.

3. Brebbia, C.A. et al.(ed.)(1985), Boundary Element Methods in Engineering. Proc. 7th Intl. Conf. on BEM., Como, Italy, 1985. Springer, Berlin.

4. Hebeker, F.K. (1985), Efficient Boundary Element Methods for 3-D Viscous Flows. Numerical Methods in PDE, 35 pp., in press.

5. Hebeker, F.K. (1985), Efficient Boundary Element Methods for 3-D Viscous Flows. In: Brebbia[3], pp. 9: 37-44.

6. Hebeker, F.K. (1986), On a New Boundary Element Spectral Method. In: Innovative Numerical Methods in Engineering (Ed. Shaw R.P. et al.), pp. 311-316, Proc. of 4th Intl. Symp., Atlanta Ga., USA., 1986. Springer, Berlin.

7. Hebeker, F.K. (1986), On the Numerical Treatment of Viscous Flows Past Bodies with Corners and Edges. In: Efficient Numerical Methods in Continuum Mechanics (Ed. Hackbusch W.), 6 pp., Proc. GAMM Seminar, Kiel, West Germany, 1986, in press. Vieweg, Braunschweig.

8. Hsiao G.C. and Porter J.F. (1986), A Hybrid Method for an Exterior Boundary Value Problem Based on Asymptotic Expansion, Boundary Integral Equation and Finite Element Approximation. In: Innovative Numerical Methods in Engineering (see Hebeker[6]), pp. 83-88.

9. Johnson C. and Nedelec J.C. (1980), On the Coupling of Boundary Integral and Finite Element Methods. Math. Comp., Vol. 35, pp. 1063-1079.

10. Varnhorn, W. (1985), Zur Numerik der Gleichungen von Navier-Stokes. PhD Thesis, U. of Paderborn.

11. Wendland, W.L. (1985), On Some Mathematical Aspects of Boundary Element Methods for Elliptic Problems. In: The Mathematics of Finite Elements and Applications V (Ed. Whiteman J.R.), pp. 193-227.

12. Wendland, W.L. (1986), On Asymptotic Error Estimates for the Combined Boundary and Finite Element Method. In: Innovative Numerical Methods in Engineering (see Hebeker[6]), pp. 55-69.

13. Zienkiewicz, O.C., Kelly D.W. and Bettess, B. (1977), The Coupling of the Finite Element Method and Boundary Solution Procedures. Intl. J. Num. Meth. Engin., Vol. 11, pp. 355-375.

Vorticity – Velocity – Pressure Boundary Integral Formulation

P. Skerget, A. Alujevic and C.A. Brebbia
Faculty of Engineering, University of Maribor, Yugoslavia
Jozef Stefan Institute, University of Ljubljana, Yugoslavia
Computational Mechanics Institute, Southampton, U.K.

INTRODUCTION

This paper deals with boundary element method applied to determine pressure distribution in steady laminar flow of an isochoric fluid governed by Navier-Stokes equations. The vorticity-
-velocity-pressure (w-v-p) boundary integral formulation of the
momentum transport equation is presented. Some specific behaviours of the mentioned formulation, which is linear integral
equation in terms of unknown pressure values, are discussed.
Two examples are studied to show the validity and applicability
of BEM to solve fluid flow problems.

GOVERNING EQUATION

Flow of isochoric viscous fluids is governed by momentum transport equation and mass conservation law, known as Navier-Stokes
and continuity equations

$$\nu \Delta \underline{v} = \underline{\nabla} p / \varrho + (\underline{v} \, \underline{\nabla}) \, \underline{v} \tag{1}$$

$$\underline{\nabla} \, \underline{v} = 0 \tag{2}$$

which are expressed with the primitive variables velocity $\underline{v}$ and
pressure p. The material properties such as viscosity ν and
density ϱ are taken constant.

In view of the vector identity

$$\underline{v} \times \underline{\nabla} \times \underline{v} = \underline{v} \times \underline{w} = \underline{\nabla} \, v^2 / 2 - (\underline{v} \, \underline{\nabla}) \, \underline{v} \tag{3}$$

where $\underline{w}$ is the vorticity vector

$$\underline{w} = \underline{\nabla} \times \underline{v} \tag{4}$$

the momentum transport equation (1) can be given in terms of
velocity-vorticity-pressure (v-w-p) variables, i.e.

$$\nu \Delta \underline{v} = \underline{\nabla} h - \underline{v} \times \underline{w} \tag{5}$$

h being the total head defined by

$$h = p/\varsigma + v^2/2 \tag{6}$$

The equation (5) can be treated as a linear equation for unknown pressure values in the case that the velocity-vorticity (v-w) formulation is used to compute velocity and vorticity vectors (Skerget et al., 1984).

INTEGRAL REPRESENTATION FOR (v-w-p) FORMULATION

One can derive the boundary integral statement for the momentum transport equation (5) using Green's theorem for vectors (Wu et al., 1973) yielding

$$c(\xi)\underline{w}(\xi)+\int(\underline{w}(S)\underline{n}(S))\underline{\nabla}u^*(\xi,S)d\Gamma(S) = \int(\underline{w}(S)x\underline{n}(S))x\underline{\nabla}u^*(\xi,S)d\Gamma(S)+$$

$$+1/\nu.\int h(S)\underline{\nabla}u^*(\xi,S)x\underline{n}(S)d\Gamma(S)+1/\nu.\int(\underline{v}(s)x\underline{w}(s))x\underline{\nabla}u^*(\xi,s)d\Omega(s) \tag{7}$$

which can also be written as (Skerget et al., 1984)

$$c(\xi)\underline{w}(\xi)+\int\underline{w}(S)q^*(\xi,S)d\Gamma(S) = \int(\underline{\nabla}u^*(\xi,S)x\underline{n}(S))x\underline{w}(S)d\Gamma(S)+$$

$$+1/\nu.\int h(S)\underline{\nabla}u^*(\xi,S)x\underline{n}(S)d\Gamma(S)+1/\nu.\int(\underline{v}(s)x\underline{w}(s))x\underline{\nabla}u^*(\xi,s)d\Omega(s) \tag{8}$$

The formulations (7) and (8) are valid for three dimensional flows while the mathematical description of the plane flows is much simpler due to $\underline{w} \to w$ and $(\nabla u^* x\underline{n})x\underline{w} = 0$. Thus, the vector equation (8) reduces to the following scalar equation

$$c(\xi)w(\xi)+\int w(S)q^*(\xi,S)d\Gamma(S) = 1/\nu.\int h(S)q_t^*(\xi,S)d\Gamma(S)+$$

$$+1/\nu.\int w(s)\underline{v}(s)\underline{\nabla}u^*(\xi,s)d\Omega(s) \tag{9}$$

For the two dimensional problems which are dealt with in this paper, the fundamental solution and its normal and tangential derivatives are

$$u^*(\xi,S) = -1/2\pi. \ln r(\xi,S)$$

$$q^*(\xi,S) = 1/2\pi. d(\xi,S)/r^2(\xi,S) \tag{10}$$

$$q_t^*(\xi,S) = 1/2\pi. d_t(\xi,S)/r^2(\xi,S)$$

with $d(\xi,S)=(x_i(\xi)-x_i(S)).n_i(S)$, $d_t(\xi,S)=(x_i(\xi)-x_i(S)).t_i(S)$, $i=1,2$ and $n_i(S)$, $t_i(S)$ are direction cosines of the unit normal and tangent at the boundary point S, while ξ is the source point.

BOUNDARY ELEMENT DISCRETIZATION

Let us divide the boundary Γ into E boundary elements with N_e nodes, and the domain Ω into C internal cells with N_c points. The functions w, h and products wv_i , $i=1,2$ are approximated within each element and cell according to space shape functions $\varnothing(\eta)$ or $\phi(\eta_1,\eta_2)$, multiplied by nodal values, i.e. for the boundary

$$w(\eta) = \underline{\varnothing}^T \underline{w}^n \quad , \quad h(\eta) = \underline{\varnothing}^T \underline{h}^n \tag{11}$$

and in the domain

$$wv_i(\eta_1,\eta_2) = \underline{\phi}^T \underline{WV}_i^n \tag{12}$$

The coordinates η and η_i are the homogeneous coordinates (Brebbia, 1978), while the index n refers to the number of the nodes within each element or cell, which are associated with the nodal values of the functions w, h and wv_i.

With the integrals

$$h_e^n = \int \underline{\varnothing}^n q^* d\Gamma_e \ , \quad h_{et}^n = \int \underline{\varnothing}^n q_t^* d\Gamma_e \ , \quad d_{ci}^n = \int \underline{\phi}^n q_i^* d\Omega_c \tag{13}$$

which are functions of the geometry only, the discretized expression of equation (9) can be formulated as

$$c(\xi)w(\xi) + \Sigma\underline{h}^T\underline{w}^n = - 1/\mathcal{J}.\Sigma\underline{h}_t^T\underline{h}^n + 1/\mathcal{J}.(\Sigma\underline{d}_{-x}^T\underline{WV}_x^n + \Sigma\underline{d}_{-y}^T\underline{WV}_y^n) \tag{14}$$

or in the matrix form of linear system of equations for all unknown boundary total head values ($\xi = 1, N_e$)

$$\underline{H}_t \ \underline{h} = \underline{D}_x\underline{WV}_x + \underline{D}_y\underline{WV}_y - \mathcal{J}\underline{H} \ \underline{W} \tag{15}$$

which can be rewritten as

$$\underline{A} \ \underline{X} = \underline{F} \tag{16}$$

It can be easily proved using Stokes theorem, that the boundary integral containing the total head in equation (9) vanishes for its constant value. As a consequence N_e-1 equations are at most independent on each other, and in the computational procedure an arbitrary value for the total head has to be specified.

The tangential derivative of the fundamental solution $q_t^*(\xi,S)$ is weak elliptic operator (Schmidt, 1983) and in order to obtain the solution of the system of equations (15) for unknown pressure values, the source points ξ have to be located in the middle of the boundary elements in the case of continuous linear boundary elements (Wendland, 1985), thus $c(\xi)=0.5$.

EXAMPLES

Couette flow

This problem represents a simple flow in a channel with the upper wall moved with a constant velocity v_h. Only one mesh has been used to compute velocity, vorticity and pressure drop in the channel. The flow region was discretized by 38 continuous linear boundary elements with 38 nodes, and 140 internal cells containing 52 points.

The exact solutions for velocity distribution and pressure drop
variation are known and given by

$$v_x = v_h \cdot y/H + S.y/H.(1-y/H).v_h \; , \; p = p_1-(p_1-p_2).x/L - \varrho.g.y$$

where H = height and L = length of the channel, while

$$S = (p_1-p_2).H^2/(2L\varrho\nu v_h)$$

and g is gravity acceleration. Results for pressure values are
depicted on Fig.1 for the following data: H = 0.05 , L = 0.2 ,
v_h = 0.01 , S = 2 , ν = o.143E-03 and 0.143E-04 , ϱ = 1 . An
excellent agreement of BEM results with the exact solution has
been achieved.

Flow over a backward facing step

This problem has been studied for various Reynolds number values and several BEM meshes. Results for velocity distribution
and recirculation zone lenth have been compared against FEM results and experimental data (Skerget et al., 1985).

Inlet and outlet boundary velocity conditions are prescribed according to the exact parabolic variation of the velocity for
fully developed laminar flow in a channel. Inlet velocity values are defined by

$$v_x = 4y.(1-y) \; , \; \bar{v}_x = 2/3$$

Pressure drop distributions were obtained for different Reynolds
numbers Re = $\bar{v}_x.H/\nu$ = 0.73 , 7.3 and 73., defined for the mean
velocity and step height. Mesh with 102 boundary elements (102
nodes) and 908 internal cells (404 points) has been used to discretize the flow region.

Results for pressure distribution along the channel are depicted
on Fig.2, clearly demonstrating the zone of positive pressure
gradient ($\partial p/\partial x$ > 0) behind the step. No comparison has been made against results from other sources of research reports.

CONCLUSION

In the above paper the boundary element technique has been used
to determine the pressure drop distribution of laminar motion in
an isochoric and viscous fluid. The vorticity-velocity-pressure
(w-v-p) boundary integral statement is presented and some of its
behaviours are discussed.

Pressure drop variation in the simple Couette flow and in a more
complex channel with the backward facing step have been analysed. The BEM results for Couette flow are compared against the
exact solution and good agreement achieved. The results for the
recirculation flow in the channel with the step also appear to
be adequate, showing the region of positive pressure gradient
in the recirculation zone behind the step.

REFERENCES

Brebbia, C.A. (1978) The Boundary Element Method for Engineers. Pentech Press, London, Halstead Press, New York

Brebbia, C.A., Telles, J. and Wrobel, L.C. (1984) Boundary Element Methods - Theory and Applications, Springer Verlag, New York

Brebbia, C.A. /ed/ (1984) Topics in Boundary Element Research, Vol.1, Basic Principles and Applications, Springer Verlag, Berlin

Brebbia, C.A. /ed/ (1985) Topics in Boundary Element Research, Vol.2, Time Dependent and Vibrational Problems, Springer Verlag, Berlin

Schmidt, G. (1983) On Spline Collocation for Singular Integral Equations. Math.Nachrichten, 111:177-196

Skerget, P., Alujevic, A., and Brebbia, C.A. (1984) The Solution of Navier-Stokes Equations in Terms of Vorticity-Velocity Variables by Boundary Elements. 6th Int.Conf.on Boundary Element Method, QE2, Southampton-New York, Springer Verlag, Berlin

Skerget, P., Alujevic, A., and Brebbia, C.A. (1985a) Analysis of Laminar Flows with Separation Using BEM. 7th Int.Conf.on Boundary Element Method, Como, Springer Verlag, Berlin

Skerget, P., Alujevic, A., and Brebbia, C.A. (1985b) Analysis of Laminar Fluid Flows by Boundary Elements. 4th Int.Conf.on Numerical Methods in Laminar and Turbulent Flows, Swansea

Wendland, W.L. (1985) Asymptotic Accuracy and Convergence for Point Collocation Methods. Topics in Boundary Element Research, 2:230-250, Springer Verlag, Berlin

Wu, J.C. and Thompson, J.F. (1973) Numerical Solution of Time Dependent Incompressible Navier-Stokes Equations Using an Integro-Differential Formulation. Computers and Fluids, 1:197-215

Wu, J.C. (1982) Problems of General Viscous Flow. Developments in Boundary Element Method, Vol.2, Ch.2, Appl.Sc.Publ., London

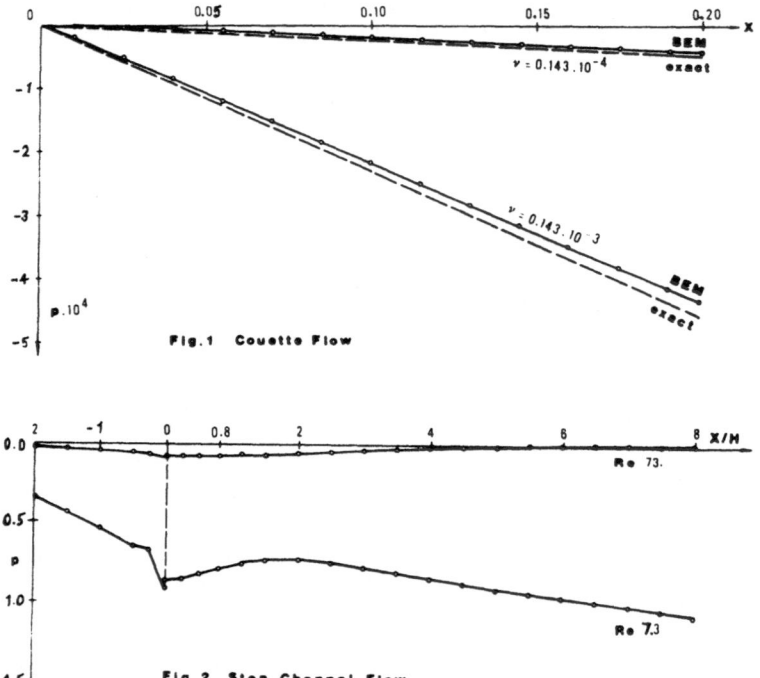

Fig.1 Couette Flow

Fig.2 Step Channel Flow

Boundary Elements Applied to Laminar Flow in Irregular Ducts

L.E. Coates

Department of Civil Engineering, University of Birmingham, Birmingham B15 2TT, U.K.

INTRODUCTION

For laminar flow of a viscous fluid along a prismatic duct, the mean axial velocity is proportional to the pressure gradient driving the flow. The constant of proportionality includes a factor K which depends only on the geometry of the cross-section. Rehme[1] has shown that the K factor for laminar flow may be used to determine the turbulent flow friction factor for non-circular sections. This has stimulated renewed interest in the computation of laminar flow. Closed form solutions are known for a number of special geometries (see e.g. Straub et al[2]). This paper is concerned with the calculation of K for arbitrary cross-sections.

To determine the K factor for an arbitrary shape the mean velocity corresponding to a given pressure gradient must be determined. This involves the solution of a Poisson equation to find the velocity distribution and then integration over the domain to obtain the mean velocity. The straight-forward application of the boundary element method (BEM) to the solution of a Poisson equation, introduces a further integral over the domain. The domain integrals may be evaluated numerically by dividing the domain into cells, but the advantages of the boundary procedure are then significantly reduced.

Salon et al[3], Danson[4] and Brebbia et al[5] have shown how, for a restricted class of forcing functions, the resulting domain integrals may be transformed into boundary integrals by using Green's theorem. Alternatively, for cases where a particular solution of Poisson's equation may be combined with a numerical solution of Laplace's equation, the domain integration arising from the right-hand side of the differential equation, may be avoided (see e.g. Jaswon and Symm[6]).

The calculation of the mean velocity would appear to involve considerable additional computation in finding values of the velocity at internal points for a numerical integration over the domain. This paper shows how this domain integral may also be obtained solely in terms of boundary integrals by a further application of Green's theorem.

The paper derives the boundary integral equations for two different approaches to the solution of the Poisson equation and the subsequent calculation of the mean velocity and K factor. One approach is then tested against the known exact solutions for a circle and a square. The rate of convergence of the computed K factor with increasing numbers of elements is shown to be good. The cross-sections corresponding to flows in symmetric and asymmetric flooded open channels are then considered. The much reduced rate of convergence is attributed to the poor handling of the singularity in the normal velocity gradient at the re-entrant corners. Symm[6] has shown how a known form of singularity may be included in the BEM solution. A version of this method which introduces only one additional unknown per corner is shown here to lead to much improved convergence.

THE K FACTOR FOR LAMINAR FLOW

For laminar flow in horizontal prismatic ducts the Navier-Stokes equations for incompressible flow reduce to the following Poisson equation (see e.g. Daily and Harleman[7])

$$\mu \nabla^2 u(y,z) = \partial p / \partial x , \qquad (1)$$

where $u(y,z)$ is the distribution of axial velocity over the cross-section, μ is the dynamic viscosity and $\partial p / \partial x$ is the pressure gradient driving the flow. Dimensional analysis suggests a relationship between the pressure gradient and the mean velocity $\bar{u}$ of the form

$$\partial p / \partial x = - \frac{f}{4m} \cdot \tfrac{1}{2} \rho \ \bar{u}^2, \qquad (2)$$

where ρ is the fluid density, m is the hydraulic radius – defined as the cross-sectional area divided by the wetted perimeter, and f is a friction factor expected to be a function of the Reynolds number Re. The linear relationship between the pressure gradient and the velocity in equation (1) should appear in equation (2) as well. Accordingly the friction factor should be inversely proportional to Re

$$f = \frac{K}{4m\rho\bar{u}} = K.Re^{-1}, \qquad (3)$$

where the K factor will take account of the geometry of the cross-section. Substituting equation (3) into equation (2) and rearranging we obtain an expression for K,

$$K = -32m^2 \left[\frac{\partial p/\partial x}{\mu \, \bar{u}} \right], \tag{4}$$

where the minus sign is because the flow is in the direction of reducing pressure. For a given pressure gradient we can solve equation (1) to find u and then integrate to obtain the mean velocity. For a given cross-section the term in square brackets in equation (4) will not change. The BEM may be used to obtain this term.

BOUNDARY INTEGRAL EQUATIONS

The numerical problem reduces to that of finding the solution of

$$\nabla^2 u = b, \tag{5}$$

in the domain Ω , subject to the ´no slip´ boundary condition

$$u = 0, \tag{6}$$

on the smooth boundary Γ . Here $b = \mu^{-1} \partial p/\partial x$ is a constant. Once the unknown boundary values have been found then the mean velocity $\bar{u}$ may be calculated from

$$\bar{u} = \int_\Omega u \, d\Omega \, / \int_\Omega d\Omega \tag{7}$$

Two related approaches to the solution of equation (5) will be presented. Both methods rely on Green´s second identity for two functions U and V whose second derivatives are continuous,

$$\int_\Omega U \, \nabla^2 V - V \, \nabla^2 U \, d\Omega = \int_\Gamma U \frac{\partial V}{\partial n} - V \frac{\partial U}{\partial n} \, d\Gamma \tag{8}$$

where n is the outward normal direction on Γ . To avoid the need for a limiting process we will define the fundamental solution of equation (5), u^*, as a solution of

$$\nabla^2 u^* = \delta_P \tag{9}$$

where δ_P is the Dirac delta function centred at point P. For two dimensions

$$u^* = (2\pi)^{-1} \ell n \, r, \tag{10}$$

where r is measured from point P.

Method 1
In equation (8), letting $U = u$ and $V = u^*$, substituting and rearranging we obtain

$$2\pi \, c_P u_P = \int_\Gamma u \frac{\partial}{\partial n}(\ell n \, r) - (\ell n \, r)\frac{\partial u}{\partial n} \, d\Gamma + \int_\Omega b(\ell n \, r)d\Omega \,, \tag{11}$$

where $c_P = 1$ for P in Ω , $c_P = 1/2$ for P on Γ and zero elsewhere. The domain integral on the right-hand side could be evaluated numerically by splitting Ω into cells (see e.g. Lafe et al[8]). Here we prefer to transform it into a boundary integral. In equation (8) let $U = b$ and

$$\nabla^2 V = \ln\ r. \tag{12}$$

Substituting and rearranging we obtain

$$\int_\Omega b(\ln\ r)\ d\Omega = \int_\Gamma b\ \frac{\partial V}{\partial n} - V\ \frac{\partial b}{\partial n}\ d\Gamma + \int_\Omega V\ \nabla^2 b\ d\Omega. \tag{13}$$

By integrating equation (12) by parts we find that

$$V = \frac{r^2}{4}\{\ \ln\ r\ -\ 1\}. \tag{14}$$

If we restrict b to be a harmonic function then the domain integral on the right-hand side of equation (13) vanishes. If b takes some other form, for example biharmonic, we may apply equation (8) a further time by integrating equation (14) again. In equation (5) b is constant so when equations (13) and (14) are substituted into equation (11) we obtain as the basis for the BEM,

$$2\pi c_P u_P = \int_\Gamma u\frac{\partial}{\partial n}(\ln\ r) - (\ln\ r)\frac{\partial u}{\partial n} d\Gamma + b\int_\Gamma \frac{\partial}{\partial n}\left\{\frac{r^2}{4}(\ln\ r - 1)\right\} d\Gamma. \tag{15}$$

Method 2

Because of the linearity of equation (5) we may choose to decompose u into two functions thus

$$u = u' + v\ , \tag{16}$$

such that

$$\nabla^2 v = b\ , \tag{17}$$

where v may be arbitrarily chosen as a particular solution of equation (17). The boundary value problem reduces to that of finding the function u' that satisfies

$$\nabla^2 u' = 0\ , \tag{18}$$

in domain Ω , subject to the boundary condition

$$u' = -\ v\ , \tag{19}$$

on Γ . For this problem equation (11) takes on the more usual form

$$2\pi\ c_P u'_P = \int_\Gamma u'\ \frac{\partial}{\partial n}(\ln\ r) - (\ln\ r)\ \frac{\partial u'}{\partial n}\ d\Gamma\ . \tag{20}$$

Equations (15) and (20) may each be used, with $c_P = 1/2$, as the basis of a BEM formulation. The unknown boundary values of $\partial u/\partial n$ and $\partial u'/\partial n$ are obtained by a process of discretisation. Once all the boundary values are known the velocity at internal points may be obtained from the discretised form of equation (15) or (20) with $c_p = 1$.

COMPUTATION OF MEAN VELOCITY

To calculate the mean velocity from equation (7) we could split the domain into cells and evaluate the velocity at selected points for numerical integration. This represents considerable additional computation if an accurate result is to be obtained. However this is unnecessary. To illustrate how Green's second identity may be used with method 1 or 2 to calculate $\bar{u}$, we use equation (8) together with the additional restrictions that

$$\nabla^2 U = b , \qquad (21)$$

and $\quad \nabla^2 V = 1 . \qquad (22)$

Upon substitution of equations (21) and (22) into (8) we obtain,

$$\int_\Omega U \, d\Omega = \int_\Gamma U \frac{\partial V}{\partial n} - V \frac{\partial U}{\partial n} \, d\Gamma + \int_\Omega V.b \, d\Omega . \qquad (23)$$

The domain integral on the right-hand side of equation (23) may be removed as before, using equation (8) to give

$$\int_\Omega U \, d\Omega = \int_\Gamma U \frac{\partial V}{\partial n} - V \frac{\partial U}{\partial n} \, d\Gamma + b \int_\Gamma \frac{\partial W}{\partial n} \, d\Gamma , \qquad (24)$$

where $\quad \nabla^2 W = V . \qquad (25)$

Of the many possible choices for V we choose $R^2/4$ and therefore $W = R^4/64$ from equation (25), where R can be measured from any convenient fixed point.

Method 1
Using the definitions for V and W and letting $U = u$ in equation (24) we obtain for the total discharge

$$\int_\Omega u \, d\Omega = \int_\Gamma u \frac{\partial}{\partial n}(\frac{R^2}{4}) - \frac{R^2}{4} \frac{\partial u}{\partial n} \, d\Gamma + b \int_\Gamma \frac{\partial}{\partial n} \frac{R^4}{64} \, d\Gamma . \qquad (26)$$

Including the boundary condition from equation (6) this reduces finally to

$$\int_\Omega u \, d\Omega = - \int_\Gamma \frac{R^2}{4} \frac{\partial u}{\partial n} \, d\Gamma + b \int_\Gamma \frac{\partial}{\partial n} \frac{R^4}{64} \, d\Gamma . \qquad (27)$$

The area of the cross-section may be obtained from equation (26) with $u = 1$ and $b = 0$. Thus the mean velocity may be obtained solely in terms of boundary integrals. The functions V and W introduce no additional singularities.

Method 2

We may substitute equation (16) into the right-hand side of equation (27) to give

$$\int_\Omega u \, d\Omega = - \int_\Gamma \frac{R^2}{4} \frac{\partial u'}{\partial n} \, d\Gamma - \int_\Gamma \frac{R^2}{4} \frac{\partial v}{\partial n} \, d\Gamma + b \int_\Gamma \frac{\partial}{\partial n} \frac{R^4}{64} \, d\Gamma \ . \qquad (28)$$

The first term on the right-hand side of both equations (27) and (28) will be evaluated numerically in the same way – by linear interpolation between point values of $\partial u / \partial n$ and $\partial u' / \partial n$ respectively. If the choice for v is left open then special integrals may need to be evaluated for the second term on the right-hand side of equation (28). If a linear interpolation between point values of $\partial v / \partial n$ on Γ is used, the resulting accuracy will be poor. For simplicity we can choose to make $v = bR^2 / 4$. For this case the last two terms in equation (28) may be combined to give for the total discharge

$$\int_\Omega u \, d\Omega = - \int_\Gamma \frac{R^2}{4} \frac{\partial u'}{\partial n} \, d\Gamma - b \int_\Gamma \frac{\partial}{\partial n} \frac{R^4}{64} \, d\Gamma \ . \qquad (29)$$

Equations (27) and (29) are not significantly different computationally. The main difference between the two methods is in the extra boundary integral in equation (15). This extra term may be calculated exactly using an integral that is used in the calculation of the other terms. Consequently very little additional computation is required.

NUMERICAL DETAILS

In equation (15), if the point P is taken onto the boundary Γ , then we obtain an integral equation relating the boundary variation of u and $\partial u / \partial n$. This equation can be approximated by a set of algebraic equations relating point values of u and $\partial u / \partial n$ on the boundary, by a process of discretisation that has been well reported in the literature (see e.g. Brebbia[9]). All the examples shown below assumed a piecewise linear interpolation for both u and $\partial u / \partial n$ between their values at specific points or nodes on the boundary.

A boundary element formulation based on method 2, using equation (20), appears, superficially, to be simpler than method 1 using equation (15). There is, however, a hidden computational difficulty in equation (20). If the boundary condition from equation (19) is substituted after the matrix equations have been formed (which would usually be the case), it will have been implicitly assumed that the function v varies linearly between the nodal values. This additional approximation will seriously affect the accuracy of the solution. To obtain the same accuracy as method 1 the choice for function v needs to be made before the matrix formation and integrated exactly to obtain the matrix coefficients. The additional integrals involved are marginally more complicated (because of the different origins for R and r) than those resulting from the last term in equation (15). Accordingly, in the examples below method 1 was used for the solution of equation (5).

The system of equations resulting from the discretisation of equation (15) may be expressed in matrix form as

$$[A]\{ u \} = [B]\{ \partial u/\partial n \} + \{ F \} \tag{30}$$

where each equation corresponds to the point P being placed at each of the boundary nodes in turn. All the integrals necessary for forming A, B and F were evaluated exactly. Since many of the boundary geometries of interest would not be smooth the values of $2\pi cp$ were calculated from the off-diagonal terms of matrix A. The vector of velocities u was retained for generality. Lines of symmetry can therefore be readily accommodated by specifying $\partial u/\partial n = 0$, although Au and Brebbia[10] have shown that this is not the most efficient way of doing this. Initially the problem of discontinuities in the boundary geometry was handled by the device of two nodes placed a small distance ε either side of the discontinuity.

To reduce the amount of input data required, the area and perimeter of the cross-section were both calculated numerically using boundary integrals.

NUMERICAL EXAMPLES

Circular cross-section
For a circular cross-section equation (1) can be integrated in the radial direction. The resulting velocity distribution is parabolic with a mean value of half its maximum value, given by

$$\bar{u} = - \frac{m^2}{2} \frac{1}{\mu} \frac{\partial p}{\partial x} \tag{31}$$

When this is substituted into equation (4) the exact value for K of 64 is obtained. The normal gradient of the velocity at the boundary is given by

$$\frac{\partial u}{\partial n} = \frac{m}{\mu} \frac{\partial p}{\partial x} \tag{32}$$

Figure (1) shows the values of K, u_{max} and $\partial u/\partial n$. The ratios of the computed values to the exact values are shown, plotted against the number of elements on the boundary. The convergence of the results with increasing number of elements is particularly good, perhaps because there are no boundary discontinuities and therefore no double nodes. It is well known that the accuracy of the results at internal points (e.g. u_{max}) is better than the point values of u and $\partial u/\partial n$ on the boundary. This is because it is the linear interpolation between the point values that satisfies the equations in a

´best fit´ sense.

Because the regular polygon used in the discretisation
inscribes the actual circle, the effective hydraulic radius to
which the results correspond is somewhat less than the

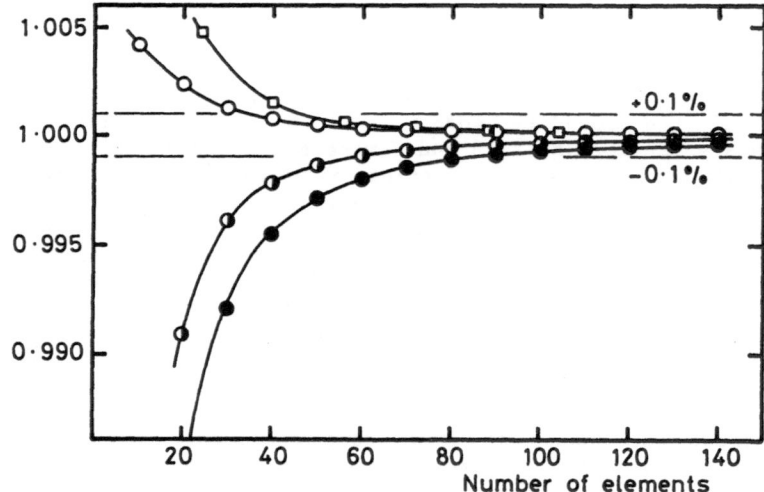

Figure 1. Ratios of computed to theoretical values for a
circle and a square. Circle – $\bullet\!-u_{max}/(m^2b)$;
$\bullet\ \partial u/\partial n/(mb)$; $\bigcirc$ K/64; Square $\square$ K/56.9084.

theoretical one. The error that this introduces into the
computation of K, fortuitously, leads to a more accurate
result. The most sensitive indicator of the performance of
the method is therefore the value of $\partial u/\partial n$ at the boundary.
The computed values are within 0.1% of the theoretical values,
using 90 elements or more.

These results were obtained with a radius of 0.5.
Theoretically the value of K does not depend on the radius.
Figure (2) shows that numerically this is not the case.
Because of the nature of the logarithmic function there will
be a particular radius for which the domain integral in
equation (11) is zero. For this radius (which happens to be
unity) all the equations become homogeneous and a solution
cannot be obtained. Figure (2) was produced using 40 elements
around the boundary. It can be seen that the error in K is
significantly larger when the radius is greater than unity.

Although for non-circular shapes the equations cannot be
completely homogeneous, there is still a possibility of errors
arising form shapes whose dimensions are close to unity.
Accordingly the dimensions of all subsequent problems were
made significantly less than unity. It may be shown that the

same difficulty affects method 2.

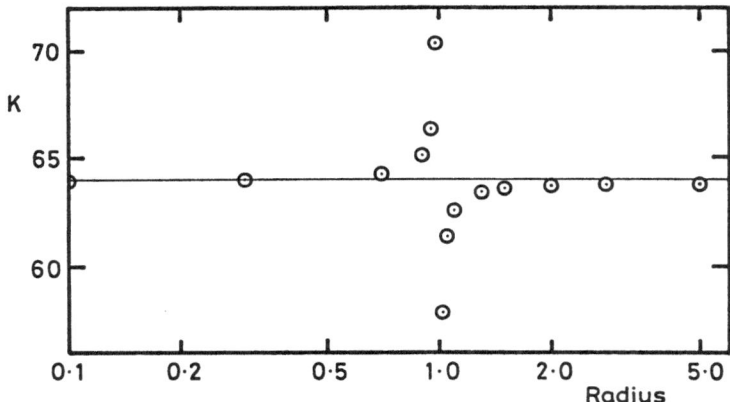

Figure 2. Computed K for a circular cross-section, plotted
against radius. Results obtained using 40 elements.

Square cross-section
Straub et al[2] gave an exact solution for a rectangular
cross-section. For the special case of a square K is 56.9084.
Figure (1) shows the ratio of computed to theoretical K
against the total number of elements. These results were
obtained for a square of side 0.8 with equal length elements
on each side. At the corners the double nodes used a value for
ϵ of 10^{-6}. This was found to be the optimum – producing K
values convergent to five significant figures, for a fixed
number of elements, whilst avoiding round-off error in the
equation-solving program.

Although the convergence is not as dramatic as for the circle,
the computed K factor is still within 0.1% using 56 elements
or more.

Compound cross-sections
The research which prompted this work involves the study of
flow in flooded open channels. Figures 3 and 4 show two of the
particular sections of interest. For laminar flow it is found
that along the free surface $\partial u/\partial n$ is zero. The problems were
solved as closed ducts by reflecting the original
cross-section about the free surface.

Figures 3 and 4 show the variation of computed K plotted
against the total number of boundary elements. Although the
exact solutions are not known, it is clear that the results
are some way from convergence even with more than 130
elements. This degree of accuracy was considered unacceptable.

The major difference between these compound channels and the

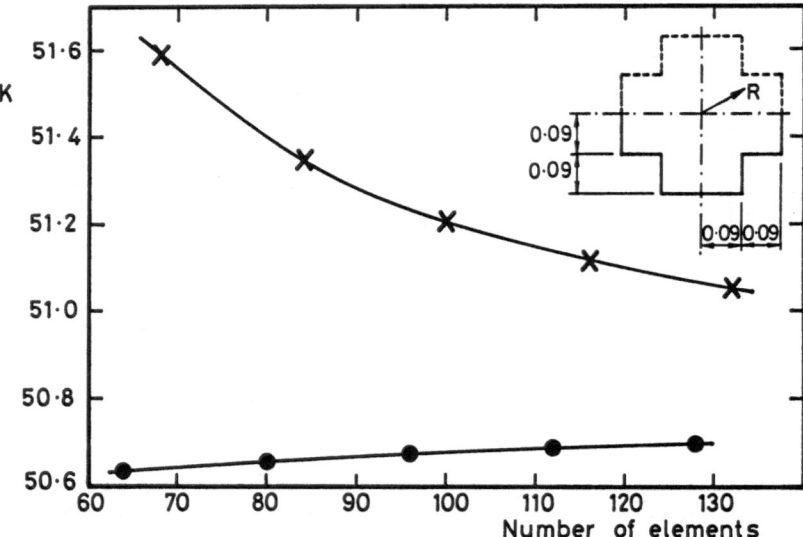

Figure 3. Computed K factors for the symmetric cross-section shown, ✕ without separation of known solutions; ● with separation of known solutions.

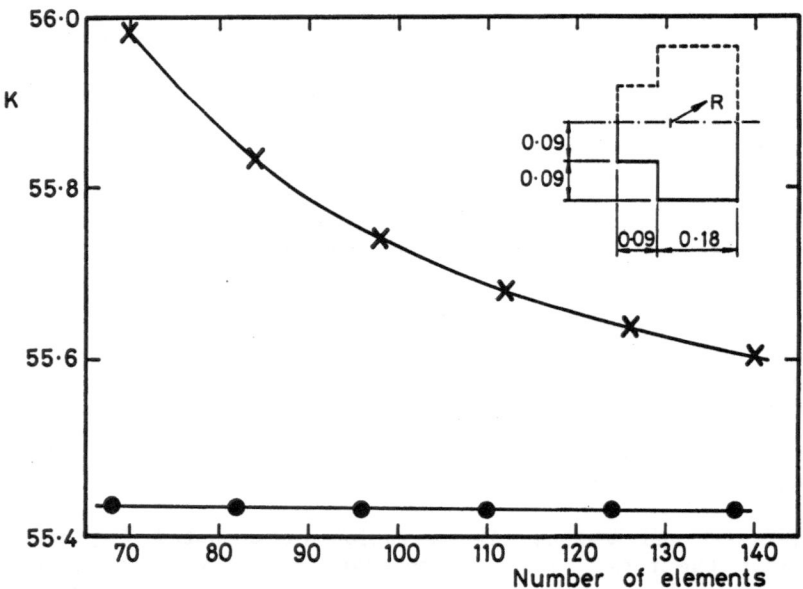

Figure 4. Computed K factors for the asymmetric cross-section shown, ✕ without separation of known solutions; ● with separation of known solutions.

square test case is that at the re-entrant corners the normal gradient of the velocity is singular. The form of the velocity close to the corner is known to contain a term proportional to $r^{(2/3)}\sin 2\theta/3$, where r is the distance from the corner and θ is the angle measured from one of the edges. The normal gradient of the velocity close to the corner is therefore dominated by a term proportional to $r^{(-1/3)}$. Figure 5 shows that the linear interpolation is unable to represent this behaviour accurately. Liggett and Liu[11] have shown how special interpolation functions based on the known form of the solution close to the corner may be used to improve the accuracy. Jaswon and Symm[6] have used an alternative technique using the known form of the singularity but with an initially unknown set of coefficients. These they obtained by matching the solution at adjacent nodes.

In order to improve the accuracy in this case it was decided to separate only that part of the known local behaviour which gives rise to the singular velocity gradient. Only one additional unknown coefficient is then introduced.

We assume that the velocity may be decomposed into known and unknown parts as

$$u = A_i u'_i + u''$$
(33)

where the repeated index implies summation and

$$u'_i = r_i^{(2/3)}\sin 2(\theta_i - \alpha_i) \quad \text{no sum}$$
(34)

is the first term of the known solution for the re-entrant corner i, and α_i is the angle between the coordinate axis and the first edge of the corner. Since u'_i is harmonic the remainder of the velocity u'' must still satisfy equation (5) but subject to the boundary condition

$$u'' = -A_i u'_i$$
(35)

on Γ. The unknown A_i may be found by recognising that at the corner i the variation of $\partial u/\partial n$ is dominated by the singular term. If the strength A_i is correct then the value of $\partial u''/\partial n$ at the corner should be close to zero. This provides an extra equation for each A_i value.

With double nodes at the corners the question arose as to which of them to choose to satisfy this condition. This was resolved when it was discovered that at the corners of the square test case, if single nodes were used, the computed value of $\partial u/\partial n$ appeared to be equal to the value obtained in the limit as $\varepsilon \to 0$. So at re-entrant corners only, single nodes were used and the value of $\partial u''/\partial n$ set equal to zero.

Equation (35) was satisfied at the nodal points only, after

Table 1. Computed K factors and coefficients A

Symmetric cross-section

Without separation		With separation		
Number of Elements.	K	Number of Elements.	K	A
68	51.58878	64	50.65321	0.0502580
84	51.35025	80	50.65838	0.0493171
100	51.20958	96	50.67631	0.0486414
116	51.11756	112	50.68929	0.0481236
132	51.05319	128	50.69875	0.0477093

Asymmetric cross-section

Without separation		With separation		
Number of Elements.	K	Number of Elements.	K	A
70	55.98535	68	55.43580	0.0384552
84	55.83376	82	55.43343	0.0382768
98	55.74177	96	55.43196	0.0381541
112	55.68084	110	55.43109	0.0380632
126	55.63798	124	55.43058	0.0379922
140	55.60643	138	55.43027	0.0379348

formation of the matrices. An opportunity to evaluate the first integral in equation (15) exactly was therefore missed. Since u´ is non-singular the error introduced is probably not large. The calculation of $\bar{u}$ now included integrals which could not be evaluated exactly. A mixture of exact integration close to the corners and three-point Gauss quadrature elsewhere was used.

The computed K values using this technique are shown by the circles in figures 3 and 4. The improvement in the results particularly for the asymmetric case is remarkable. The figures show where R in equation (27) was measured from. The results were not at all sensitive to this choice. The two sets of results are also shown in table 1 together with the computed values of A_i, which for both cases were identical at each re-entrant corner. From this table it is possible to extrapolate to obtain, for the symmetric and asymmetric cases, values for K of 50.72 and 55.430 respectively. As before the values of A_i are slightly further from their converged values than are the K values.

Figure 5 shows the variation of $\partial u/\partial n$ close to a re-entrant corner in the symmetric section. These results correspond to the 128 element discretisation. The oscillations in the original solution are confined quite close to the corner. They have been removed by the separation technique. The behaviour of $\partial u''/\partial n$ is not entirely as expected. The value at the node nearest the corner seems too high.

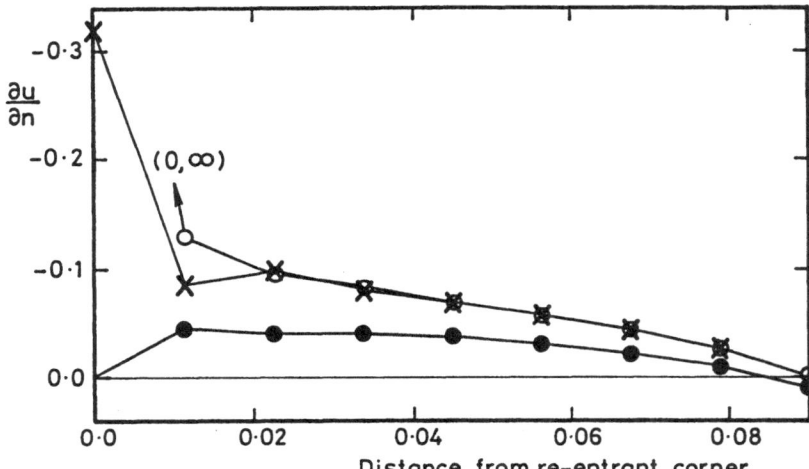

Figure 5. Variation of the normal gradient of velocity between a re-entrant corner and the next corner for the symmetric cross-section. ✕ Without separation; ◯ With separation; ● the resulting $\partial u''/\partial n$.

Figures 3 and 4 demonstrate the importance of proper examination of the convergence of computed results. It is not enough to verify a formulation against simple analytical solutions, and then solve an actual problem once.

CONCLUSIONS

In this paper the BEM has been successfully applied to the solution of problems of laminar flow in irregular ducts. In particular the shape factor K has been obtained accurately without the need to resort to numerical integration over the domain. This has been achieved by repeated application of Green's second identity.

Two methods of solving the problem have been derived. One of the methods has been tested against the known exact solutions for circular and square cross-sections. It has been shown that if the dimensions of the cross-section are close to unity the solution may be extremely inaccurate.

It has been shown that for geometries involving re-entrant corners accurate results can only be obtained by subtracting from the total, that part of the solution that gives rise to the singularity in the velocity gradient.

REFERENCES

1. Rehme K. (1973), Simple Method of Predicting Friction
    Factors of Turbulent Flow in Non-Circular Channels, Int.
    J. Heat Mass Transfer, Vol 16, pp. 933-950.

2. Straub L.G. Silberman E. and Nelson H.C. (1958), Open
    Channel Flow at Small Reynolds Numbers, Trans. ASCE,
    Vol 123, pp. 685-714.

3. Salon S.J. Schneider J.M. and Uda S. (1981), Boundary
    Element Solutions to the Eddy Current Problem, Boundary
    Element Methods, Proc. of the third Int. Seminar,
    pp. 14-25.

4. Danson D.J. (1981), A Boundary Element Formulation of
    Problems in Linear Isotropic Elasticity with Body Forces,
    Boundary Element Methods, Proc. of the third Int.
    Seminar, pp. 105-122.

5. Brebbia C.A. Telles J.C.F. and Wrobel L.C. (1984),
    Boundary Element Techniques, Springer-Verlag, Berlin and
    New York.

6. Jaswon M.A. and Symm G.T. (1977), Integral Equation
    Methods in Potential Theory and Elastostatics, Academic
    Press, London and New York.

7. Daily J.W. and Harleman D.R.F. (1966), Fluid Dynamics,
    Addison-Wesley USA.

8. Lafe E.O. Liggett J.A. and Liu P.L-F. (1981), BIEM
    Solutions to a Combination of Leaky, Layered, Confined,
    Unconfined, Anisotropic Aquifers, Water Resources
    Research, Vol 17, pp. 1431-1444.

9. Brebbia C.A.(ed.),(1983), Topics in Boundary Element
    Research, Volume 1: Basic Principles and Applications
    Springer-Verlag, New York,

10. Au M.C. and Brebbia C.A. (1983), A Three Dimensional
    Boundary Element Approach for Rotationally Symmetrical
    Problems, Boundary Elements, Proc. of the fifth Int.
    Conf. Hiroshima, Japan, Springer-Verlag, New York.

11. Liggett J.A. and Liu P.L-F. (1983), The Boundary Integral
    Equation Method for Porous Media Flow, George Allen and
    Unwin, London.

Application of Boundary Element Method in Transient Flow Problems with Moving Boundaries

C.S. Chang
Department of Civil Engineering, University of Massachusetts, Amherst, MA 01003, U.S.A.

INTRODUCTION

The problem under consideration is classified as a unsteady state flow with moving boundaries. Due to the complex nature of this type of problem, only very few mathematical solutions are available with simplified assumptions. To obtain a general solution for such problem, numerical methods would be a more practical approach. Among them, one of the earliest is a finite difference method proposed by Kirkham and Gaskell[1]. They treated the transient condition as a succession of steady state conditions and used a finite difference technique for solving Laplace equation for each successive position of the water table in tile and open ditch. Desai[2] and Neuman and Witherspoon[3] used finite element techniques to determine the free surface in river banks due to drawdown. More recently, concepts of the boundary element method have been used for solving problems involving free surface (Liggett and Liu[4], Chang[5]).

Among the various numerical techniques, the boundary element method perhaps is the most flexible and efficient tool for solving this type of problem. However due to the large variation of gradients along the moving boundary surface in some cases of transient flow in an earth dam, the oscillating shape of moving boundary often occurs, resulting in inaccurate solutions. This paper reports a solution procedure using averaged flow velocity to minimize this problem of oscillating shape of the moving boundary. Applicability of the boundary element method in the analysis of a transient flow in an earth dam due to drawdown of the water level in reservoir is then evaluated with observed experimental results.

STATEMENT OF THE PROBLEM

A typical two-dimensional free surface problem of flow through an earth dam is shown in Fig. 1. Neglecting the capillary and surface tension, the flow domain in R has the following four types of boundaries: (1) a prescribed head boundary B_1, (2) a prescribed flux boundary B_2, (3) a seepage face S, and (4) a free surface F. This boundary value problem can be described by the following set of equations:

$$\frac{\partial}{\partial x_i}\left(K_{ij}\frac{\partial h}{\partial x_j}\right) = 0 \qquad\qquad \text{on R} \qquad\qquad (1)$$

$$h = H \qquad\qquad \text{on } B_1 \qquad\qquad (2)$$

$$K_{ij}\frac{\partial h}{\partial x_j}\, n_i = v_n \qquad\qquad \text{on } B_2 \qquad\qquad (3)$$

$$h = y \qquad\qquad \text{on S} \qquad\qquad (4)$$

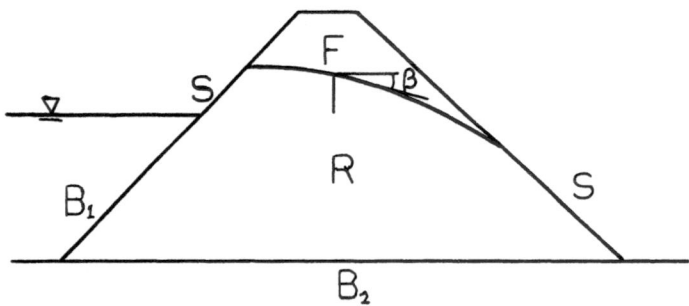

Figure 1. Schematic Illustration of a Typical Free Surface Seepage Problem

On F, the free surface, water pressure is zero. Since velocity of water seeping through soil is negligibly small, according to Bernoulli equation, the potential head h is equal to the elevation head z.

$$h = z(x,t) \qquad\qquad (5)$$

The kinematic boundary condition on the free surface is

$$\frac{\partial h}{\partial t} = -\frac{1}{f\,\cos\beta}\,v_n \qquad\qquad (6)$$

where f is the porosity of soil; β is the angle between the free surface and the horizontal; v_n is the flux across the free surface which can be obtained by Eq. (3). In a steady state condition, the flux across the free surface boundary is zero.

The above equations are based on the assumption that no lag in flow or water pressure is caused by inertia or consolidation, and the changes in boundary conditions are transmitted quickly throughout the soil mass. Capillarity induced force in fine soil and negative pore pressure stresses are also neglected.

In many real situations location of this free surface is not known a priori. Therefore the difficulty in solving the problem lies in the determination of the free surface boundary rather than solving the differential equation. One of the solutions which comes out of the numerical method is the location of free surface.

BOUNDARY ELEMENT FORMULATION

A detailed formulation of the Boundary Element Method can be found in many publications (for example, Refs. 4,8), therefore only a brief description of the procedure is presented. Based on the method of boundary integration, relationship between potential heads h_j and gradients $\left(\frac{\partial h}{\partial n}\right)_j$ at nodal points on the boundary surface can be derived as:

$$H_{ij}h_j = G_{ij}\left(\frac{\partial h}{\partial n}\right)_j \qquad (7)$$

In Eq.(7), matrices H_{ij} and G_{ij} are obtained by

$$H_{ij} = \delta_{ij}\alpha - \int_S N_i \frac{\partial}{\partial n}(\ln r_j)dS \qquad (8)$$

$$G_{ij} = \int_S N_i(\ln r_j)dS \qquad (9)$$

where δ_{ij} is the kronecker delta matrix, r_j is the distance from node j to any other point on the boundary surface. The value of constant α depends on the shape of boundary surface. For example, α is 1/2 if the boundary surface is smooth at node j. N_i is the shape function. In the following procedure linear shape functions are used and the boundary is represented by straight line segments.

In a flow problem, either flux or potential head is known at a node on the boundary. Therefore Eq. (7) represents a set of simultaneous equations with the same number of equations and unknowns. The unknown potential heads and gradients in Eq. (7) can be readily solved.

Once the potential heads and gradients on the boundary are solved, the potential head at any point within the flow domain can be calculated by the following equation.

$$h = \int_S N_i \frac{\partial}{\partial n}(\ln r)dS\, h_i + \int_S N_i(\ln r)dS \left(\frac{\partial h}{\partial n}\right)_i \qquad (10)$$

where h_i and $\left(\frac{\partial h}{\partial n}\right)_i$ are the potential heads and gradients at nodal points on the boundary surface, r is the distance from the point of interest to any point on the boundary surface.

BOUNDARY MOVING PROCEDURE

In an earth dam, the transient condition caused by a decrease of pool level is treated as a succession of steady state conditions. The solution procedure includes the following steps:

Step 1: For any given time, specify the boundary conditions in accordance with the pool level as shown in Fig. 1 and Eqs. (1-4). The potential heads of the nodal points on free surface are specified to be equal to their elevation. Then Eq. (7) is solved to obtain the values of gradients, $\frac{\partial h}{\partial n}$, for all nodes on the free surface. If the problem is indeed transient, the condition of null normal gradients across the free surface is not satisfied.

Step 2: Calculate the water particle velocity normal to the free surface at node i by

$$v_i = \frac{1}{f} \ k \ \left(\frac{\partial h}{\partial n}\right)_i \tag{11}$$

where f is the porosity of soil, $\left(\frac{\partial h}{\partial n}\right)_i$ is the gradient at node i, positive is outward, negative inward.

Step 3: The movement of any node on the free surface is computed by multiplying the normal velocity by a given time increment Δt. The new location of the free surface is obtained by connecting the new locations of all the free surface nodes.

Step 4: The process from Step 1 to step 3 is repeated again with the new location of the free surface. By updating the coordinates of nodal points on the free surface at the end of each time step, the successive movement of free surface can thus be obtained.

However, based on the author's experience, this procedure may give problems of oscillating shapes of free surface and often the degree of oscillation increases with the successive steps, leading to an unstable solution. A schematic example is shown in Fig. 2 for the case of water movement in a thin horizontal layer. The oscillation problem exists in both space and time, since the discretization technique is used in both space and time to obtain the solution. This problem is often more severe in the case of earth dams, perhaps due to the larger variation or discontinuity of gradients along the free surface in certain more complex configurations.

A solution scheme is therefore proposed to overcome such problems. The new scheme involves an averaging process for the velocities at nodes in both space and time.

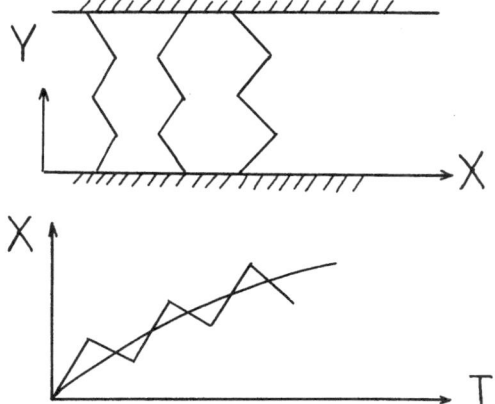

Figure 2. Schematic Illustration of Oscillation in
both Space and Time

Velocity Averaging Spacewise

More accurate and stable solutions can be achieved by averaging the velocity spacewise. In a discretized boundary as in Fig. 3, the representative velocity at a node C should be the averaged value of velocities on the adjacent area ACB.

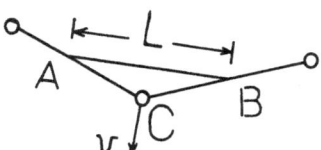

Figure 3. Averaged Velocity at a Node in Spacewise

The flow quantity across the boundary surface ACB is

$$q = \int_{ACB} k \, \frac{\partial h}{\partial n} \, dS \tag{12}$$

Using the shape functions, the flow quantity q_i for nodes on the boundary can be obtained by the following equation.

$$q_i = A_{ij} \left(\frac{\partial h}{\partial n} \right)_j \tag{13}$$

where

$$A_{ij} = \int_S N_i N_j k \, dS \tag{14}$$

Once the flow quantities on the nodes are solved using Eq. (13), the velocities at the nodes on free surface are computed in the following way,

$$v_i = q_i / L f \qquad\qquad (15)$$

L is the length from point A to point B, which are midpoints of adjacent segments as shown in Fig. 3. The velocity is thus an averaged value along the surface length from point A to point B as shown in Fig. 3. The direction of flow is perpendicular to the line AB, positive sign means outward flow and negative sign means inward flow.

Velocity Averaging Timewise

Timewise, the velocity is averaged between the velocity at time t, v^t, and the velocity at time t+Δt, $v^{t+\Delta t}$, by the Wilson-Θ method.

$$v = (1-\Theta) v^t + \Theta v^{t+\Delta t} \qquad\qquad (16)$$

where Θ is a weighting factor. When Θ=0 (backward difference), the velocity is v^t. When Θ=1 (forward difference), the velocity is $v^{t+\Delta t}$. And when Θ=0.5 (central difference), the velocity is an average of that at time t and time t+Δt. The effects of Θ on the stability and accuracy of solution have been discussed by Liggett[4] for a linear case. The oscillation can be damped or amplified to unstable solutions depending greatly on the magnitude of time steps used. From Leggett's study, for the same magnitude of time steps, the optimum value for Θ is between 0.5 and 0.6, and the time steps should be as small as possible to minimize the numerical distortion. For the drawdown case, these guidelines are still generally valid.

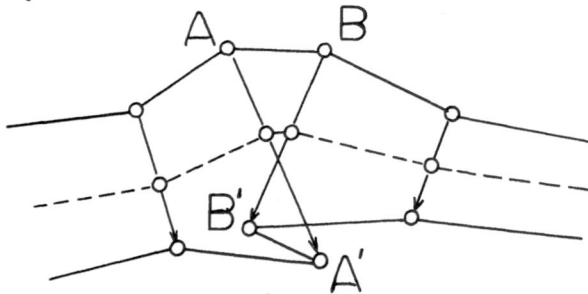

Figure 4. Illustration of Crossing Flow Line Problem

It is noted that, a problem of flowline crossing frequently occurs when the choice of time step is inappropriate. This problem is illustrated in Fig. 4. In a given time step Δt, two points A and B on the free surface tend to move to their new position A' and B'. If the time step is relatively large and the curvature of the free surface is relatively high, the two flow paths may cross each other and form a distorted front of free surface as the solid line in Fig. 4. However if the time step is small enough, the flow path crossing can be avoided and a reasonably smooth front of free surface can be achieved as the dashed line shown in Fig. 4.

As a rule of thumb, in a drawdown problem, the beginning time step may be selected by the following equation,

$$\Delta t = 0.02 \ h \ / \ v_{max} \qquad (17)$$

where h is the total potential head difference, v_{max} is the maximum averaged velocity among the nodes on free surface, calculated by Eq. (16). However, the time step used in the analysis should always be dictated by the consideration of avoiding crossing flow lines.

COMPARISONS WITH EXPERIMENTAL RESULTS

The analyses using boundary element method are evaluated by using experimental results measured in a laboratory model embankment. The variation of pore pressures within the embankment involving the change of reservoir water level are observed from pore pressure cells embedded inside model embankment (Kellogg[6]).

This sand embankment model is illustrated in Fig. 5. Six pore pressure cells were embedded in the model at various places. The height of the model is 15 in. and is made of standard Ottawa sand. The reservoir was filled to a height of 15 in. and the level maintained there long enough to permit the attainment of a steady state of seepage, followed by an unsteady state seepage induced by drawdown. The rate of drawdown of the reservoir level was 2.4 in. per minute.

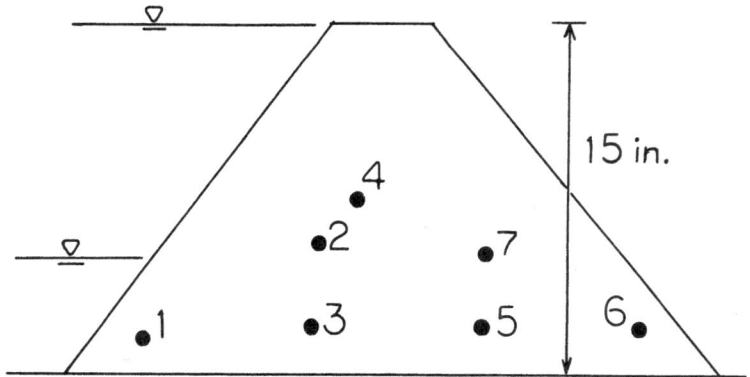

Figure 5. Soil Dam Model

Typical values of permeability for clean coarse sand such as the sand used in the model tests ranges from 0.2 in./min. to 10 in./min. This computation uses permeability equal to 2.4 in./min. Porosity of the sand is 0.5. The Boundary Element

model and the successive positions of free surfaces are shown
in Fig. 6. The observed free surface shape shown as the
dashed curve matches well with the calculated shape.

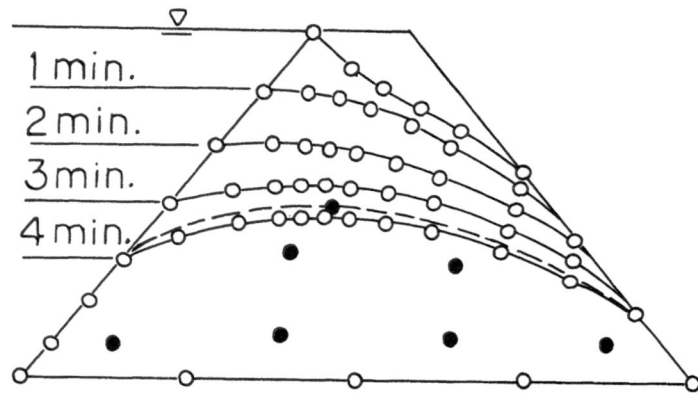

Figure 6. Free Surfaces Positions versus Time due to
Drawdown of Impounding Water

Fig. 7 shows the comparisons of observed and calculated potential head at various locations in the embankment model due to
drawdown of impounding water. The values of potential head at
zero time represent the potential head during the initial
steady state. In general, computed values of potential head
show very good agreement with those observed in the model
tests.

Some discrepancies between computed and measured results are
summarized below. Cell no. 3 shows appreciably different
initial value than that calculated from the boundary element
method. The predicted drainage rate on the upstream side
(cell nos. 1,3) is slightly faster than that measured. The
predicted drainage rate on the downstream side (cell nos. 5,6)
shows opposite trend. These discrepancies, however, are
reasonably small considering the possible inaccuracies that
might be involved in the laboratory measurements.

The time integration scheme used here with $\theta=0.5$ is similar to
the central finite difference procedure. The rate of draw
down is approximated by several step-wise sudden drawdown
conditions. The proposed procedure provides acceptable accuracy for the solutions. It seems that for a practical range
of discretizations (25 nodes, 10 time steps in this case), the
class of problem is not significantly affected by the time
steps nor by the spacing of the nodes.

CONCLUSION

Numerical problem of oscillation can be overcome by the
averaged velocity scheme which leads to a stable solution.

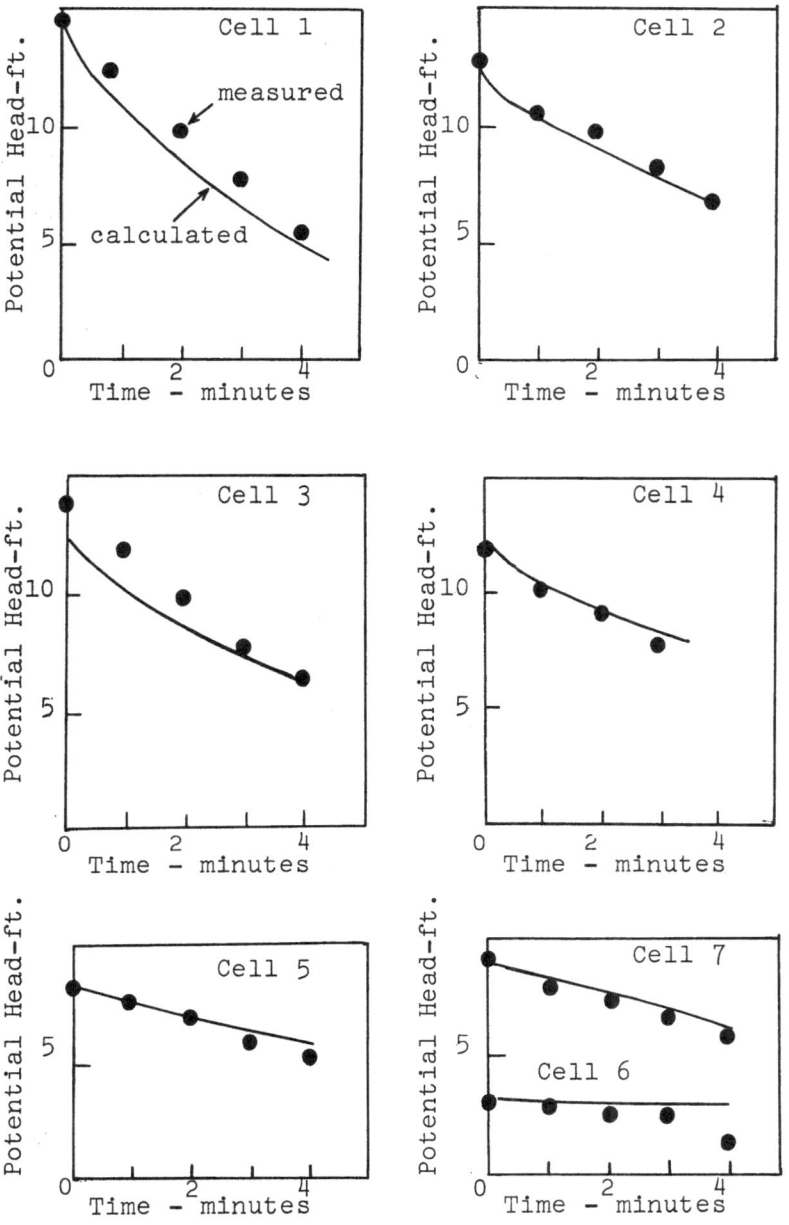

Figure 7. Potential Head Change due to Drawdown of
Impounding Water

The satisfactory correlation between the numerical solutions and laboratory model observations is encouraging. Although the method outlined above has neglected the effect of capillarity and neglected the change of soil permeability due to drainage, refinements of the analyses method may not be necessary, especially when the controlling factor, permeability, in most cases can only be determined approximately.

The method is helpful in establishing probable limits within which the prototype can be expected to perform. It would enhance the usefulness of the boundary method for practical analysis of banks and dams.

ACKNOWLEDGEMENTS

The author would like to thank his graduate students, Mr. E. A. Delis, Mr. A. Misra and Mr. J. H. Xue for their substantial help in computation and in preparation of this manuscript.

REFERENCES

1. Kirkham, D., and Gaskell R.E. (1951), The Falling Water Table in Tile and Ditch Drainage, Proceedings, Soil Science Soc. of America, Vol. 15, 1951, pp. 37-42.
2. Desai, C.S. (1972), Seepage Analysis of Earth Banks under Draw Down, Journal of Soil Mechanics and Foundation Division, American Society of Civil Engineers, Vol. 98, No. SM1, Nov. 1972.
3. Neuman, S.P. and Witherspoon, P.A. (1971), Analysis of Nonsteady Flow with a Free Surface Using the Finite Element Method, Journal of Water Resources Research, Vol. 7, No. 3, June, 1971, pp. 611-623.
4. Liggett J.A. and Liu, P.L. (1983). The Boundary Integral Equation Method for Porous Media Flow, George Allen & Unwin Ltd, USA.
5. Chang, C.S. (1981), Boundary Element Method in Seepage Analysis with a Free Surface, Proceedings of Implementation of computer Procedures and Stress-Strain Laws in Geotechnical Engineering, Chicago, Illinois.
6. Kellogg, F.H. (1941), Investigation of Drainage Rates Affecting Stability of Earth Dams, American Society of Civil Engineers, Transactions, Vol. 113, pp. 1261-1309.
7. Delis E.A. (1985), Evaluation of The Boundary Element Method for Steady and Transient Flow in Earth Dams, Master of Science Project Report, Department of Civil Engineering, University of Massachusetts, Amherst, Massachusetts.
8. Brebbia C.A. (1978), The Boundary Element Method for Engineers, Pentech Press Limited, Plymouth Devon.

Numerical Simulation of Separated Flow by Discrete Vortex Method

T. Miyamoto, M. Hashiguchi
Aerodynamics Research Department, Mazda Motor Corporation 3-1, Shinchi, Fuchu-cho, Aki-gun, Hiroshima, Japan

INTRODUCTION

Flow around a moving automobile is the unsteady high Reynolds number flow, and shows complex separated flow phenomena. The investigation and control of such a separated flow has a very important role in the production of automobile with high aero-dynamic performance, such as low drag, good cross-wind stability, low wind-noise level, etc.

The flow past an automobile is of course three dimensional, but at some local region of the automobile including front corners, hood, and middle of wind shield, the flow may be regarded as two-dimensional. Our experiences show that the control of the local separated flow results in a significant drag reduction or improvement of total aerodynamic performance.

At a high Reynolds number, the effect of viscosity in the flow field is localized within a thin boundary layer along the surface of the car body, and the vorticity is assumed to be generated only in that thin layer. Therefore, a method that can treat the vortex motion is desirable. One such method is the discrete vortex method. By this method the flow field is represented by a number of dis-crete vortices in inviscid fluid and the motion of each vortex is computed. This method is found to be very effective because it is particularly suitable for computer simulation, whereas the original flow problem governed by Navier-Stokes equations is in most cases too difficult to be computed directly.

The discrete vortex method has been discussed by a variety of au-thors. Sakata et al.[1] have used the panel method in the represen-tation of the potential flow field around a rectangular cylinder. This method distributes a number of bound vortices along the sur-face of the body. Another approach to represent the potential flow is the conformal mapping method. In calculation around a body of arbitrary shape, the panel method has advantage over the conformal mapping method because the former method does not need the mapping function. We have calculated the flow past a trapezoidal cylinder[2] by using the discrete vortex method combined with the panel method. We met some difficulties when a separated shear layer was close to or reattached the body surface. Owing to the singularity of the bound vortex, the calculation of the velocity distribution near the surface become erroneous and resulted in the penetration of separated shear layer into the inside of the body.

In this paper, we represent some results of numerical experiments carried out by the discrete vortex method combined with the panel method. For practical simulation, two models which treat the behavior of the separated shear layer near the body surface, are considered here. The flows around various trapezoidal cylinders were calculated to simulate critical phenomenon which shows an extremum of the drag or the Strouhal number when its geometrical shape is changed. These critical phenomenon has been reported by Hayashi et al.[3] The calculated results were compared with the experimental results obtained by them. In this calculation, we also used the continuously distributed vortex in the representation of bound vortex to improve the calculation accuracy near the body surface.

As other applications of this method, the flow patterns around a car-shape body and a truck were computed.

COMPUTATIONAL METHOD

Let us consider the flow past a bluff body as shown in Fig.1. The flow is assumed to be a two-dimensional, incompressible, unsteady high Reynolds number flow. Under this assumption, the boundary layer along the body surface has negligible thickness and can be replaced by a sheet of bound vortex. When flow separation occurs, the bound vortex is shed from the separation point as a free vortex into the outer inviscid flow field and a lump of free vortices forms the separated shear layer or the wake region. In this flow field, the velocity potential exists and satisfies the Laplace equation:

$$\nabla^2 \phi = 0 \quad , \tag{1}$$

where ϕ is the velocity potential.

The discrete vortex method combined with the panel method which is a type of boundary element method, describes the entire flow field with the sum of three velocity potentials, i.e., the potential of main inviscid flow, bound vortices, and the free vortices shed from the separation points, which are denoted by $\phi_U, \phi_{\Gamma B}, \phi_{\Gamma W}$, respectively:

$$\phi = \phi_U + \phi_{\Gamma B} + \phi_{\Gamma W} \quad , \tag{2}$$

where

$$\phi_U = U_\infty \cos\alpha \; x + U_\infty \sin\alpha \; y \quad , \tag{3}$$

$$\phi_{\Gamma B} = -\oint \frac{\gamma_B}{2\pi} \tan^{-1} \frac{y - y_B}{x - x_B} \, ds \quad , \tag{4}$$

and

$$\phi_{\Gamma W} = -\sum_{k=1}^{N} \frac{\Gamma_{Wk}}{2\pi} \tan^{-1} \frac{y - y_{Wk}}{x - x_{Wk}} \, ds \quad , \tag{5}$$

where x and y are the Cartesian coordinates, U_∞ the flow velocity at infinite upstream, α the angle of attack, and Γ_W the circulation of the bound vortex and free vortex. The positive sign is given to the vorticity of clockwise rotation.

The potential must satisfy the boundary condition on the body surface such that

$$\partial \phi / \partial n = 0 , \tag{6}$$

where n is the normal vector to the body surface. This condition means that the flow does not penetrate into the inside of the body.

Two velocity components u and v are obtained by the differentiation of the velocity potential:

$$(u , v) = (\partial\phi/ \partial x, \partial\phi/ \partial y)$$

In order to proceed with our calculation, we first assume that the bound vortex is distributed in unknown manner. The usage of the concentrated vortex shown in Fig.2 gives the induced flow velocity (u_B, v_B) due to all bound vortices as follows:

$$u_B = \frac{1}{2\pi} \sum_{j=1}^{M} \frac{y - y_j}{(x-x_j)^2 + (y-y_j)^2} \Gamma_{Bj} ,$$

and

$$v_B = \frac{-1}{2\pi} \sum_{j=1}^{M} \frac{x - x_j}{(x-x_j)^2 + (y-y_j)^2} \Gamma_{Bj} , \tag{7}$$

where Γ_{Bj} is the circulation of the j'th bound vortex. This approximation of the bound vortex corresponds to collocation method in BEM.

Moreover the present vortex system must satisfy Kelvin's theorem:

$$\sum_{j=1}^{M} \Gamma_{Bj} + \sum_{k=1}^{N} \Gamma_{Wk} = 0 \tag{8}$$

The induced velocity (u_W, v_W) due to all free vortices can be given by the following equations:

$$u_W = \frac{1}{2\pi} \sum_{k=1}^{N} \frac{y - y_{Wk}}{(x-x_{WK})^2 + (y-y_{Wk})^2} \Gamma_{Wk} ,$$

and

$$v_W = \frac{-1}{2\pi} \sum_{k=1}^{N} \frac{x - x_{Wk}}{(x-x_{WK})^2 + (y-y_{Wk})^2} \Gamma_{Wk} . \tag{9}$$

The rate of the shedding of the free vortex at the separation point is assumed to be

$$|\Gamma_W| = 0.5 u_s^2 \Delta T , \tag{10}$$

where ΔT is the time interval of the vortex shedding and u_s denotes the surface velocity induced at the separation point. According to this assumption, the flow problem is now reduced to the determination of Γ_{Bj} such that the resulting flow is parallel to the body surface at each control point where the' boundary condition Eq.(6)

is applied.

The position of the i th control point is defined as located in the middle of the i th and i+1 th bound vortex, as shown in Fig.2.

The substitution of Eqs.(2)-(5) for Eq.(6) gives a system of linear algebraic equations for M unknown Γ_B's as follows:

$$\sum_{j=1}^{M} C_{ij} \Gamma_{Bj} = d_j \quad \text{for } i=1, M, \tag{11}$$

where C_{ij} is the coefficient matrix.

Since there are M+1 equations for M unknown Γ_B's, one of the control points must be disregarded. Its position can be selected so that its elimination has only a small influence on the resulting flow. This procedure has been verified by Sakata et al[1].

The calculation procedure is as follows. Consider an unsteady flow impulsively started from rest to a constant velocity U_∞ . First, Γ_B's are determined from Eq.(11) at time t=0. Then u_s can be calculated at the prescribed separation points and Γ_w can be determined by Eq.(10). Secondly, positions of free vortices after vorticity shedding are computed by the Euler scheme:

$$x_w(t+ \Delta t)= x_w(t) + u_w(t) \Delta t ,$$

$$y_w(t+ \Delta t)= y_w(t) + v_w(t) \Delta t ,$$

where (x_w,y_w) are the position of the free vortex, Δt the time step for the time integration. At that time, the resulting flow may not satisfy the boundary condition Eq.(6), therefore Γ_B's must be calculated at the next time step. The repetition of this procedure gives a result for the unsteady vortex movement.

MODELLING OF SEPARATED SHEAR LAYER NEAR THE BODY SURFACE

In order to simulate the behavior of the separated shear layer existing near the body surface, two different phenomenological models were considered.

Model A   As shown in Fig.3.a, the circulation of the free vortex is set at zero when it enters inside the body. The physical meaning of this model may be interpreted from Kelvin's theorem so that the boundary layer absorbs the shear layer.

Model B   In some cases, the flow calculated with Model A procedure shows an abrupt change when a free vortex is suddenly eliminated at the body surface. Thus Model B as shown in Fig.3.b is proposed. In this model, the shear layer is allowed to slip along the body surface without vanishing as in Model A. In actual algorithm, the position of the free vortex was reset to outside position near the surface when it enters inside of the body.

RESULTS AND DISCUSSION

(I) Flow past a trapezoidal cylinder
Let us consider two-dimensional unsteady flow past a trapezoidal
cylinder as shown in Fig.4.

Experimental results for this flow has been reported by Hayashi
et al[3]. They have found that the flow shows a critical phenomenon
such that the drag and the Strouhal number have peaks at a certain
geometrical parameter b/h, where b and h are the size of the trapezoidal cylinder as shown in Fig.4. They state that the dominant
factor of this phenomenon is the interaction of the separated shear
layer and the body surface.

In the calculation, thirty-six vortices were used to represent the
body. The separation points were assumed to be located at the points
A,B,C and D, from which the free vortices were shed in the wake with
time interval of   T. Nondimensional time interval $\Delta \overline{T}$ ($=U_{\infty}\Delta T/h$)
was chosen as 0.2. We chose the time interval of the Euler scheme $\Delta \overline{t}$
as $\Delta \overline{T}/2$ to increase the accuracy of time integration. The calculation
conditions are as follows: U_{∞} =10 (m/s), d=0.03 (m), h=0.05 (m).
The shape parameter was changed as b/h=0.25,0.5,0.75,1.0.
the vortex blob concept used by Sakata et al[1]. was also used to
avoid any unreasonable vortex motion. The vortex clustering far
downstream was not performed.

Time averaged drag coefficient Fig.5 shows the results of the
calculation of the time averaged drag coefficient $\overline{C_d}$. The drag
was calculated by the Blausius equation.

It is shown that both Model A and Model B predict a peak of $\overline{C_d}$
for 0<b/h<1, however, the value of b/h at the peak is slightly
different from the experimental value obtained by Hayashi et al.
This discrepancy in b/h seems to be caused by the calculation
accuracy of the velocity distribution near the body surface.
The result for Model A is close to that for Model B.

In order to improve the accuracy of the velocity distribution
calculation, we used the continuously distributed bound vortex,
which is defined by the following equation:

$$\gamma_B(x,y) = \gamma_{Bj} + \frac{\gamma_{Bj+1} - \gamma_{Bj}}{\sqrt{(x_{j+1}-x_j)^2 + (y_{j+1}-y_j)^2}}\sqrt{(x-x_j)^2+(y-y_j)^2} \quad (12)$$

instead of the concentrated bound vortex. The calculation was
performed only for b/h=0.75 because the discrepancy between the
calculation and the experiment was great in that case. The number
of unknowns was not changed. Only Model A was used.

In this calculation the drag was calculated by the integration
of time-averaged surface pressure instead of Blausius equation
since we now discuss the time averaged drag only. We used the
following formula which is derivable from Ref.4, that is

$$\overline{C}_p = 1 - (\frac{\gamma_B}{U_\infty})^2 - \sum_{j=1}^{N_s}(\frac{u_{sj}}{U_\infty})^2 \quad , \qquad (13)$$

Where N_s is the number of the separation points upstream of the
calculation point, and u_{sj} the j th surface velocity counted
upstream.
This formula is more useful than Bernoulli's equation because
there is no need of calculating the time derivative of the velocity potential, although exactly speaking, the formula can be only
applied to the steady state pressure calculation.

The results of C_d and pressure calculation are shown in Fig.5
and Fig.6, respectively. These results prove the usefulness of
this calculation method.

Time history of drag and lift   Fig.7 and Fig.8 show the calculated
time histories of drag and lift respectively. As is shown, there
was no noticeable difference between Model A and Model B for C_d
calculation, but a remarkable difference in lift coefficient C_l
time history. As expected, there was no abrupt change in C_l when
Model B was used.

Strouhal number   The strouhal number calculated from C_l time
history shown in Fig.8 is plotted in Fig.9. The agreement of the
calculation and the experimental result is considerably good.
there is no remarkable difference between Model A and B.

Flow pattern   Various flow patterns are displayed by plotting
the free vortices. It is found that the present calculation can
simulate the characteristic flow pattern that Hayashi et al.
reported: the vortices are from the trailing edges C and D at
b/h=0.25; then the separation points were moved to the leading
edges A and B from C and D at b/h=0.75. See Fig.10.
This movement of the separation points has a dominant role on the
rolling up of the separated shear layers behind the body.

In particular, the usage of the distributed bound vortex gives
the flow pattern quite similar to the result of the smoke wire
flow visulization done by Hayashi et al.

(II) Another flow calculation   Fig.11 shows an application to
the three dimensional flow which may be assumed to be locally
two dimensional. This flow pattern agrees with the smoke lines
observed in the wind tunnel.

Fig.12 shows the flow around the full scale truck cabin. The flow
in the symmetry plane was calculated. The calculated result agrees

with the smoke line observed in the wind tunnel.

These results show that the present method is applicable to simulate the flow regarded to be locally two-dimensional flow.

CONCLUDING REMARKS

Some results of the discrete vortex simulation were presented. These computations show that the present method can simulate the critical phenomenon of the trapezoidal cylinder and the flow locally two dimensional flow.

The introduction of the phenomenological free vortex- wall interaction model has an important role on the simulation of the shear layer near the body.

The present calculations were carried out on IBM 3081. Computation times for a typical calculation up to nondimensional time T=45 are 30 minutes. Therefore the discrete vortex method can keep the CPU time within reasonable limits.

ACKNOWLEDGMENT

The authors would like to thank Professor Akira Sakurai of the Aeronautical Department of Kyushu University for his suggestions on the calculations presented here.

REFERENCES

1. Sakata H.et al.(1983),Numerical calculation of unsteady separated flow by discrete vortex method,J.of Japan Soc. of Mech.Eng., Vol.49,No.440,p.801 (in Japanese).

2. Miyamoto T.et al.(1984),Numerical Simulation of separated flow by discrete vortex method,Japan Soc. of Aeronautical and Space Sciences,p.25 (in Japanese).

3. Hayashi M. et al.(1981),Experimental investigation of critical phenomenon of flow around trapezoidal cylinder, Technical Report, Kyushu University, Vol.54, No.2, p.151 (in Japanese).

4. Lewis R.I.(1981), Surface vorticity modeling of separated flows from two-dimensional bluff bodies of arbitrary shape, J. of Mechanical Engineering Science, Vol.23, No.1, p.1.

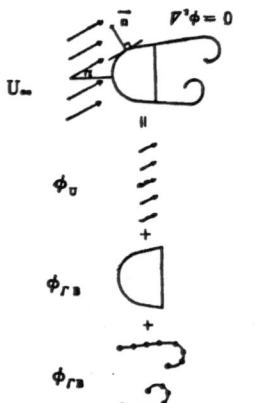

Figure 1. High Reynolds number
flow.

Figure 2. Concentrated
bound vortex.

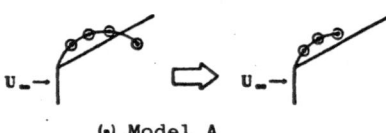

(a) Model A

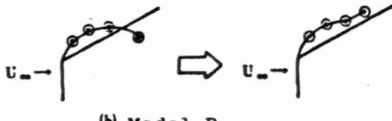

(b) Model B

Figure 3. Separated shear layer-
wall interaction model.

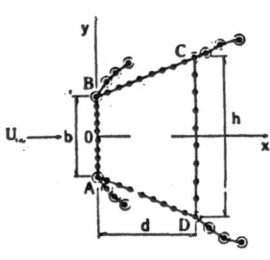

Figure 4. Trapezoidal cylinder
and flow model.

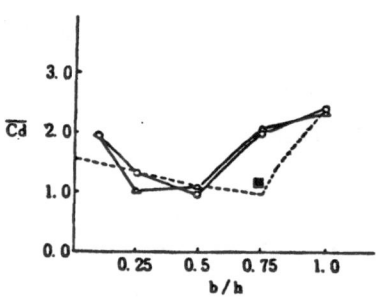

Figure 5. Time-averaged drag
coefficient,
-:calculation(O:Model A,
Δ:Model B),---:Ref.3.
■:continuously distribute
and Model A.

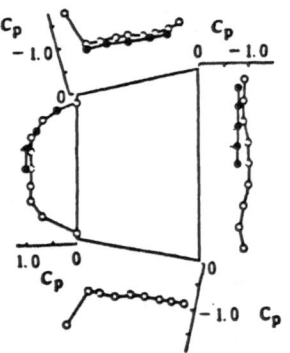

Figure 6. Surface pressure distribution
of b/h=0.75,-O-:calculation,
-●-: Ref.3.

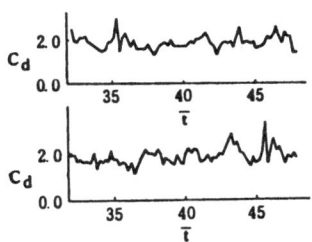

Figure 7. Time history of drag
for b/h=0.75.

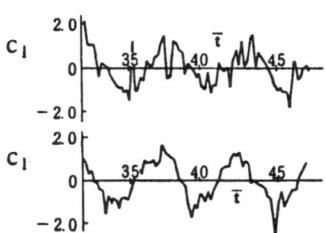

Figure 8. Time history of lift
for b/h=0.75.

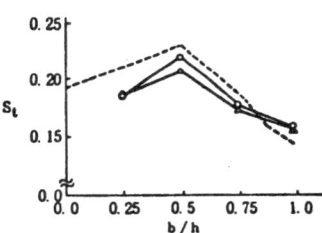

Figure 9. Strouhal number ,
  -:calculation(o:Model A,
  △:Model B),---:Ref.3.

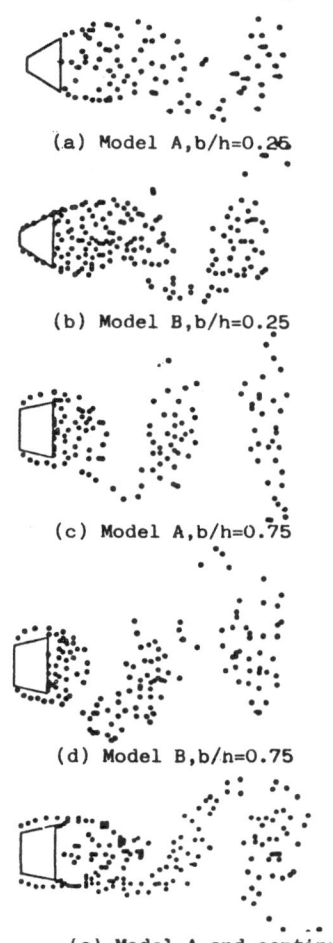

(a) Model A,b/h=0.25

(b) Model B,b/h=0.25

(c) Model A,b/h=0.75

(d) Model B,b/n=0.75

(e) Model A and continuously distributed vortex,b/h=0.75

(f) Smoke visualization by Ref.3.

Figure 10. Flow pattern, O:clockwise,
●:anti-clockwise.

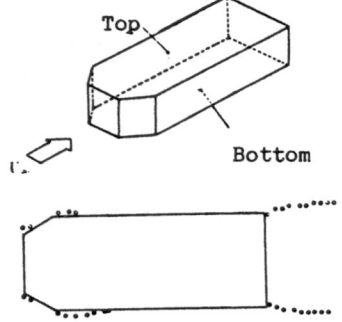

Figure 11. Plane flow around
a one-box car shape.

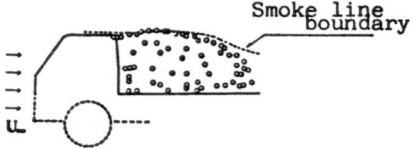

Figure 12. Flow behind the cabin
of a truck.

Analyses of Fresh – Salt Water Interface in Coastal Aquifers

T. Fukuhara, T. Fukui
Department of General Construction Engineering, University of Fukui, Fukui, Japan

INTRODUCTION

Analyses of ground water flows in coastal aquifers are required for managing coastal aquifers, in view of the prevention of the obstacles relating to the sea water intrusion into a fresh water aquifer. In the case of an unconfined aquifer, especially, there are two free boundaries: one is a phreatic surface and the other is a fresh–salt water interface(f–s interface).

We propose two different schemes to decide the shape of movable boundaries: one is useful for steady states and the other is useful for unsteady states. For steady states, we analyze only a fresh water region, because the salt water is immobile in a aquifer, i.e. the piezometric head of salt water is constant throughout the salt region. The immobile condition of the sea water is reduced to a potential problem in the fresh water region with two free surface. These shapes can also be determined by the method, proposed by one of authors. In the latter half of this paper, we present the numerical scheme of the unsteady analysis to solve the hydraulic problems which are related to the fresh water lens in islands and the salt water intrusion in unconfined aquifers. Furthermore, some geohydraulic problems in islands are discussed based on the computed results.

SEAWATER INTRUSION PROBLEMS

In this section we will define the problems mathematically and represent briefly the basic equations and the boundary conditions.

Basic Equation

The present analysis is applied to the seepage flow through the unconfined aquifers as shown in Fig.1. If it is assumed that Darcy's law is valid in a homogeneous porous

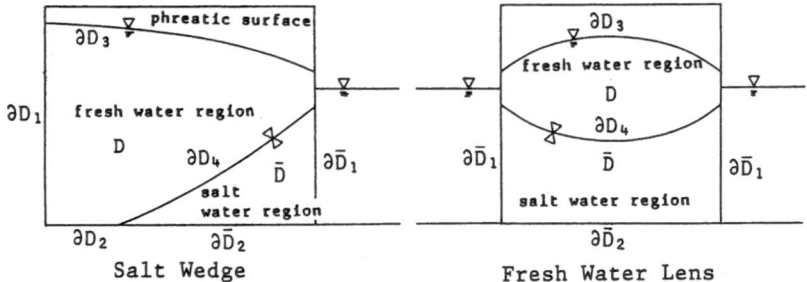

Fig.1 Typical boundaries in unconfined aquifers

media, the discharge velocity vector components, $v_i(x)$ can be expressed by

$$v_i(x)=-K\cdot\partial h(x)/\partial x_i \qquad , \qquad (1)$$

where K is the coefficient of permeability and $h(x)$ is the piezometric head.
The continuity equation is given by

$$\partial v_i(x)/\partial x_i+Q(x)=0 \qquad , \qquad (2)$$

where $Q(x)$ is the specific source intensity. Substituting Eq.(1) into Eq.(2), the continuity equation is rewritten as

$$-K\cdot\partial^2 h(x)/\partial x_i{}^2+Q(x)=0 \qquad (3)$$

for $h(x)$ in a flow domain D.

Boundary Conditions

(i) $\qquad h(x)=H_1$ or $h(x)=H$ $\qquad$ **on ∂D_1** , $\qquad$ (4)

where H_1 is the freshwater level at the upstream boundary and H is the seawater level.

(ii) $\qquad q(x)\equiv n\cdot v=-K\partial h/\partial n$ $\qquad$ on ∂D_2 , $\qquad$ (5)

where **n** is the unit outward normal vector. If ∂D_2 is impervious, $q(x)$ becomes to be zero.

(iii) $\qquad h(x)=z_3(x)$ $\qquad$ on ∂D_3 , $\qquad$ (6)

where z_3 is the vertical distance from the base face to the phreatic surface or the seepage face. Especially, the phreatic surface must be a material surface(kinematic condition). If **k** is an appropriately defined unit direction vector, the rate of change of the phreatic surface along **k** is given by

$$\partial \xi_1(x)/\partial t = q(x)/kn+\omega \qquad \text{on } \partial D_3 \qquad , \qquad (7)$$

where ξ_1 is the distance along k from the base face to the phreatic surface. and ω is the rate of acceration. In the case of steady flows, this condition becomes

$$q(x)=0 \qquad\qquad \text{on} \quad \partial D_4 \qquad , \qquad (8)$$

(iv)

$$\overline{\sigma h}_4(x)-h_4(x)=(\sigma-1)z_4(x) \quad \text{on} \quad \partial D_4 \qquad , \qquad (9)$$

$$q_4(x)+\overline{q}_4(x)=0 \qquad\qquad \text{on} \quad \partial D_4 \qquad , \qquad (10)$$

where a symbol $(-)$ indicates the seawater region and z_4 is the vertical distance from the base face to the f-s interface. The value of σ is defined as ρ_2/ρ_1, where ρ_2 is the density of the salt water and ρ_1 is that of the fresh water.

The kinematic condition is given by

$$\partial \xi_2(x)/\partial t = \overline{q}_4(x)/k\overline{n}_4 = q_4(x)/kn_4 \qquad \text{on } \partial D_4 \qquad , \qquad (11)$$

where ξ_2 and k are defined in the same manner as on ∂D_3 . In the steady problem, this is also reduced to

$$\overline{q}_4(x)=q_4(x)=0 \qquad\qquad \text{on } \partial D_4 \qquad . \qquad (12)$$

BOUNDARY INTEGRAL EQUATION

The boundary integral equation of Eq.(3) can be obtained by using the Green's third theorem as follows ;

$$F(y)h(y)=\int_D G(x;y)Q(z)dDx-\int_{\partial D}\left[G(x;y)q(x)-S(x;y)h(y)\right]dSx \qquad , \qquad (13)$$

where y is a source point. $G(x;y)$ and $S(x;y)$ are the fundamental singular solutions of Eq.(3) and can be written as

$$G(x;y)=\frac{-1}{2\pi k}\ln(1/|x-y|) \quad , \quad S(x;y)=\frac{1}{2\pi}\mathbf{n}(\mathbf{x-y})/|x-y|^2 \qquad . \qquad (14),(15)$$

$F(y)$ takes the value $0, 1/2$ or 1 when $y \in D+\partial D, y \in \partial D$ or $y \in D$, respectively, if the boundary ∂D is smooth. Subdividing the given boundaries into N small linear segments (constant element), ΔS_i $(i=1,N)$ and choosing boundary points $,x_i,$ such that $x_i \in \Delta S_i$, Eq.(13) is transformed into the following from :

$$\sum_{j=1}^{N}[\,G(x_i;x_j)q(x_i)\,-\,(\,S(x_i;x_j)\,-\,\tfrac{1}{2}\delta_{ij}\,)h(x_j)\,]\Delta S_j$$
$$=\sum_{j=1}^{N}\frac{Q}{2\pi k}\ln(1/|x_j-y|) \qquad (16)$$

where δ is Dirac's delta function.
Consequently, Eq.(16) can be also expressed in the following matrix form as

$$[G]\{q\} - [S]\{h\} = [F]Q \tag{17}$$

ANALYSIS STEADY PROBLEM

Solution Procedure

In the case of steady flow, the value of the piezometric head in the seawater region is constant everywhere, i.e. $h = H$. The problem can be concentrated only about the fresh water region. This can be attributed to a problem with two kind of free surface boundaries. We use here the iteration technique that is used to determine one free boundary , i.e. the phreatic surface in Niwa et.al.(1974)[1]. The algorithm is summarized as follows ;

(i) Firstly, $h_3 = z_3$ on ∂D_3 and $h_4 = \sigma H - (\sigma - 1) z_4$ on ∂D_4 are set and the unknown boundary values q_1, h_2, q_3 and q_4 are decided by resolving the following matrix,

$$[G_1 - S_2 G_3 G_4] \begin{Bmatrix} q_1 \\ h_2 \\ q_3 \\ q_4 \end{Bmatrix} = [S_1 - G_2 S_3 S_4] \begin{Bmatrix} h_1 \\ q_2 \\ z_3 \\ \sigma H - (\sigma - 1) z_4 \end{Bmatrix} + \begin{bmatrix} F_1 \\ F_2 \\ F_3 \\ F_4 \end{bmatrix} Q \quad . \tag{18}$$

(ii) The values of q_3 and q_4 are used as the boundary conditions on ∂D_3 and ∂D_4 , respectively and Eq.(17) is transformed as

$$[G_1 - S_2 - S_3 - S_4] \begin{Bmatrix} q_1 \\ h_2 \\ h_3 \\ h_4 \end{Bmatrix} = [S_1 - G_2 - G_3 - G_4] \begin{Bmatrix} h_1 \\ q_2 \\ \alpha q_3 \\ \alpha q_4 \end{Bmatrix} + \begin{bmatrix} F_1 \\ F_2 \\ F_3 \\ F_4 \end{bmatrix} Q \quad , \tag{19}$$

where α is a control factor and the value of α is from 0 to 1.
(iii) The new locations of free surfaces are predicted by using h_3 and h_4 calculated above and

$$z_3(x) = h_3(x) \quad \text{and} \quad z_4(x) = \sigma H - h_4(x)/(\sigma - 1) \tag{20}$$

are reset. These steps are repeated until q_3 and q_4 become small enough.

Text Example

In order to check the applicability the present method, a typical example is tested. An iterative process to determine the steady solution is shown in Fig.2. The steady solution can be obtained after several iterations and agrees well with the analytic solution based on Dupuit assumption. The appropriate value of α is approximatly from 0.8 to 0.9 and gives the stable convergence of computation.

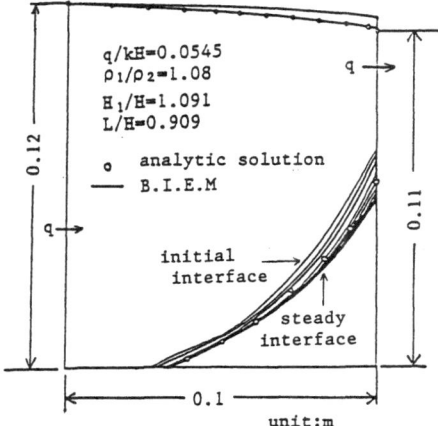

Fig.2 Comparison of computation results with analytic
results based on Dupuit assumption

ANALYSIS OF UNSTEADY PROBLEM

Solution Procedure

In the case of unsteady flows, it is needed to analyze
both freshwater region and seawater region, according to
interfacial conditions, Eq.(9) and Eq.(11). The discretized
equation corresponding to Eq.(17) can be described as

$$
\begin{bmatrix} G_1-S_2G_3 & 0 & 0 & -S_4\sigma_4 \\ 0 & 0 & 0 & G_1-S_2-\frac{S_4}{\sigma}-\frac{k_1}{k_2}G_4 \end{bmatrix}
\begin{Bmatrix} q_1 \\ h_2 \\ q_3 \\ \overline{q}_1 \\ \overline{h}_2 \\ h_4 \\ q_4 \end{Bmatrix}
=
\begin{bmatrix} S_1-G_2S_3 & 0 & 0 & 0 \\ 0 & 0 & 0 & \overline{S}_1-\overline{G}_2\overline{S}_4 \end{bmatrix}
\begin{Bmatrix} h_1 \\ q_2 \\ h_3 \\ \overline{h}_1 \\ \overline{q}_2 \\ \frac{(\sigma-1)z_4}{\sigma} \end{Bmatrix}
+
\begin{Bmatrix} F_1 \\ F_2 \\ F_3 \\ F_4 \end{Bmatrix} Q \quad . \quad (21)
$$

On solving the system of linear algebraic equation
above, $q(x)$ at the nodal points on the free surfaces, i.e.
$q_3(x)$ and $q_4(x)$ become known.
Finally, the new locations of free boundaries are predicted by

$$
\xi_1^{m+1}(x)=\xi_1^m(x)+\frac{\Delta t}{k n}\, q_3^m(x)+\Delta t\,[\theta\omega^{m+1}+(1-\theta)\omega^m] \quad , \quad (22)
$$

$$
\xi_2^{m+1}(x)=\xi_2^m(x)+\frac{\Delta t}{k n}\, q_4^m(x) \quad , \quad (23)
$$

in which Δt is a time interval, θ is a weighting factor and
the subscript m denotes the mth time level.

Text Example

As shown in Fig.3 (a), a rectangular unconfined coastal aguifer is fully saturated up to a constant level,$z=h_1$. At the initial time, $t=0$, the surface water level at the downstream end,i.e. the sea level is suddenly lowered to a lower elevation, $Z=H < H_1$ and it remains for all time. The obtained phreatic surface and the salt wedge under the steady condition is shown in Fig.3 (b). Fig.3 (b) also shows the analytic results which are calculated based on Dupuit assumption. The computed results agree well with the analytic results and with the experimental results by the present authors[2].

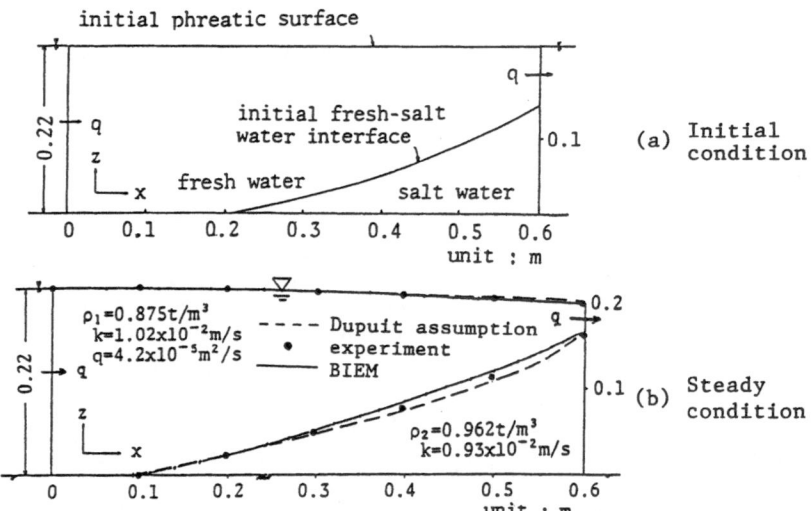

Fig.3 Comparison of computational results with experimental ones of salt wedge

Fig.4 shows a swelling process of the fresh water lens due to a natural replenishment from precipitation when $\omega/k=0.07$, where is the ratio of the infiltration of the fresh water. It was found that an initial configuration of the f-s interface does not affect the steady configuration of the f-s interface. Furthermore, a swelling process of the fresh water lens is illustrated in detail in Fig.5, which describes the temporal variation of the height of the fresh water table above the sea level, h_1max-H and one of the depth of the f-s interface below it, $H-h_2min$ at the center of island,i.e. $x=L/2$, where L is a length of island . Furthermore, the temporal variation of the fresh water discharge, q to the sea is shown in Fig.6. It is seen from a series of figures related to the fresh water lens that (i) in early stage of infiltration, a thickness of fresh water lens swells suddenly, (ii)

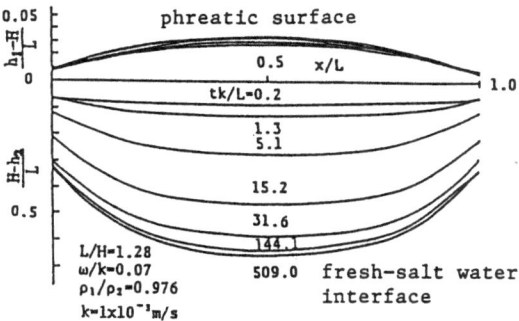

Fig.4 Change of fresh-salt water interface with time

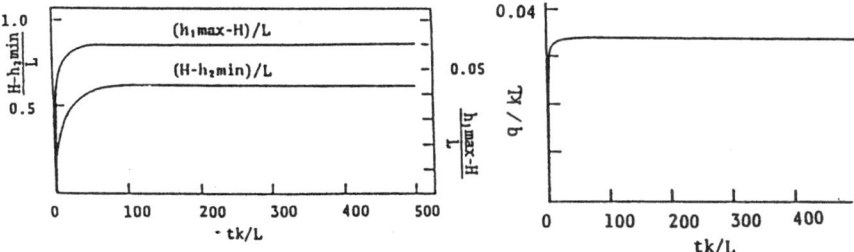

Fig.5 Change of locations of free
surface and interface in the
middle of island

Fig.6 Change of groundwater
run-off rate with time

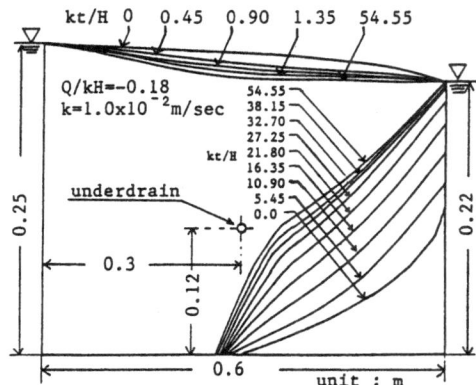

Fig.7 Change of free surface and interface under a drain

subsequently, as the discharge q is increased the intrusion of fresh water in the salt water region gets weaker, and (iii) at time elapsed enough, the fresh water intrusion ceases and the discharge q remains constant.

Fig.7 shows the temporal variation of the fresh water table and that of the salt wedge due to the withdrawal installation, in the form of a underdrain which is located at a relatively small distance from the coast. The fresh water table falls suddenly towards an underdrain from the beginning of drainage and falls gradually with time. Finally, this drawdown ceases and the fresh water table keeps a constant level at the dimensionless time, tk/H= 5.0 , where H is the sea level. The salt wedge , however , still moves towards the underdrain and the phenomenon of upconing appears at some distance from the underdrain.

CONCLUDING REMARKS

Two kinds of BIEM for the transient analysis on the underground density flows have been proposed. The validity of the proposed methods were studied by comparing the numerical results with the experimental results and with the analytic solution for the steady salt wedge. These methods can be used to solve the problems for the underground density flows, regardless the complexity of boundary shape and boundary conditions. BIEM are much more efficient than other methods, since BIEM can decide the temporal variation of flow domain more accurately and easily than other numerical methods such as the finite difference method and the finite element method.

REFERENCES

1) NIWA, Y., KOBAYASHI, S. and FUKUI, T. (1974). An Application of the Integral Equation Method to Seepage Problems Theor. Appl. Mech.,24, Univ. of Tokyo Press, 479-486.

2) FUKUHARA, T. (1986). Unsteady Dynamic Analysis of Underground Salt WEdge by Boundary Integral Equation Method, Ann. Conf. Civil Eng. (in Japanese).

Three-Dimensional Simulation of Hydro-Mass Transportation in Nearly Closed Basins

J. Matsunashi
Department of Civil Engineering, Kobe University, Kobe, Japan
T. Teratani, H. Okuda
Undergraduate School of Kobe University, Kobe, Japan

1. INTRODUCTION

In recent years the increase in the amount of waste deposited in estuary from industry and urban development has led to concern regarding the quality of water in these areas. Such environmental problems have been getting more serious in nearly closed water areas, for example, the Sea of Harima, the Bay of Osaka in Japan. It has therefore become important for planners to be able to predict the effects of any proposed development on water quality.

In this paper, the mass transport equation is assumed to govern the pollutant distribution process. In general, the process is a three-dimensional phenomenon. The problem is however used to treat as a two-dimensional one by assuming that the water body is well mixed in the vertical direction. In order to predict the vertical diffusion mechanism of the dissolvable pollutants eluting into a basin through its bottom, it is essential that the problem is treated as a three-dimensional phenomenon. The boundary element method (BEM) may be specially well suited to solving three-dimensional problems. The main advantage of this method for generating numerical solutions to the mass transport equation is that the dimension of the problem is reduced by one.

Here, it is assumed that the contribution of the tidal currents in the Bay of Osaka on the fundamental relationships may be evaluated by introducing a turbulent flow basin with the corresponding large scale eddies to the tidal currents. The hydro-mass balance in the bay may be therefore analyzed as a three-dimensional diffusion problem in a homogeneous and anisotropic turbulent flow basin without mean velocity.

2. BOUNDARY ELEMENT FORMULATION

The governing equation for the steady state hydro-mass transport can be written as, (see Fig.1)

$$\frac{\partial}{\partial X}(D_X\frac{\partial C}{\partial X})+ \frac{\partial}{\partial Y}(D_Y\frac{\partial C}{\partial Y})+ \frac{\partial}{\partial Z}(D_Z\frac{\partial C}{\partial Z})=0 \quad \text{in } \Omega \tag{1}$$

where, C is the concentration of pollutants, D_X, D_Y and D_Z are the eddy diffusivities in the directions of the coordinate axis, X,Y and Z respectively. The boundary conditions relating to this problems are of two types,

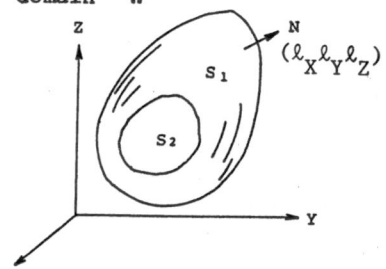

domain Ω

Fig. 1 Definitions

$$C=\overline{C} \qquad \text{on } S_1 \tag{2}$$

$$q_N=\overline{q}_N \qquad \text{on } S_2 \tag{3}$$

where,

$$q_N=D_X\frac{\partial C}{\partial X}\ell_X+D_Y\frac{\partial C}{\partial Y}\ell_Y+D_Z\frac{\partial C}{\partial Z}\ell_Z \tag{4}$$

and $N(\ell_X,\ell_Y,\ell_Z)$ is the normal to the boundary. The total boundary is $S=S_1+S_2$, $\overline{C}$ is the concentration prescribed and $\overline{q}_N$ is the concentration flux prescribed. A weighting fuction C* can now be introduced such that it has continuous first derivatives. The following weighted residual statement for this problem can be written as,

$$\iiint_\Omega\{\frac{\partial}{\partial X}(D_X\frac{\partial C}{\partial X})+ \frac{\partial}{\partial Y}(D_Y\frac{\partial C}{\partial Y})+ \frac{\partial}{\partial Z}(D_Z\frac{\partial C}{\partial Z})\}C*d\Omega =$$

$$\iint_{S_2}(q_N-\overline{q}_N)C*ds - \iint_{S_1}(C-\overline{C})q_N*ds \tag{5}$$

where,

$$q_N*=D_X\frac{\partial C*}{\partial X}\ell_X+D_Y\frac{\partial C*}{\partial Y}\ell_Y+D_Z\frac{\partial C*}{\partial Z}\ell_Z \tag{6}$$

The function C* is now assumed to be the fundamental solution of the equation, representing a concentrated mass source at a point i, that is, the solution of

$$\frac{\partial}{\partial X}(D_X\frac{\partial C*}{\partial X})+ \frac{\partial}{\partial Y}(D_Y\frac{\partial C*}{\partial Y})+ \frac{\partial}{\partial Z}(D_Z\frac{\partial C*}{\partial Z})+ \Delta_i = 0 \tag{7}$$

where, Δ_i is the Dirac delta function.
The fundamental solution of Eq.(7) is given as,

$$C*= \frac{1}{4\pi r_0\sqrt{D_X D_Y D_Z}}, \tag{8}$$

$$r_0=\{\frac{(X-Xi)^2}{D_X}+\frac{(Y-Yi)^2}{D_Y}+ \frac{(Z-Zi)^2}{D_Z}\}^{1/2} \tag{9}$$

where, (Xi, Yi, Zi) and (X,Y,Z) are respectively the coordinate of the point of application of the concentrated mass source and of the point under consideration.
Integrating by parts the left hand side of Eq.(5) and using the relation of Eq.(7), Eq.(5) becomes

$$C_i+\iint_S Cq_N*ds = \iint_S q_N C*ds \tag{10}$$

where, C_i represents the value of the unknown function C at the

point of application of the charge.
Eq.(10) is valid for any point in the domain Ω, but in order to formulate this problem as the boundary technique one needs to take the point of application of the charge to the boundary. For this case Eq.(10) can be written as,

$$k_i C_i + \iint_S C q_N^* ds = \iint_S q_N C^* ds \qquad (11)$$

The value of k_i is :

$$k_i = 1 \quad \text{if the point i is in the domain}$$
$$k_i = 1/2 \quad \text{if the point i is on an smooth boundary}$$

3. APPLICATIONS

3.1 Two-dimensional long basin (Model-A)
As an illustration of the application of the fundamental relationships to the Bay of Osaka, let one consider a hydro-mass transport in the very long two-dimensinal basin shown in Fig.2.(a=60km,h=50 m) The governing equation of the phenomena in the basin and the boundary conditions are written as,

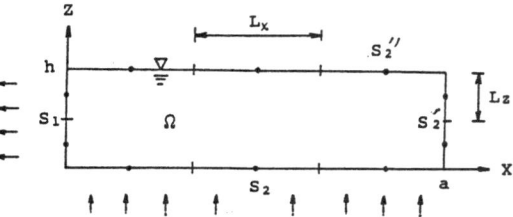

Fig.2 Two-dimensional long basin (Model-A)

$$\frac{\partial}{\partial X}(D_X \frac{\partial C}{\partial X}) + \frac{\partial}{\partial Z}(D_Z \frac{\partial C}{\partial Z}) = 0 \qquad \text{in } \Omega \qquad (12)$$

$$C = \overline{C} = 0 \qquad \text{on } S_1$$

$$q_N = \overline{q}_N = 50 mg/m^2/ day = f(X) \qquad \text{on } S_2 \qquad (13)$$

$$q_N = \overline{q}_N = 0 \quad \text{on the other boundaries } S_2' \text{ and } S_2''$$

where, S_2 denotes the boundary through which the dissolvable pollutants elute in to the basin.
For the two-dimensional case, the fundamental solution is given as, $C^* = \ell_n (1/r_0)/2\pi(D_X D_Z)^{1/2}$, $r_0 = \{(X-Xi)^2/D_X + (Z-Zi)^2/D_Z\}^{1/2}$,
which correspond to Eqs.(8) and (9) respectively. In order to apply Eq.(11) to the basin boundary, the boundaries S_2 and S_2'' are divided into n equal length elements and the boundaries S_1 and S_2' divided into m equal length elements respectively, as shown in Fig.2. L_x and L_z denote the length of one elemnt in the direction X and the direction Z respectively. In this case, the so-called" constant elements" are applied as the types of boundary elements. Numerical results are shown in Fig.3 for assumed values of L_x=1km, L_z=25m, D_z=10,20,100m^2/day. In this case an analytical solution is written as,

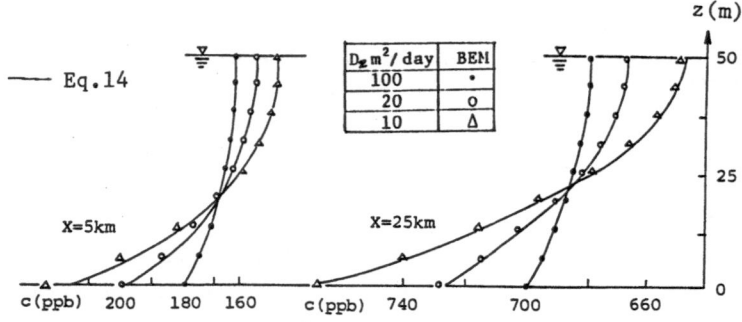

Fig. 3 Vertical distribution of C

$$C(X,Z) = \frac{2}{a} \sum_{n=0}^{\infty} \frac{\cosh\{(2n+1)\pi(Z_1 - Z\sqrt{D_X/D_Z})/2a\}}{\sqrt{D_X D_Z}\{(2n+1)\pi/2a\}\sinh\{(2n+1)\pi Z_1/2a\}} \cdot$$

$$\sin\frac{(2n+1)\pi X}{2a} \int_0^a f(X)\frac{(2n+1)\pi X}{2a} \, dX \qquad (14)$$

where, $Z_1 = h\sqrt{D_X/D_Y}$.

The concentration values computed for each value of the eddy diffusivities D_Z and the corresponding analytical results are in close agreement. It is clear that the smaller the value of D_Z, the smaller the value of concentration at the points near the water surface of the basin and the greater the value at the points near the bottom of the basin.

Fig.4 shows the equiconcentration profile in the basin for assumed values of $D_X = 20 m^2/s$, $D_Z = 100 m^2/day$. According to

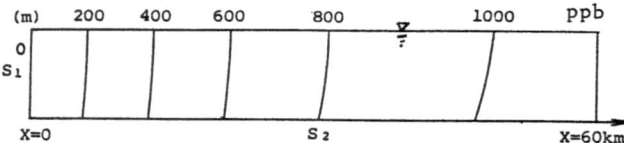

Fig.4 Equiconcentration profiles

Fig.4, it is known that the dissolvable polutants eluting into the basin through its bottom may be removed toward the boundary S_1.

Next, in view of the fact that it is necessary to compare the accuracy of the computed values against the different size of the element lengths L_X and L_Z for the assumed values of $D_X = 20$ m^2/s, $D_Z = 100 m^2/day$, the summation of the computed values of the concentration flux through the total boundary S, that is,

$$W = \int_{S=S_1+S_2+S_2'+S_2''} q_N \, dS \qquad (15)$$

is introduced as the scale of the accuracy for the solutions by the BEM.

It is shown in Fig.5 that the values of W decrease with an decrease in the value of L_X for the equal values of L_Z. As a result, it is obvious that the smaller the values of L_X, the higher the accuracy.

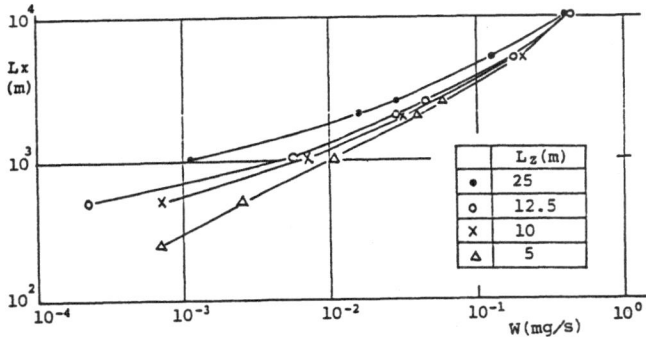

Fig.5 Comparison of the accuracy by the value of W

3.2 Three-dimensional flat basin (Model-B)

As an other illustration of the application of the fundamental relationships to the Bay of Osaka, let one consider a hydro-mass transport in the three-dimensional flat rectangular parallele-piped basin shown in Fig.6. (a_X=30 km, a_Y=50 km, a_Z=50 m) The governing equation in the basin is given by Eq.(1), and the boundary conditions are given as,

$C=\overline{C}=0$ on S_1

$q_N=\overline{q}_N$=50mg/m^7/ day   on S_2

$q_N=\overline{q}_N$=0   on the other boundary

(16)

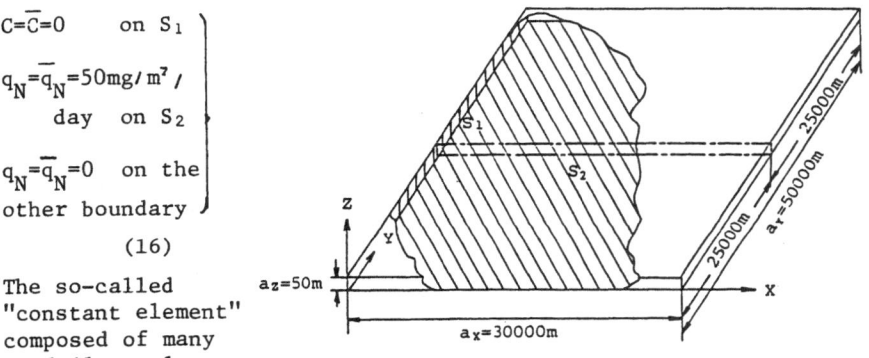

Fig.6 Flat rectangular parallelepiped basins

The so-called "constant element" composed of many quadrilaterals are applied as the types of the boundary elements. For the assumed case of $\overline{q}_N$ =F(X,Y)=const., an analytical solution is obtained as,

$$C(X,Y,Z)=\sum_{n=0}^{\infty} \frac{8a_X\overline{q}_N}{(D_XD_Z)^{1/2}(2n+1)^2\pi^2} \cdot \frac{\cosh\{(2n+1)\pi(Z_1-Z\sqrt{D_X/D_Z})/2a_X\}}{\sinh\{(2n+1)\pi Z_1/2a_X\}} \cdot$$

$$\sin\frac{(2n+1)\pi X}{2a_X} \qquad (17)$$

where, $Z_1=a_Z\sqrt{D_X/D_Z}$.

In general, the total accuracy of the BEM solution depends remarkably on the accuracy of the integration computation of each elements. In this case treating the diffusion equations, it is necessary to

integrate the functions C* and q_N* revealing special properties
on some elements. For the reason that it is difficult to get
the analytical values of the surface integrations of C* and q_N*
in the three-dimensional anisotropic mass transport problems,
these values may be generally evaluated by using numerical integrations. In the case where the distance between the point of
application of the concentrated mass source and the point under
consideration is very small and these points are on a same element,
it is difficult to get the high accuracy by using the numerical
integration, because the fuctions C* and q_N* stand up extremely
in the vicinity of the source point. Here, in order to get the
high accuracy, two methods, (1) the boundary of basin is divided
into a large number of elements, (2) the order of the numerical
integration on each elements is taken in high grade as much as
possible, are applied. In addition, the same methods stated above
are applied in the computation of unknown values at the internal
points of the basin. (see Fig.7)

The computed results
are compared with
the analytical
solution (17) as
shown in Fig.8.
In this figure the
values H_{ij} and G_{ij}
represent the second
term of the left hand
side and the right
hand side of Eq.(11)
respectively, that is,

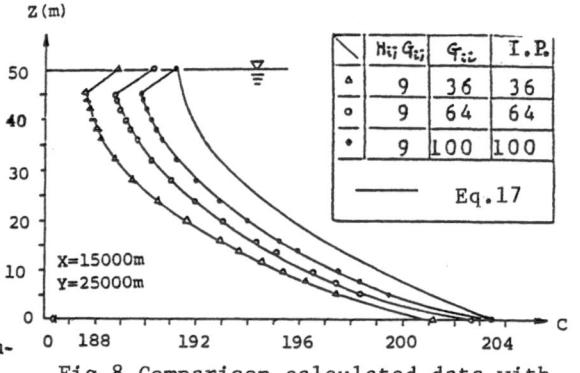

Fig.7 High order integration elements

$$H_{ij} = \iint_{S_j} q_N^* ds = \iint_{S_j} \frac{1}{4\pi (D_X D_Y D_Z)^{1/2}} \frac{\gamma}{r_0^{3}} \frac{\partial r}{\partial N} ds, \qquad (18)$$

$$G_{ij} = \iint_{S_j} C^* ds = \iint_{S_j} \frac{1}{4\pi (D_X D_Y D_Z)^{1/2}} \frac{1}{r_0} ds \qquad (19)$$

where, $r^2 = (X-Xi)^2 + (Y-Yi)^2 + (Z-Zi)^2$.

The Gauss-Legendre's formula is used on the numerical integrations.
According to Fig.8,
it is shown that
the higher the order
of the numerical
integration, the
higher the accuracy
of computed values.
Therefore, it may
be consider that
the methods stated
above are reasonable.
However, the detailed
concentration dis-
tribution in the neigh-
borhood of the water

Fig.8 Comparison calculated data with
solution (17); unit ppb

surface and the
bottom of the basin
is obtained as shown
in Fig. 9. Under
this figure it may
be clear that the
computation accuracy
is not sufficiently
high in the vicinity
of the water surface
and bottom. There-
fore, it is seemed
that the efficiency
of the method stated
above has its limit.

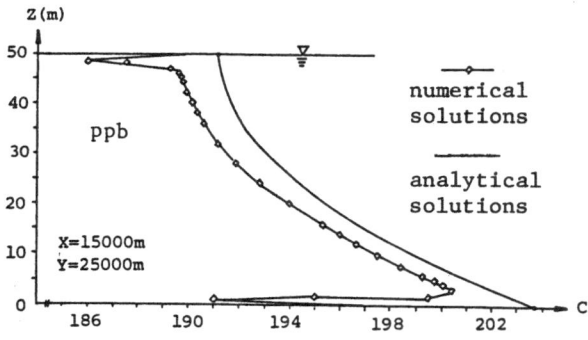

Fig.9 Detailed concentration distributions

The element number applied is equal to 830, and the Gauss Seidel's
method is used as the numerical solution of simultaneous linear
equations. Fig. 10 shows the concentration distribution on a
vertical section, Y=25 km. In this figure, the Gauss-Legendre's
100 points inte-
gration is used
in the integration
of C* and q_N*,
and the computed
values in the
vicinity of the
water surface and
bottom are so
incorrect as shown
in Fig. 9 that

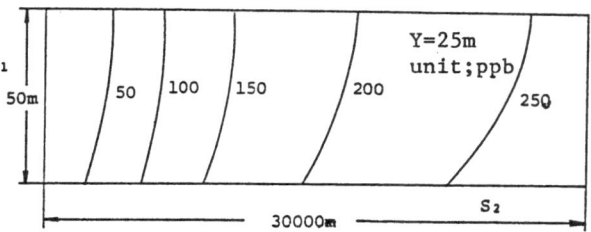

Fig.10 Vertical profiles of equiconcentration

these values are neglected. The results shown in Fig.10 may be
agree with the analytical solution in the accuracy of errors within
plus or minus 1 percent. Therefore, it is obvious that the BEM
is able to be applied to the hydro-mass transport problems in
such a vast flat plane model with a thin thickness.

3.3   Model of the Bay of Osaka (Model-C)
Fig. 11 shows the boundary element network in the   Bay of Osaka.
The governing equation is given by Eq.(1), and the boundary condi-
tions are assumed as follows,

$$C=4 \text{ ppb} \qquad\qquad \text{on } \Gamma_1$$
$$C=1 \text{ ppb} \qquad\qquad \text{on } \Gamma_1{}'$$
$$q_N=50 \text{ mg/m}^2\text{/ day} \qquad \text{on } \Gamma_2$$
$$q_N=0 \qquad \text{on the other boundary } \Gamma_2{}' \qquad\qquad (20)$$

Defining the maximum depth of the bay, hm, as a representative
length and the water density ρ , as a representative concentration
respectively, and introducing non-dimensional quantities as
follows,

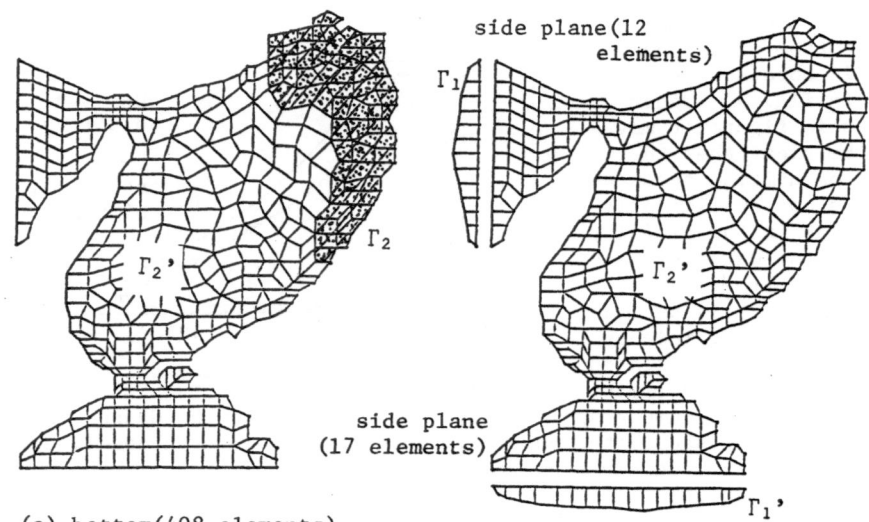

(a) bottom(408 elements)

(b) water surface(408 elements)

Fig.11 Boundary element network

$$x=\sqrt{\frac{D_Z}{D_X}}\frac{X}{hm}, \quad y=\sqrt{\frac{D_Z}{D_Y}}\frac{Y}{hm}, \quad z=\frac{Z}{hm}, \quad c=\frac{C}{\rho} \tag{21}$$

and substituting these quantities into Eqs. (1), (2) and (3) yield

$$\frac{\partial^2 c}{\partial x^2} + \frac{\partial^2 c}{\partial y^2} + \frac{\partial^2 c}{\partial z^2} = 0 \tag{22}$$

$$
\begin{aligned}
c=\bar{c}=4.0 \cdot 10^{-9} & \qquad \text{on } \Gamma_1 \\
c=\bar{c}=1.0 \cdot 10^{-9} & \qquad \text{on } \Gamma_1{}' \\
q_n=\bar{q}_n=4.652 \cdot 10^{-8} & \qquad \text{on } \Gamma_2 \\
q_n=\bar{q}_n=0 & \qquad \text{on } \Gamma_2{}'
\end{aligned}
\tag{23}
$$

In general, the boundary of a basin constitutes a three-dimensional curved surface. However, dividing the surface into many fine elements as shown in Fig. 11, it may be assumed that each of such fine elements constitutes a flat plane in the three-dimensional space respectively. Introducing two equations, $F(X,Y,Z)=0$ and $f(x,y,z)=0$, as the equations of the corresponding fine element planes on the old coordinate system (X,Y,Z) and on the new non-dimensional system (x,y,z) respectively, and using Eq.(21) yield

$$F(X,Y,Z)=hm \ f(x,y,z) \tag{24}$$

With the aid of Eq.(24), the flux q_N is related to the flux q_n as follows,

- ● point on element
- ○ G_1: G.C. of Δ124
- ○ G_2: G.C. of Δ234
- ○ G_3: G.C. of Δ134
- ◣ representative plane

Fig.12 Representative planes of elements

$$q_N = D_X \frac{\partial C}{\partial X} \frac{\partial F}{\partial X} + D_Y \frac{\partial C}{\partial Y} \frac{\partial F}{\partial Y} + D_Z \frac{\partial C}{\partial Z} \frac{\partial F}{\partial Z}$$

$$= \frac{D_2 \rho}{hm} \left(\frac{\partial c}{\partial x} \frac{\partial f}{\partial x} + \frac{\partial c}{\partial y} \frac{\partial f}{\partial y} + \frac{\partial c}{\partial z} \frac{\partial f}{\partial z} \right) = \frac{D_2 \rho}{hm} q_n \tag{25}$$

where, hm=93 m.
In this case, the quadrilateral constant elements of 845 are applied as the types of elements. Owing to calculate the value H_{ij}, it is necessary to evaluate the distance between the element plane and the point of application of the concentrated mass source. However, for the reason that the four points composing a element are not on one plane generally, it is necessary to introduce some techniques. It is here assumed that as shown in Fig.12, the representative plane of the element constituting of four points 1,2,3 and 4 may be given by the triangular plane $G_1 G_2 G_3$, where three points G_1, G_2 and G_3 are the center of gravity of Δ124, Δ234 and Δ134 respectively. Fig.13 shows the numerical results in this case. It may be recognized that the dissolvable pollutants eluted in to the basin through its bottom in the interior of the

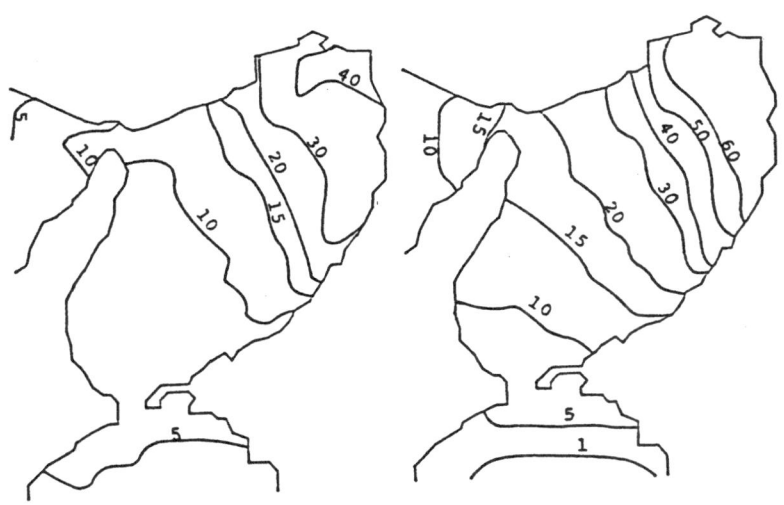

(a) bottom   (b) water surface
Fig.13 Equiconcentration distribution (unit: ppb)

Osaka Bay, may be removed toward the two open boundaries Γ_1 and Γ_1'. These results are however uncertain for the reason that the concentration values in the vicinity of the water surface are considerably greater than the values in the neighborhood of the basin bottom, and these facts have been not verified by the observation data in the Bay of Osaka as yet.

4. CONCLUSIONS

In order to predict the vertical diffusion mechanism of the dissolvable pollutants eluting into a basin through its bottom, it is essential that the problem are treated as a three-dimensional phenomenon. In this paper, the hydro-mass balance in the Bay of Osaka is analyzed as a three-dimensional diffusion problem in a homogeneous and an-isotropic turbulent flow basin without mean velocity. For this a few models are used, and the results obtained are as follows,

(1) Using the very long two-dimensional rectangular model (a = 60 km, h=50 m), the accuracy of the computed values are compared against the different size of the elements lengths L_x and L_z . According to the results, the computation accuracy of this problem is mainly dependent upon the size of L_x, and the order of the accuracy, $W=10^{-3}$mg/s is obtained for the values of L_x=1 km and L_z=25 m .

(2) Using the three-dimensional flat rectangular parallelepiped model (a_x=30km, a_y=50km, a_z=50m), and applying the constant quadrilateral elements of 830 and the element size of 2km x 2km, the solution coinciding with the analytical solution in the accuracy of errors within plus or minus 1 percent is obtained except for the points in the vicinity of the water surface and the bottom. It is therefore obvious that the BEM is able to be applied to the hydro-mass transport problems in such a vast flat plane model with the thin thickness.

(3) The governing equation is reduced to the non-dimensional form, and the diffusion problems in the Bay of Osaka are analyzed, using the constant quadrilateral elements of 845. The results obtained are reasonable on the whole. However, several problems concerning the numerical integration on the bottom elements of the Bay of Osaka are pending.

5. REFERENCES

(1) Matsunashi J., Nakago Y. and Yoshikawa M. (1983), Applications of Boundary Element Method to Hydro-Mass Transport Problems, the Construction Engineering Research Institute Foundation, No.25, pp.137-150.

(2) Brebbia C. A. (1978), The Boundary Element Method for Engineers, Pentech Press, London.

SECTION XV GROUNDWATER FLOW

Computation of the Fluid Flow in Zoned Anisotropic Porous Media and Determination of the Free Surface Seepage

E. Bruch, S. Grilli and A. Lejeune
Laboratories of Hydrodynamic, Applied Hydraulics and Hydraulic Constructions (L.H.C.H.), University of Liège, 4000 Liege, Belgium

1. INTRODUCTION

In this paper, we present new developments concerning the application of the Boundary Element Method, to the solution of the steady state two-dimensional flow in zoned anisotropic porous media, including the iterative determination of the free surface seepage position.

Since 1978, many authors have contributed to the solution of the seepage problem. In their paper, (Lu, Brebbia and Adey[1]) give a detailed description of the chronological evolution in this field. Particularly (Liggett and Liu[2]) present, in their book on boundary integral equations for porous media flow, their numerous contributions to the seepage problem.

In the paper, we first introduce the classical BEM equations for the porous media potential flow. We then use, for the discretization, linear, quadratic or cubic elements and therefore, we develop singular numerical integration methods, well suited to the considered singularities. After that, we improve the possibilities by including a division in sub-regions, having different permeabilities and subdividing, the boundaries and the domain into several parts. The potential gradient continuity is imposed on the interfaces between the regions and on the multiple nodes of their intersection, by special compatibility equations. After that, we check the method validity by a comparison with analytical and numerical (FEM) solutions. In each case, we study the discretization effect (number of nodes, degree of the elements) on the convergence of the iterative determination of the free surface position. At last, we present the real case of a zoned anisotropic dam.

2. APPLICATION OF THE BOUNDARY ELEMENT METHOD TO THE POTENTIAL FLOWS

2.1. Potential Theory Equations

The continuity equation gives, in two dimensions :

isotropic medium $\qquad\qquad \nabla^2 u = 0$ $\qquad\qquad$ on Ω $\quad$ (1)

anisotropic medium $\qquad k_x \dfrac{\partial^2 u}{\partial x^2} + k_y \dfrac{\partial^2 u}{\partial y^2}$ $\qquad$ on Ω $\quad$ (2)

k_x, k_y : the permeability coefficients in the directions x and y of
$\qquad\qquad$ orthotropy
u : the velocity potential

The equation (1) or (2), coupled with the appropriate boundary conditions (Dirichlet or Neuman) allows to resolve any particular potential problem.

In case of the BEM (Brebbia and Walker) have established :

$$\left[\begin{array}{l} c_i u_i + \displaystyle\int_{\Gamma_q} u q^*_i \ d\Gamma - \int_{\Gamma_u} q u^*_i \ d\Gamma = \int_{\Gamma_q} \bar{q} u^*_i d\Gamma - \int_{\Gamma_u} \bar{u} q^*_i \ d\Gamma \\[4mm] c_i = 1 \text{ on } \Omega \qquad\qquad \Gamma = \Gamma_u \cup \Gamma_q \\[2mm] c_i = 1/2 \text{ on } \Gamma \ \ (\text{smooth}), \qquad c_i = (1 - \dfrac{\omega}{2\pi}) \ \ \text{on } \Gamma \ (\text{non smooth}) \end{array}\right. \quad (3)$$

Γ_u, Γ_q : parts of the boundary Γ where u or q is imposed
ω : exterior angle of the tangents to the boundary at i

with, by determining the fundamental solution u^*_i of (2) :

$$\left[\begin{array}{l} u^*_i = -\dfrac{1}{2\pi\sqrt{k_x k_y}} \ \ln \left[\dfrac{(x-x_i)^2}{k_x} + \dfrac{(y-y_i)^2}{k_y} \right]^{\frac{1}{2}} \\[6mm] q^*_i = -\dfrac{1}{2\pi\sqrt{k_x k_y}} \ \dfrac{n_x(x-x_i) + n_y(y-y_i)}{\dfrac{(x-x_i)^2}{k_x} + \dfrac{(y-y_i)^2}{k_y}} = \dfrac{\partial u^*_i}{\partial n} \end{array}\right. \qquad (4)$$

(n_x, n_y) : the normal outwards unitary vector at the point (x,y) of Γ

$q = k_x \dfrac{\partial u}{\partial x} n_x + k_y \dfrac{\partial u}{\partial y} n_y$: the normal velocity or potential gradient on Γ

2.2. Discretization, Numerical Integration

We introduce a discretization of the boundary using linear, quadratic or cubic elements, for the potential and for the potential gradient. Therefore, to compute the terms of (3), we need to develop singular numerical integration methods in order to evaluate the integrals on the elements containing a singularity of order ln r_i or $1/r_i$ (see (4)) at the point i, when the distance r_i to i tends to zero. For the other cases, we use the classical regular Gauss formula.

2.2.1. <u>Integrals containing $q^*{}_i$</u>. We pose :

$$K^e_{qij} = \int_{\Gamma_{qe}} N_j \, q^*{}_i \, d\Gamma, \qquad N_j = N_j(\xi) = P_{n-1}(\xi)$$

n : number of nodes of the considered element
$P_{n-1}(\xi)$: polynomial of degree n-1 in ξ
N_j : shape function corresponding to the node j   (5)

After changing of variable from the considered element e, to the reference element (coordinate ξ) and by expressing the terms of (5) analytically in function of ξ, we can integrate by parts and we obtain (Grilli[4], Bruch[5]) :

$$K^e_{qij} = \begin{cases} -\dfrac{\alpha}{2\pi}\,[\text{arctg } v(-1) + I_j] & \text{if } j = 1 \\[2ex] -\dfrac{\alpha}{2\pi}\, I_j & \text{if } 1 < j < n \\[2ex] \dfrac{\alpha}{2\pi}\,[\text{arctg } v(+1) - I_j] & \text{if } j = n \end{cases}$$

with $\quad I_j = \displaystyle\int_{-1}^{1} \dfrac{dN_j(\xi)}{d\xi} \text{ arctg } v(\xi)\, d\xi$

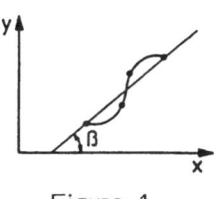

for $0 \leqslant |\beta| \leqslant \dfrac{\pi}{4}$ $\quad \begin{bmatrix} \alpha = 1 \\[1ex] v(\xi) = \sqrt{\dfrac{k_x}{k_y}}\, \dfrac{y(\xi)-y_i}{x(\xi)-x_i} \end{bmatrix}$

Figure 1

for $\dfrac{\pi}{4} \leqslant |\beta| \leqslant \dfrac{\pi}{2}$ $\quad \begin{bmatrix} \alpha = -1 \\[1ex] v(\xi) = \sqrt{\dfrac{k_y}{k_x}}\, \dfrac{x(\xi)-x_i}{y(\xi)-y_i} \end{bmatrix}$ (6)

In (6), we have :

$$x(\xi) = \sum_{l=1}^{n} N_l(\xi)x_l, \qquad y(\xi) = \sum_{l=1}^{n} N_l(\xi)y_l$$

(x_l, y_l) : the coordinates of the n element nodes.

Thus, to compute $v(\xi)$ for $\xi = -1$ or $+1$ we obtain, for the numerator and the denominator, terms like (y_1-y_i) or (y_n-y_i). Therefore, for i = 1 or n, we need to apply the l'Hospital theorem which performs the derivation of the numerator and the denominator.

2.2.2. <u>Integrals Containing $u^*{}_i$</u>. We pose :

$$K^e_{uij} = \int_{\Gamma ue} N_j \, u^*_i \, d\Gamma \tag{7}$$

In this case, we apply the classical technique of the kernel trans-formation, generalized for the cubic elements, which subdivides (7) into a regular part (first, second and fourth integrals of (8)) for which we apply the Gauss formula and a singular part where the logarithmic singularity is exactly integrated using the Berthod-Zabo-rowisky formula :

$$\begin{aligned}
\Bigg[K^e_{uij} = - \frac{1}{2\pi\sqrt{k_x k_y}} \Bigg[& \int_{-1}^{1} \ln \frac{2r_i(\xi)}{|\xi - \xi_i|} \, N_j(\xi) \, J(\xi) \, d\xi \\
& + \frac{1+\xi_i}{2} \ln \frac{1+\xi_i}{2} \int_{-1}^{1} N_j(\eta_1) \, J(\eta_1) \, d\eta_1 \\
& - (1+\xi_i) \int_{0}^{1} \ln \frac{1}{\eta_2} N_j(\eta_2) \, J(\eta_2) \, d\eta_2 \\
& + \frac{1-\xi_i}{2} \ln \frac{1-\xi_i}{2} \int_{-1}^{1} N_j(\eta_3) \, J(\eta_3) \, d\eta_3 \\
& - (1-\xi_i) \int_{0}^{1} \ln \frac{1}{\eta_4} N_j(\eta_4) \, J(\eta_4) \, d\eta_4 \Bigg]
\end{aligned} \tag{8}$$

with

$$r_i(\xi) = \left[\frac{(x(\xi)-x_i)^2}{k_x} + \frac{(y(\xi)-y_i)^2}{k_y} \right]^{1/2} ; \quad J(\xi) = \frac{ds(\xi)}{d\xi}$$

the jacobian

$$\eta_1 = \frac{\xi_i - 2\xi - 1}{\xi_i + 1} , \quad \eta_2 = \frac{\xi_i - \xi}{\xi_i + 1} , \quad \eta_3 = \frac{\xi_i - 2\xi + 1}{\xi_i - 1} , \quad \eta_4 = \frac{\xi_i - \xi}{\xi_i - 1}$$

2.3. Matric Equations

Introducing the discretization in (3), we directly obtain :

$$[(\mathrm{diag}\, c + \underline{\underline{K}}_q)\underline{u} - \underline{\underline{K}}_u \, \underline{q}] = \underline{P}_q - \underline{P}_u$$

$$\Bigg[\begin{aligned}
& K_{qij} = \sum_{e=1}^{Ne} K^e_{qij}, \quad K_{uij} = \sum_{e=1}^{Ne} K^e_{uij} \\
& P_{ui} = \sum_{e=1}^{Ne} \int_{\Gamma ue} \bar{u} \, q^*_i \, d\Gamma , \quad P_{qi} = \sum_{e=1}^{Ne} \int_{\Gamma qe} \bar{q} \, u^*_i \, d\Gamma
\end{aligned} \tag{9}$$

Ne : Number of boundary elements.

By solving the linear system (9), we obtain the unknown values of u or q, depending on the case, on Γ. Then, we explicitly can compute the potential and its derivaties, by application and derivation of (3).

2.4. Division in sub-regions

This technique is very usefull in two cases. The first one is the case of a zoned domain, i.e. composed by several regions having different permeabilities. The second one is for purely numerical purposes. Indeed, if the domain geometry is quite complicated or narrowed in some parts, it will be more accurate to use several sub-regions (see the example of § 4.1.2.). Besides, we will obtain, in each case a system matrix containing null blocks which could be taken into account to optimize the system resolution.

Using sub-regions, we impose the potential and normal gradient continuity on the region interfaces (figure 2) :

$$u_A = u_B \quad \text{and} \quad q_A = - q_B \tag{10}$$

However, on the multiple intersection nodes between the regions and between their interfaces and the exterior boundary, special problems will occur, and will require special compatibility equations (see § 3.).

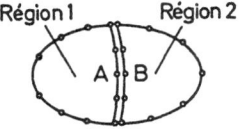

Figure 2.

2.5. Iterative Determination of the Free Surface Position

The phreatic free surface position (AB, see figure 3) in the porous medium is one of the problem unknowns. On the same way, the seepage surface (BC, see figure 3) is determined by the point B of the free surface. This last is a stream line, and is at the constant atmospheric pression. Hence, the boundary conditions are :

$$\frac{\partial u}{\partial n} = 0 \quad \text{and} \quad u = y \ , \quad \text{on AB} \tag{11}$$

Therefore, we use an iterative procedure, starting from a given linear or polygonal, approximation of AB and solving (9) for the first condition (11). After each iteration, AB is modified by (12) in order to satisfy the second condition (11), until the convergence is reached, for a given relative precision ε_o, using the criterion (13) :

$$y_i^{k+1} = \theta u_i^{k+1} + (1-\theta) y_i^k \ ; \qquad \begin{bmatrix} i = 1, \ldots, n_f \\ \theta \in]0, 1[\end{bmatrix} \tag{12}$$

$$\text{Max}_i (\varepsilon) = \text{Max}_i \ |[y_i^{k+1} - y_i^k] / y_i^{k+1} | \leqslant \varepsilon_o \tag{13}$$

Θ : a coefficient improving the procedure convergence
k : the number of iterations
y_i : the y coordinate of the free surface nodes
n_f^i : the number of nodes of AB

However, to determinate the new position of B, we perform the intersection between a straight line, built on the two previous nodes of AB, and the seepage surface. Doing so, we found a better convergence of the procedure.

Other seepage surfaces can occur inside the domain. For instance, if we consider a zoned dam (figure 4) for which the kernel (zone 2) has a very low permeability compared to the other regions (zone 1) and if we allow the possibility of obtaining a seeping on RS, we find the results shown on the figure 4. In the computation, the seeping flow on RS is introduced as a hydraulic head for the element ST (see the example of the § 4.2.).

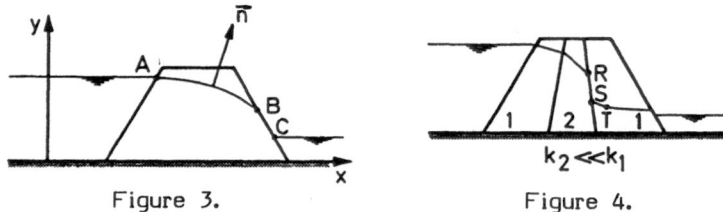

Figure 3.   Figure 4.

3. GENERAL COMPATIBILITY EQUATIONS ON MULTIPLE NODES

3.1. Introduction
On the multiple intersection nodes between the regions and between their interfaces and the exterior boundary, the potential values, obtained after resolution, are in good concordance. However, the normal gradients (velocities) are in bad concordance, mainly near those nodes. Indeed, the normal and tangential velocities are not independant on the interfaces, but linked by some relations.

(Lafe, Montes, Cheng, Liggett and Liu[6]) proposed a first solution to this problem by using the analytical solution of the Laplace equation (1), in polar coordinates (for regions having the same permeability), in the nearness of the multiple intersection node. However it is necessary to compute the smallest positive eigenvalue of a linear system, and this method becomes quickly useless.

As we already said before, we developed special compatibility equations, taking into account the relations between the velocity vectors, on the interface intersection nodes. We will first present the case of an interior multiple node and, afterwards, the one of an intersection node between interfaces and the exterior boundary.

3.2. Interior Multiple Intersection Node
We consider several sub-regions intersecting at the point O (figure 5), and say t and q, the tangential and normal interface velocities at O. For the angle α of the region A, at O, we draw the bisectrix OR, and the perpendicular line OS to this one (figure 6). We

directly obtain :

$$\left[\begin{array}{ll} Q_1 = t_1 \cos \frac{\alpha}{2} + q_1 \sin \frac{\alpha}{2}, & Q_2 = -t_2 \cos \frac{\alpha}{2} + q_2 \sin \frac{\alpha}{2} \\ T_1 = t_1 \sin \frac{\alpha}{2} - q_1 \cos \frac{\alpha}{2}, & T_2 = t_2 \sin \frac{\alpha}{2} + q_2 \cos \frac{\alpha}{2} \end{array}\right. \quad (14)$$

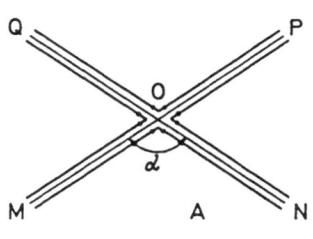

Figure 5.   Figure 6.

Now as, at O, we have a unique velocity, we have :

$$Q = Q_1 = Q_2 \text{ and } T = T_1 = T_2$$

Hence, we have, by (14)

$$2Q \sin \frac{\alpha}{2} = q_1 + q_2, \quad 2T \cos \frac{\alpha}{2} = q_2 - q_1 \quad (15)$$

If we express relations such (15) for the N regions intersecting at O, we obtain, by summation, the general compatibility relations between all the velocity vectors at O :

$$\left[\begin{array}{l} \dfrac{1}{2} \sum_{r=1}^{N} (t_2^r - t_1^r) \sin \alpha_r = \sum_{r=1}^{N} (q_1^r + q_2^r) \sin^2 \dfrac{\alpha_r}{2} \\[2ex] \dfrac{1}{2} \sum_{r=1}^{N} (t_1^r + t_2^r) \sin \alpha_r = \sum_{r=1}^{N} (q_2^r - q_1^r) \sin^2 \dfrac{\alpha_r}{2} \end{array}\right. \quad (16)$$

In the particular case of only 2 regions intersecting at O, the first equation (16) vanishes and we only use the second equation.

3.3. Multiple Intersection Node between one Interface and a Non Smooth Exterior Boundary

In this case, we must distinguish between interfaces intersecting the Dirichlet (Γu) or Neuman (Γq) boundary, at a point O where it is not smooth.

3.3.1. Neuman condition. At O, the velocity vector should instanta-neously change of direction. However, in the reality, such singula-rities does not occur, and the change in velocity always is gradual. Therefore we suppose, in case of a homogeneous Neuman conditions

($\bar{q}=0$), that the velocity vector is in the direction PP' at O (figure 7.) with :

$$\gamma = \frac{\alpha+\beta-\pi}{2} \ , \quad \lambda_A = \frac{\alpha-\beta}{2} \ , \quad \lambda_B = \frac{\alpha-\beta}{2} \tag{17}$$

We then express the following relation between the velocity vectors at O (figure 8).

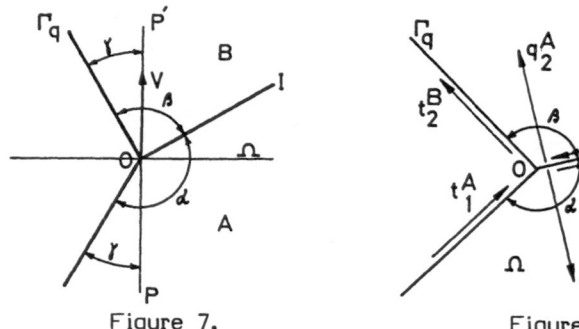

Figure 7. Figure 8.

$$\frac{1}{\cos \gamma} (t_1^A + t_2^B) - (t_2^A \sin \lambda_A + t_1^B \sin \lambda_B) = q_2^A \cos \lambda_A - q_1^B \cos \lambda_B \tag{18}$$

3.3.2. Dirichlet condition. Doing on the same way, we express that at O, the velocity vector V is acting on the direction perpendicular to PP' (figure 9) in case of a homogeneous Dirichlet condition ($\bar{u}=0$) i.e. for all the Dirichlet boundaries, except the seepage surface. We again obtain, with (17), the following relation (figure 10) :

$$t_2^A \cos \lambda_A - t_1^B \cos \lambda_B = q_1^B \sin \lambda_B + q_2^A \sin \lambda_A - \frac{1}{\cos \gamma} (q_2^B + q_1^A) \tag{19}$$

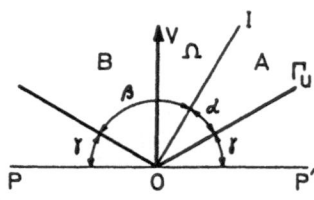 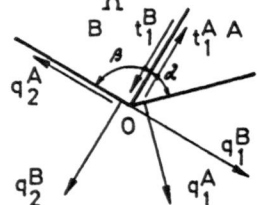

Figure 9. Figure 10.

3.4. Introduction of the Compatibility Equations in the Linear System.

If we add equations like (16), (18) or (19) to the linear system (9), it becomes overdeterminate which would necessitate its solving by methods such that the least square method. Therefore, in each case we eliminate a number, of multiple node equations, equal to the number of compatibility equations we add at this node. For instance, in case of the § 3.3.2., 4 nodes are located at O and give 4 equations, whose one is suppressed, to add (19).

3.5. Computational Examples for the Compatibility Equations Checking.

3.5.1. <u>Solution compared to an analytic one.</u> The problem is described by the figure 11. Its analytic solution is given by the following relation.

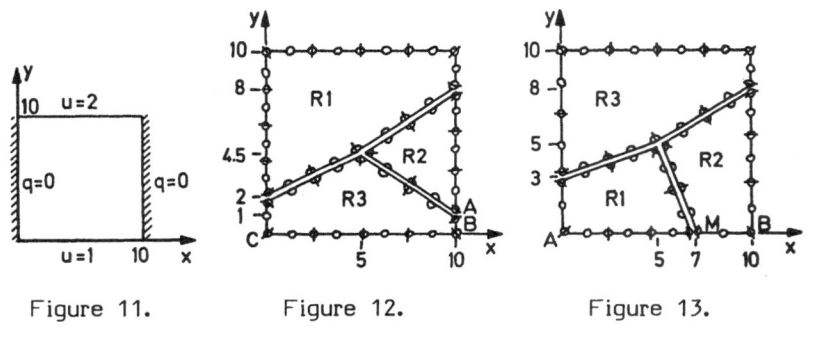

Figure 11.   Figure 12.   Figure 13.

$$u = y/10 + 1 \qquad\qquad (20)$$

To check the sub-region division and the multiple node treatment, we divide the square domain in 3 sub-regions (R1, R2, R3, see figure 12) using quadratic 3 nodes elements, except between A and B which gives 70 nodes.

After resolution, without using the compatibility equations, we obtain for the potential, deviations of 2.7 % to the analytic solution. But for the normal gradient the results are very bad, especially along the interfaces (see table 1, case A). Now, if we use the compatibility equations (16) at the point P the potential values and the normal gradient along the interfaces are in concordance with (20) at 0.1 % (see table 1, case B).

Without the compatibility equations the total C.P.U. time was of 10.35 s, and with them, of 10.43 s. Thus, the difference is neglictable.

After that, we tested a second sub-region division (figure 13) using quadratic 3 nodes elements, for a total of 65 nodes.
In a first computation, we directly introduce the compatibility equations at P. We obtain quite good results, except for the gradient on PM (see table 2, case A).

However, after introducing the compatibility equation (19) at M, we obtain results of the same accuracy than in the previous example of the figure 12 (see table 2, case B).

Gradient along BC			Gradient along AP		
x	case A	case B	x	case A	case B
0	- 1.461	- 1.000	10.00	0.4444	- 0.8192
1.25	- 0.907	- 1.000	8.75	- 3.110	- 0.8192
2.50	- 1.002	- 1.000	7.50	- 0.6971	- 0.8192
3.75	- 1.409	- 1.000	6.25	- 4.907	- 0.8192
5.00	- 1.101	- 1.000	5.00	15.86	- 0.8192
6.25	- 0.787	- 1.000			
7.50	- 1.048	- 1.000			
8.75	- 1.066	- 1.000			
10.00	- 0.862	- 1.000			

Table 1.

Gradient along AB			Gradient along PM		
x	case B	case A	x	case B	case A
0	- 1.000	- 1.010	5.00	0.371	0.335
1.17	- 1.000	- 1.000	5.50	0.371	0.351
2.33	- 1.000	- 0.975	6.00	0.371	0.369
3.50	- 1.000	- 0.973	6.50	0.371	0.532
4.67	- 1.000	- 0.993	7.00	0.371	-0.264
5.83	- 1.000	- 1.190			
7.00	- 1.000	- 0.249)$_M$			
7.00	- 1.000	- 1.610)			
8.50	- 1.000	- 0.896			
10.00	- 1.000	- 0.934			

Table 2.

3.5.2. Example with sub-regions of different permeabilities. We now present the example of the figure 12, but with the following permeabilities :

$$k_1 = 1, \qquad k_2 = 0.1, \qquad k_3 = 10$$

We consider two cases. In the first one, we do not use any compatibility equation (see table 3, case A). In the second one, we use compatibility equations at the 4 nodes A, D, E and P (see table 3, case B). The previous conclusions about the potential values are again valid, and for the gradients, we see, in the comparison of the table 3, that the results are clearly improved in the case B. However, in this case, we have no analytic solution to exactly check the results.

Gradient along DP			Gradient along AP			Gradient along EP		
x	case A	case B	x	case A	case B	x	case A	case B
0	0.0116	0.3263	5.00	-0.1288	0.0186	5.00	-0.1436	0.0405
1.25	0.3998	0.2968	6.25	0.0646	0.0262	6.25	0.0758	0.0274
2.50	0.3586	0.3498	7.50	0.0183	0.0197	7.50	0.0219	0.0212
3.75	0.2586	0.3013	8.75	0.0066	0.0133	8.75	0.0001	0.0050
5.00	0.1624	0.0818	10.00	0.0143	0.0184	10.00	0.0064	0.0205

Table 3.

4. COMPUTATIONAL EXAMPLES

4.1. Comparison with the Finite Element Method

4.1.1. The dam case. We consider the simple case of the figure 14
for which we have the FEM results computed by Cheng and Li[7].
They use 702 elements and 421 nodes. We compare the BEM results to their solution, for several discretizations and relative pre-
cisions. As the seeping surface is quite small compared to the dam
dimensions, we use two sub-regions, to have a better accuracy. The
seepage free surface is discretized, using 2 nodes linear elements.
Indeed, with a higher degree, we obtain perturbating oscillations.
On the other boundaries we use high order elements (quadratic and
cubic). We found they improve the method convergence and accuracy.

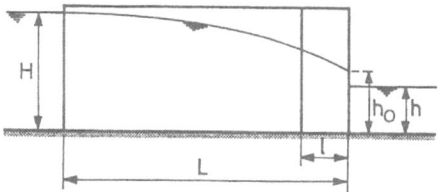

Figure 14.

L = 20 m   l = 4 m
H = 10 m   h = 4 m
k = 0.01 m/s

NDE : Number of nodes
NLE : Number of elements
NIT : Number of iterations
T : CPU time (IBM 4341)
Q : Seepage discharge

NDE	NEL	θ	ε_o	NIT	T(s)	h_o(m)	Q (m²/s/m)
52	37	0.5	0.01	5	25.20	3.32	0.02259
52	37	0.5	0.001	8	35.22	3.34	0.02277
59	44	0.5	0.01	6	37.57	3.19	0.02260
59	44	0.5	0.001	8	44.38	3.23	0.02271
73	56	0.5	0.01	6	55.99	3.21	0.02260
73	56	0.5	0.001	10	84.83	3.21	0.02270

Table 4.

Cheng and Li[7] give for h_o, a value of 3.32 m. Polubarinov-Kochine[9]
give 3.23 m. All the results of the table 4 are in this field. The
seepage discharge can be easily computed by the formula :

$$Q = k \ (H^2 - h^2) \ / \ 2L \tag{21}$$

Which gives in this case $Q = 0.02275$ m³/s/m. Therefore all the presented results lead to a very good accuracy, for a low computation time. The figure 15 shows the discretization and the free surface position, in the last case of the table 4.

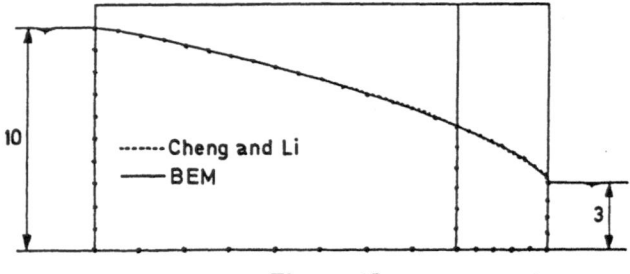

Figure 15.

4.1.2. The sheet-pile wall case. We now consider the case of the figure 16, which has been studied by Bottiau and Roose at the Free University of Brussels (Belgium), and which is still not published. They computed the free surface position for different holds of a sheet-pile wall, tending to the depth H_2 of the impervious medium.

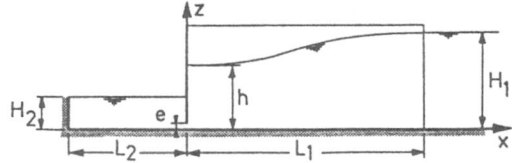

$H_1 = 15$ m, $H_2 = 5$ m, $L_1 = 100$ m, $L_2 = 50$ m.

Figure 16.

They used a very refined finite element mesh but we obtain, quite exactly, the same results using a very low number of boundary elements. We computed two cases with two different discretizations Their comparison is presented at the table 5. The two softwares were implemented on a IBM 4341.

In our computations, the domain is divided into two sub-regions by the sheet-pile wall. However, in the case B (133 nodes), we use 4 sub-regions (figure 17), in order to improve the results accuracy in the bottom area (see table 5).

Thus, the previous comparison shows a decrease in the computation time of a factor 2 to 6, depending on the case, for the same results accuracy, by using the Boundary Element Method. Moreover, the amount of data to introduce is highly reduced in our case, and their automatic generation is easily performed.

A. $e = 2m$, $\varepsilon_o = 0.001$, $\theta = 0.5$

B. $e = 1\ cm$, $\varepsilon_o = 0.001$, $\theta = 0.5$

A.	BEM		FEM	B.	BEM		FEM
NDE	72	97	3225	NDE	119	133	3241
NEL	52	66	771	NEL	96	91	771
NIT	7	7	6	NIT	7	7	5
T(s)	57.72	99.28	317.63	T(s)	119.81	146.84	293.54
h	7.08	7.04	6.92	h	10.55	10.47	10.50
Q	0.895k	0.903k	0.912k	Q	0.595k	0.601k	0.603k

Table 5

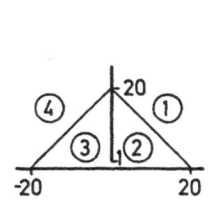

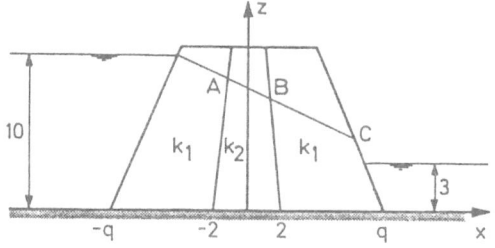

Figure 17.   Figure 18.   (m)

4.2. The Real Zoned Dam Case

We now consider the dam of the figure 18, already studied by Liu and Liggett[8] for the case $k_1 = k_2$. We computed the results of the table 6, for different decreasing permeabilities k_2 ($k_1 = 10^{-3}$ m/s, $\alpha = k_2/k_1$).

α	ε_o	θ	NDE	NEL	NIT	T	H_A	H_{B1}	H_{B2}	H_C	Q
1	10^{-3}	0.5	91	52	9	78.22	8.79	7.70	7.70	4.61	$3.31\ 10^{-3}$
0.5	10^{-3}	0.5	85	46	12	91.43	9.12	7.44	7.44	4.18	$2.67\ 10^{-3}$
0.1	10^{-3}	0.5	100	51	10	97.66	9.71	7.90	5.48	3.11	$1.05\ 10^{-3}$
0.05	10^{-3}	0.5	97	50	10	92.74	9.84	8.02	4.54	3.00	$5.98\ 10^{-4}$
0.01	4.10^{-3}	0.5	97	50	20	182.12	9.97	8.11	3.37	3.00	$1.34\ 10^{-4}$

Table 6.

The figure 19 shows the free surface discretization and position. The concordance is very good with the Liu and Liggett's results for $\alpha = 1$. If the kernel permeability becomes quite low, a seeping surface occurs inside the dam, like in its downstream part.

This phenomenon can be physically explained by the following observations. Due to the permeability k_2 lower than k_1, the free
surface level must go up in the upstream region. Then, in the kernel, its level can not decrease abruptly because of the small k_2.
At last, for a low discharge Q, the level must go down in the
downstream part of the dam. Therefore, we implemented in our
software, the possibility of the occurence of interior seeping surfaces.

The last line of the table 6 shows that for a small α, the convergence becomes very slow and the total CPU time increases a lot.

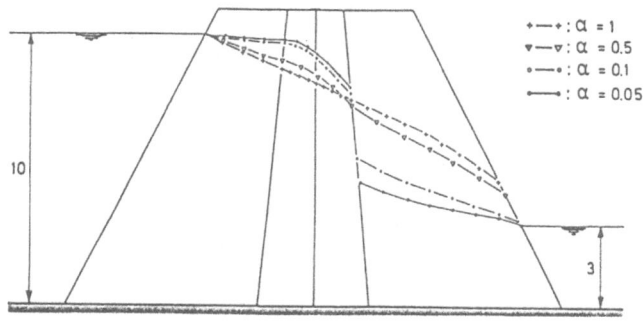

Figure 19.

5. CONCLUSIONS

We thus developed a very powerfull tool (by comparison with the
other numerical methods) allowing to resolve any type of bidimen-
sional flow in porous medium. To optimize its use, we implemented an automatic data generation, giving a great ease to resolve
industrial applications. We are now working on the method genera-
lization to the three-dimensional and unsteady flows, and we hope
that the application of the techniques described in this paper will
lead, in those cases, to a good behaviour of our algorithms.

6. REFERENCES

1. Lu, Z.K., Brebbia, C.A. and Adey, R.A. (1985). Calculation of
   Free Surface Seepage through Zoned Anisotropic Dams. Procee-
   dings of the 7th Boundary Element Conference, Villa Olmo,
   Springer-Verlag. Berlin and New-York.

2. Liggett, J.A. and Liu, P.L.F. (1983). The Boundary Integral
   Equation Method for Porous Media Flow. George Allen and Unwin. London.

3. Brebbia, C.A. and Walker, M.A. (1980). Boundary Element Techniques in Engineering. Newnes-Butterworths London-Boston.

4. Grilli, S. (1985). Etude expérimentale et numérique du comportement hydrodynamique de grandes portes flottantes automotrices pour écluses et barrières marées-tempêtes. Thèse de doctorat, Université de Liège. Belgique.

5. Bruch, E. (1985). Résolution par éléments frontières des écoulements permanents en milieu poreux, à surface libre éventuellement indéterminée. Travail de fin d'études, Université de Liège Belgique.

6. Lafe, O.E., Montes, J.S., Cheng, A.H.D., Liggett, J.A. and Liu, P.L-F. (1980). Singularities in Darcy Flow through Porous Media J. Hyd. Div., ASCE 106 (HYG), 977-97.

7. Cheng, R.T. and Li, C. (1973). On the Solution of Transient Free-Surface Flow Problems by the Finite Element Method. J. Hydrol., 20, 49-63.

8. Liu, P. L-F. and Liggett, J.A. (1979). Boundary Solutions to two Problems in Porous Media. J. Hyd. Div., A.S.C.E. 105 (HY3), 171-83.

9. Polubarinov-Kochine P.Y. (1962). Theory of Groundwater Movement. Translated from the Russian by J.M. Roger de Wiest. Princeton University Press.

EXTRA PAPER

The Calculation of Transient Eddy Currents by Means of the Boundary Element Method

A. Krawczyk
Department of Fundamental Research, Electrotechnical Institute, Warsaw, Poland

INTRODUCTION

The computation of transient eddy currents, very important in designing of electrical devices is very difficult when using the so called "domain me- thods", like finite element or difference method (Krawczyk[5]). Therefore, an effort was made to find a new approach and it is just the boundary element method BEM which seems to be far more appropriate to this problem. The eddy current problem can be described in various terms and in this paper the magnetic vector potential was employed.

Let us consider the conducting region V, in which the source current of density 6_s excited by an external source varies non-sinusoidally in time. Then, the process is described as follows

$$\nabla^2 A = \mu \delta \frac{\partial A}{\partial t} - \mu 6_s (\underline{x}, t) \qquad \forall \underline{x} \in V, t > 0 \qquad (1)$$

and

$$\left. \begin{array}{l} A (\underline{x}, t) = \overline{A} (\underline{x}, t) \qquad \forall \underline{x} \in S_1 \\[2mm] B_t (\underline{x}, t) = \frac{\partial A}{\partial n} = \overline{B}_t (\underline{x}, t) \qquad \forall \underline{x} \in S_2 \end{array} \right\} t > 0 \qquad (2)$$

$$A (\underline{x}, 0) = A_0 (\underline{x}) \qquad \forall \underline{x} \in V \qquad (3)$$

where

A - the MVP
B_t - tangential component of magnetic flux density
σ_s - the density of source current
$S_1 \cup S_2 = S$ - the boundary of V
μ - magnetic permeability
γ - electric conductivity

The induced eddy currents are determined with the well-known formula

$$\sigma_e = - \gamma \frac{\partial A}{\partial t} \tag{4}$$

and then the total current density is as follows

$$\sigma = \sigma_e + \sigma_s \tag{5}$$

The formula (5) lacks the component arising due to electrical field, but it can be so, because this component is eliminated in the one-conductor problem. The problem is considered as two-dimensional, so that the vector $\underline{x}$ has two components x_1 and x_2. We also assume the material parameters to be constant. The boundary condition result from the physical properties of the medium and its neighbourhood. The initial condition is implied by the electromagnetic process in the object of interest just before transient phenomenon has begun.

BOUNDARY INTEGRAL APPROACH

To obtain the boundary integral formulation of the described differential problem one can use a few ways, e.g. the Green's formulae, the integration by parts or the weighed residuals method.
All details are well-described in numerous papers and books   e.g.(Brebbia[1], John[2], Krawczyk[4]).
Thus this formulation is given without derivation and it has the following form

$$\int\limits_0^T\int\limits_S A(\underline{x},t)\frac{\partial \vartheta}{\partial n}\, dSdt \; - \int\limits_0^T\int\limits_S B_t(\underline{x},t)\vartheta dSdt \; + \; c_i\alpha^2 A(\underline{x}_i,T) =$$

$$\alpha^2\int\limits_V A_0(\underline{x})\,\vartheta\,(t=0)\;dV - \mu\int\limits_0^T\int\limits_V \sigma_s(\underline{x},t)\,\vartheta\, dVdt \qquad (6)$$

where

T is the time under consideration

$$c_i = \begin{cases} 1 & \underline{x}_i \in V \\ .5 & \underline{x}_i \in S \\ 0 & \underline{x}_i \notin \overline{V} \end{cases}$$

$$\alpha^2 = \mu\tau$$

The function ϑ is the fundamental solution and for 2-D space the function itself and its normal derivative take the form of

$$\vartheta = \frac{\alpha^2}{4\pi(T-t)}\, \exp\left(-\alpha^2\,r^2\,/\,4\,(T-t)\right) \qquad (7)$$

$$\frac{\partial \vartheta}{\partial n} = \frac{\alpha^4\,r}{8\pi\,(T-t)^2}\, \exp\left(-\alpha^2\,r^2\,/\,4(T-t)\right)\frac{\partial r}{\partial n} \qquad (8)$$

$r = |\underline{x} - \underline{x}_i|$ — the distance between the observation and source points.

Approximating the functions B_t and A by the stair-case functions in time and changing order of integration one obtains the following boundary integral equation (Krawczyk[4])

$$\sum\limits_{m=1}^M \int\limits_S A(\underline{x},\,t_m)\,\beta\, dS \; + \; \int\limits_S B_t\,(\underline{x},t_m)\,\varphi\, dS \; +$$

$$+ c_i \alpha^2 A (\underline{x}_i, T) = \int_V A_0(\underline{x}) \, \mathcal{V} \, (t=0) dS -$$

$$-\mu \int_0^T \int_S \, \xi_s \, (\underline{x}, t) \, \mathcal{V} \, dSdt \tag{9}$$

The functions β and φ are analytically established time integrals and have the following forms

$$\beta(\underline{x}_i; \underline{x}, T; t_m; t_{m-1}) = - \frac{\alpha^2}{2\pi r} [\exp(-w_{m-1}) -$$

$$\exp(-w_m)] \frac{\partial r}{\partial n} \tag{10}$$

$$\varphi(\underline{x}_i; \underline{x}, T; t_m; t_{m-1}) = \frac{\alpha^2}{4\pi} [E_1(w_{m-1}) - E_1(w_m)] \tag{11}$$

$$w_m = \frac{\alpha^2 r^2}{4(T-t_m)} \quad , \quad E_1(w_m) = \int_{w_m}^\infty \frac{e^{-z}}{z} \, dz$$

$(E_1$ - exponent integral$)$

The equation (11) is the starting point for further
numerical implementation based on the BEM.

BOUNDARY ELEMENT TECHNIQUE

Since the boundary element technique in respect to
the spatial variables is well-described in many
books and papers (Brebbia[1]) we do not feel obliged
to repeat it here. Thus we shall focus our attention on the time problems. Taking the straight line
elements and the constant approximation on each of
them, one obtains the following matrix equation

$$\sum_{m=1}^{M} [H_{mM}] \{A_m\} = \sum_{m=1}^{M} [G_{mM}]\{B_{tm}\} + \{F_M\} \qquad (12)$$

The coefficients of the matrices are as follows

$$h_{mM}^{ij} = \int_{e_j} \beta (\underline{x}_i, \underline{x}_j, T, t_m, t_{m-1}) \, dS + c_i \alpha^2 \delta_{ij} \, \delta_{mM} \qquad (13)$$

$$g_{mM}^{ij} = \int_{e_j} \varphi (\underline{x}_i, \underline{x}_j, T, t_m, t_{m-1}) \, dS \qquad (14)$$

The vector $\{F_M\}$ contains both the integral of initial condition and the integral of source cur‑ rent, so its "i"-th component has the form of

$$f_{Mi} = \int_{V} A_o (\underline{x}) \vartheta (\underline{x}_i, \underline{x}, T, 0) \, dV -$$

$$\mu \int_{0}^{T} \int_{V} \sigma_s (\underline{x}, t) \vartheta (\underline{x}_i, x, T, t) \, dV dt \qquad (15)$$

Of course, the above matrix equation must be re-written because only parts of the vectors $\{A\}$ and $\{B_t\}$ are to be found, while the remaining parts are known. There are two possibilities of the ana-lysis of the time. One of them is the so called "step by step method" and the second one was named the collection method (Krawczyk[4]) . In the author's programme the latter one has been employed but it was done without any suggestion as to its quality. Both methods have advantages and disadvantages and the choice of one them depends finally on the ana-lyst's needs and possibilities. The other problem appearing in the time analysis is how to calculate

the integrals of a component of the vector $\{F_M\}$
(Equation 15). The first integral containing the
initial condition is, in many electromagnetic cases
equal to zero since the source current before the
transient process is equal to zero. But, if it is
not so, one has to solve the stationary problem to
find the function A_o. It is the solution of the
Poisson's equation. And, then, approximating somehow the function A_o one can calculate the integral.
In the author's programme the whole region of interest is divided into N_r elementary rectangles
$Q_r = (x_{II}, x_{12}) \times (x_{21}, x_{22})$ and on each of them
the constant value of A_{og} is prescribed. It enables
getting the results of integration in an analytical
form:

$$I_1 (\underline{x}_i, T) = \sum_{q=1}^{N_r} A_{og} \left\{ \left[erf\left(\frac{\alpha (x_{11}-x_{1i})}{2\sqrt{T}} \right) - \right. \right.$$

$$\left. erf\left(\frac{\alpha (x_{12}-x_{1i})}{2\sqrt{T}}\right)\right] \left[erf\left(\frac{\alpha (x_{21}-x_{2i})}{2\sqrt{T}}\right) - erf\left(\frac{\alpha(x_{22}-x_{2i})}{2\sqrt{T}}\right)\right]\right\}$$

(16)

The same way can be used in the source current integral. But it is only half of work because the
time integral remains. However this integral can be
relatively easily calculated numerically, as the
one-dimensional quadrature can be, then, applied.
We used here the six-point Gaussian quadrature at
each time interval (t_{m-1}, t_m). It is obvious that
if the domain cannot be covered by set of rectangles
we are not allowed to use the formula (16) and then
it is necessary to adopt some 2-D quadrature.
In time dependent analysis the problem of singularity is more acute because the functions β and φ
vary dramatically in respect to space coordination.
Like in stationary problem, the coefficient
h_{MM}^{ii} is equal to c_i as the radial and normal directions are orthogonal. For coefficient g_{MM}^{ii} one can
use the analytical representation of the exponent
integral and integrating it one obtains

$$\int_0^R E_1 (ar^2) \, dr = [2-C- \ln (ar^2) +$$

$$\sum_{k=1}^{\infty} (-1)^{k-1} \frac{(ar^2)^k}{k \, k! \, (2k+1)}]_R \qquad (17)$$

$$a = \frac{\mu \, \upsilon}{4 \, (T-t)}$$

However the series appearing in the formula (17) converges very slowly, expecially when the argu - ment is not small enough. That is why the formula (17) should be used only if $aR^2 \leqslant 1$ and for $aR_1^2 = 1$ we have (18) with 8-digital accuracy

$$\int_0^{R_1} E_1 (ar^2) \, dr = .29024786 \, R_1 \qquad (18)$$

Thus the whole integral when $R > R_1$ is as follows

$$\int_0^R E_1 (ar^2) \, dr = .29024786 \, R_1 + \int_{R_1}^R E_1 (ar^2) dr \qquad (19)$$

The integral on the right-hand side can be calcula- ted using the Gaussian quadrature without any fear. If $R < R_1$ one can use the formula (17) taking the same number of series components as in formula (18) i.e. $k=10$.
But there is also the problem of the so called "quasi-singularity". When the internal point under consideration is very close to the boundary, the numerical integration fails again because of the very fast variation of the function β and φ .
In this case the following way of integration has

been applied:

$$\int_{e_j} E_1 (ar^2)dS = \int_{e_j} [E_1 (ar^2) + \ln(ar^2)]dS -$$

$$- \int_{e_j} \ln (ar^2) dS \qquad (20)$$

$$\int_{e_j} \frac{e^{-ar^2}}{r^2} dS = \int_{e_j} (\frac{e^{-ar^2}}{r^2} - \frac{1}{r^2})dS + \int_{e_j} \frac{1}{r^2} dS \quad (21)$$

The first integrals on the right-hand sides are
without singularity and the other ones are calculated analytically. Of course, this procedure needs
some additional computer time, so it is used for
special cases only, when the average distance between internal point and the ends of an element is
less than its length.

Eddy current calculation

The analysts who deal with the finite element method know how it is difficult to calculate the eddy
currents by means of this method. The BEM enables
to avoid many numerical problems as it is based on
analytical rather than numerical concept of time
treatment. Thus, using the formula (4) one can
calculate the eddy current density by analytical
differentiation of (11) in respect to the T-varia-
ble. Then, of course $c_j = 1$ and we use the same
procedure as it was described for the calculation
of the MVP. The differentiation is done in accordance with the theory of convolution, so one obtains

$$\beta_{\overline{T}}' = \frac{\alpha^4}{8\pi}[\exp(-w_{m-1})/(T-t_{m-1})^2 - \exp(-w_m)/(T-t_m)^2]$$

$$(22)$$

$$\varphi'_T = \frac{\alpha^2}{4\pi} \left[\exp(-w_{m-1})/(T-t_{m-1}) - \exp(-w_m)/(T-t_m) \right]$$

$$(23)$$

$$\vartheta'_T = \frac{\alpha^2}{4\pi(T-t)^2} \exp(-w_t)(w_t-1)$$ (24)

Introducing (22) - (24) to the formula (11) instead of β, φ, v we can compute the eddy currents den-situ at any desired time-space point. It is incal-culable advantage of the BEM.

THE RESULTS AND DISCUSSION

To illustrate our theoretical discussion we shall solve the simple problem. Consider the following physical model : the long conductor of the square cross-section carrying the time-dependent current is surrounded by highly permeable walls from three sides and the fourth one is open. It represents a slot of electrical machine. From the mathematical point of view it is the parabolic non-homogenous Neuman's problem. The boundary conditions are homo-genous on the highly permeable walls and on the fourth side the boundary condition results from the Ampere's law. It means that the eddy currents are to be confined to the cross-section.
In Fig. 1 the domain of interest and the input data are given.
The values of material parameters and the form of the source current density are as follows:

$$\gamma = 56 \cdot 10^6 \frac{S}{m} , \quad \mu = .4\pi \ 10^{-6} \ H/m,$$

$$\delta_S = \delta_c \ \eta(t), \quad \delta_0 = 1.10^6 \frac{A}{m^2}$$

The boundary has been divided into 16 equal ele-ments. It was checked that the greater number of element did not cause any considerable difference. Two variants of time division have been examined: the one with uniform time-step ($\Delta t = .01$) and the other with non-uniform one. For the uniform time division the results were compared with the ones obtained with the finite element method. The re-

sults of both methods are shown in Tab. 1.
For some time-space points in Tab. 1 the two values of the BEM results are quoted. The first comes out from the uniform time division and the second from the non-uniform time division.
All the results show that, in spite of methodologi-
cal difference between both methods, the results
are somehow close. It confirms the correctness of
both methods, but the results of the BEM are far
more likely and compatible with the intuitive expectations. Since the method enables to calculate the
eddy current density at the points placed very near
the boundary, it was possible to draw the distributions of eddy currents densities along the
x_2 - axis emphasizing the neighbourhood of the top
and the bottom of the conductor (Fig. 2). It should be stressed that such curves are of the great
interest in the eddy current analysis and it would
be very difficult to obtain them with the "domain
methods".

GENERAL REMARKS

The theoretical discussion and the quoted calculation show that the BEM is of great usefulness in
the analysis of transient eddy currents. There are
three main reasons of this conclusion.
Firstly, one can compute the eddy currents exactly
at the time-space point one wishes. What is more,
the value of eddy current density is given analytically rather then numerically, so that one avoids
the difficult and inaccurate numerical differentiation. Secondly due to analytical nature of the fundamental solution one can adjust the time step to
the course of exciting quantities /source current
or/and boundary conditions/. Thirdly, all advantages appearing in the stationary BEM last here.
There are also some difficulties. One of them and
in our opinion, the most important one is how to
accelerate the calculation of the function β and
φ . On the basis of our experiences it takes the
major part of computer time. The author's programme bases on the standard procedure for exponent
integral and probably it can be improved by some
approximation approaches. Yet the problems of singularity or quasi-singularity suffer the lack of
general solution. And the problem of non-lineari-
ty, remains still unsolved, even when dealing with
stationary case. Roughly speaking, in magnetic
problems one uses mostly the MVP formulation.
It causes that the coefficients of differential
equation depend on the derivative of unknown fun-

-ction, so that the Kirchoff's transformation emplo
-yed succesfully in temperature field analysis
(Ingham[6]) cannot be applied here. We recognize this
problem as the main limitation of the common use
of the BEM in magnetic field analysis.

TABLE 1.

$(x_1 = .05,\ x_2 = .09\ )$

METHOD ——— TIME sec.	MVP $x10^{-3}$		TOTAL CURRENT DENSITY $x10^6$	
	BEM	FEM	BEM	FEM
.01	-.633 -.582	-.434	3.114 3.463	4.114
.02	-1.058 -1.009	-.862	2.491 2.536	3.296
.03	-1.367 -1.317	-1.192	2.144 2.161	2.757
.04	-1.607	-1.454	1.815	2.4
.05	-1.801 -1.733	-1.667	1.751 1.695	2.148
.06	-1.961	-1.884	1.627	1.96
.07	-2.096 -2.018	-1.994	1.529 1.473	1.813
.08	-2.209	-2.122	1.449	1.696
.09	-2.306	-2.232	1.384	1.599

REFERENCES

1. Brebbia C. A., Telles J. C. F., Wrobel L. C.
   (1984). Boundary Element Techniques, Springer
   Verlag, Berlin and New York.

2. John F. (1978). Partial Differential Equations,
   Springer Verlag, Berlin and New York.

3. Krawczyk A. (1985). The Solution of Transient
   Electromagnetic Problems by Means of the Boundary Element Method, in ISEF'85, Proceedings of
   the International Symposium on Electromagnetic
   Fields, Warsaw, Poland.

4. Krawczyk A. (1984). Application of the Boundary Element Method to Transient Electromagnetic
   Problems. Prace Instytutu Elektrotechniki 132.

5. Krawczyk A. (1983). The Application of Finite
   Element Method to the Transient Problems of
   Electrodynamics, Rozprawy Elektrotechniczne
   2/1983 in Polish .

6. Ingham D. B., Kelmanson M. A. (1984). Boundary
   Integral Analyses of Singular, Potential and
   Biharmonic Problems, Springer Verlag, Berlin
   and New York.

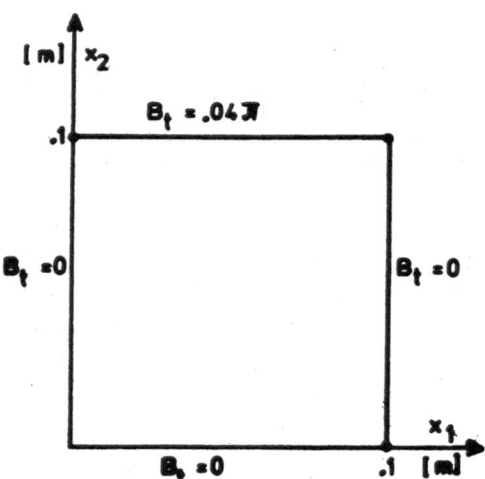

Fig. 1. The sample problem

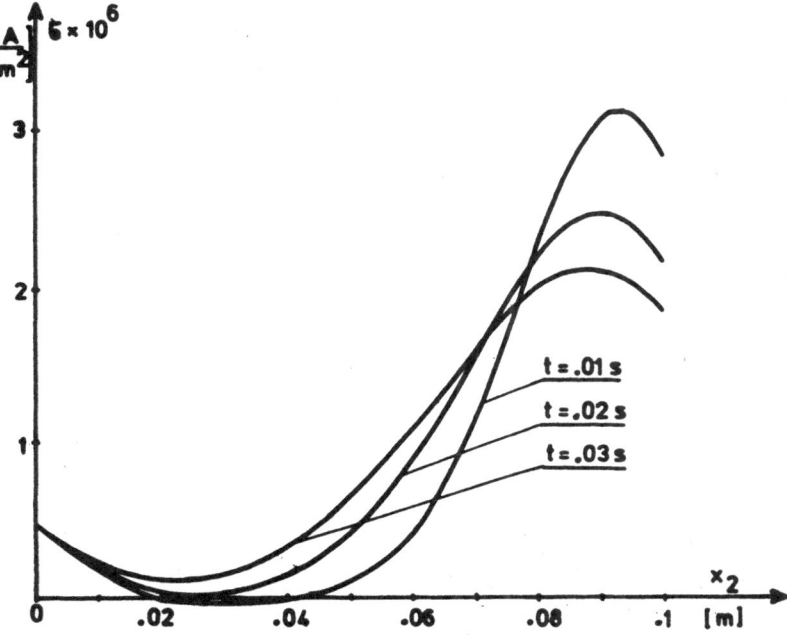

Fig. 2. The distributions of current densities